METHODS OF BIOCHEMICAL ANALYSIS

Volume 37

METHODS OF
BIOCHEMICAL ANALYSIS

Series Editor

Clarence H. Suelter

Volume 37

BIOANALYTICAL INSTRUMENTATION

Edited by
CLARENCE H. SUELTER
Department of Biochemistry
Michigan State University
East Lansing, Michigan

An Interscience Publication
JOHN WILEY & SONS, INC.
New York • Chichester • Brisbane • Toronto • Singapore

This text is printed on acid-free paper.

Copyright © 1994 by John Wiley & Sons, Inc.

All rights reserved. Published simultaneously in Canada.

Reproduction or translation of any part of this work beyond
that permitted by Section 107 or 108 of the 1976
United States Copyright Act without the permission of the copyright
owner is unlawful. Requests for permission or further
information should be addressed to the Permissions Department,
John Wiley & Sons, Inc., 605 Third Avenue, New York, NY
10158-0012.

This publication is designed to provide accurate and
authoritative information in regard to the subject
matter covered. It is sold with the understanding that
the publisher is not engaged in rendering legal, accounting,
or other professional services. If legal advice or other
expert assistance is required, the services of a competent
professional person should be sought.

Library of Congress Catalog Card Number: 54-7232

ISBN 0-471-58260-3

Printed in the United States of America

10 9 8 7 6 5 4 3 2 1

SERIES PREFACE

Methods of Biochemical Analysis was established in 1954 with the publication of Volume 1 and has continued to the present, with volumes appearing on a more or less yearly basis. Each volume deals with biochemical methods and techniques used in different areas of science. Professor David Glick, the series' originator and editor for the first 33 volumes, sensed the need for a book series that focused on methods and instrumentation. Already in 1954, he noted that it was becoming increasingly difficult to keep abreast of the development of new techniques and the improvement of well-established methods. This difficulty often constituted the limiting factor for the growth of experimental sciences. Professor Glick's foresight marked the creation of a unique set of methods volumes which have set the standard for many other reviews.

With Professor Glick's retirement from the series and beginning with Volume 34, I have assumed editorship. Because the rationale used in 1954 for the establishment of the series is even more cogent today, I hope to maintain the excellent traditions developed earlier. The format of Volume 34 and later volumes, however, is changed. Rather than cover a variety of topics as previous volumes did, each volume will now focus on a specific method or the application of a variety of methods to solve a specific biological or biomedical problem.

CLARENCE H. SUELTER

East Lansing, Michigan

PREFACE

Volume 37 of *Methods of Biochemical Analysis* focuses on the application of special instrumental techniques to problems in biology. Focusing selected volumes of this series on instrumentation is particularly appropriate because advances in instrumentation often times provide new vistas in science. The development of the pH meter, spectrophotometer, radioactivity and the associated measuring devices early in this century had a significant impact on research in many areas of biology and biochemistry. Likewise the instrumentation reviewed in this volume have and will continue to have a significant affect on the rate in which we solve problems.

The first chapter in this volume describes the application of x-ray crystallography in biochemistry. The application of this methodology is often thought to be beyond the reach of most biochemists to apply this technique in their research. The next chapter reviews the application of both transmission microscopy and scanning probe microscopy to problems in Biology. The newest instrumentation reviewed makes use of fluorescence and fluorescent probes to study cell function. The method offers many unique and perhaps yet to be explored approaches to the study of biology. Chapter Four reviews well-developed techniques for assaying enzymes in tissues of importance to clinical laboratories. The application of these techniques in biochemical laboratories particularly in studying the steady state levels of enzymes and metabolites in respiring tissues has yet to be exploited. Finally, the last chapter presents a review of rapid-scanning spectroscopy. This powerful technique makes it possible to study chemical reactions or processes which have short-lived intermediates with spectral properties significantly different from those of reactants and products. Repetitive scans of the appropriate spectral region provides information on the number and types of species present as well as the kinetics of the formation and decay of intermediates.

CLARENCE H. SUELTER
Michigan State University

CONTENTS

Abbreviations	xi
X-ray Crystallography of Proteins *J. P. Glusker*	1
Transmission Electron Microscopy and Scanning Probe Microscopy *Karen L. Klomparens and John W. Heckman, Jr.*	73
Quantitative Fluorescence Imaging Techniques for the Study of Organization and Signaling Mechanisms in Cells *Margaret H. Wade, Adriaan W. de Feijter, Melinda K. Frame and Melvin Schindler*	117
Automated Enzyme Assays *John A. Lott and Daniel A. Nealon*	143
Rapid-Scanning Stopped-Flow Spectroscopy *Peter S. Brzović and Michael F. Dunn*	191
Author Index	275
Subject Index	291
Cumulative Author Index, Volumes 1–37 and Supplemental Volume	299
Cumulative Subject Index, Volumes 1–37 and Supplemental Volume	313

ABBREVIATIONS

a-GPO	a-Glycerophosphate oxidase
ACP	Acid phosphatase
ADP	Adenosine diphosphate
AFM	Atomic force microscope
ALD	Aldolase
ALP	Alkaline phosphatase
α-Ct	Alpha chymotrypsin
ALT	Alanine aminotransferase
AMP	Adenosine phosphate
AOM	Acousto optic modulator
AST	Aspartate aminotransferase
ATP	Adenosine triphosphate
BCECF	2,7-bis-(2-carboxyethyl)-5-(and-6)-carboxyfluorescein
BCR	Community Bureau of Reference
BSA	Bovine serum albumin
BSO	Buthionine disulfoxide
BZ	Benzimidazole
CCD	Charge coupled devices
CD	Circular dichroism
CGS	Cystathionine-γ-synthase
ChE hydrolase	Pseudocholinesterase
CID	Charge injection devices
CK	Creatine kinase
CM	Colorimetric
CME	Carboxylmethylated cysteine-46 LADH
CPD	Critical point drying
Creatine-P	Creatine phosphate
CRM	Certified Reference Material
CV	Coefficient of variation
DACA	4-trans-N,N-dimethylaminocinnamaldehyde
DMEM	Dulbecco's Modified Eagle Medium

EDS	Energy dispersive spectroscopy
EDTA	Ethylene diammine tetraacetic acid
ELISA	Enzyme linked immuno sorbent assay
EM	Electron microscopy
FATME	N-furylacryloyl L-tryptophan methyl ester
FISH	Fluorescence in situ hybridization
FITC	Fluorescein isothiocyanate
FRAP	Fluorescence redistribution after photobleaching
FT-IR	Fourier transform infra red
G-6-PDH	Glucose-6-phosphate dehydrogenase
GDH	Glutamate dehydrogenase
GGT	γ-Glutamyl transferase
GK	Glycerol kinase
GLC	Gas liquid chromatography
GPO	Glycerol phosphate oxidase
GSH	Glutathione
GST	Glutathione-S-transferase
H_2O_2	Hydrogen peroxide
HK	Hexokinase
HPLC	High-performance liquid chromatography
HRP	Horseradish peroxidase
HSA	Human serum albumin
ICES	International Clinical Enzyme Scale
IFCC	International Federation of Clinical Chemistry
IgG	Immunoglobulin G
IgM	Immunoglobulin M
L	Lactate
LADH	Liver alcohol dehydrogenase
LC	Liquid chromatography
LCI-CCD	Lens coupled intensifed charge-coupled device
LD	Lactate dehydrogenase
LH2	Dihydroluciferin
LM	Light microscopy
LMCT	Ligand-to-metal charge transfer
LSM	Laser scanning confocal microscopy
MCB	Monochlorobimane
MCP	Microchannel plate
MCP-SPD	Microchannel plate image intensified-coupled silicon photodiode
MDH	Malate dehydrogenase

MPH	5,10-dihydro-5-methyl phenazine
MT	Metallothionein
NAD	Nicotinamide adenine dinucleotide
NADH	Dihydronicotinamide adenine dinucleotide
NADP	Nicotinamide adenine dinucleotide phosphate
NADPH	Dihydronicotinamide adenine dinucleotide phosphate
NBD-PC	1-acyl-2-(N-nitrobenz-2-oxa-1,3-diazole)-aminocaproylphosphatidylcholine
NCCLS	National Committee for Clinical Laboratory Standards
NDMA	p-nitroso-N,N-dimethylaniline
NIST	National Institute of Standards and Technology
NPA	p-nitroso-N-phenylalanine
NRSCL	National Reference System for the Clinical Laboratory
OASS	O-acetyl-*L*-serine sulfhydrylase
OSHS	O-succinyl-*L*-homoserine
PAA	Polyacrylamide
PAN	1-(2-pyridylazo)-2-naphthol
PCA	Principal component analysis
PCR	Polymerase chain reaction
PI	Propidium iodide
PK	Pyruvate kinase
PLP	Pyridoxal phosphate
PM	Potentiometric
RSSF	Rapid scanning stopped flow
RT	Rate
SAV	Supplementary assigned value
SBC	S-benzyl-*L*-cysteine
SFP	Simulated fluorescence process
SPD	Silicon photodiode
SPM	Scanning probe microscopy
SRM	Standard Reference Manual
STEM	Scanning transmission electron microscopy
STM	Scanning tunneling microscope
SWSF	Single wavelength stopped flow
TEM	Transmission electron microscopy
TEM-AR	Transmission electron microscopy-autoradiography
TMV	Tobacco mosaic virus
TRITC	Rhodamine isothiocyanate
UA	Uranyl acetate
WDS	Wavelength dispersive spectroscopy

METHODS OF BIOCHEMICAL ANALYSIS

Volume 37

BIOANALYTICAL INSTRUMENTATION   VOLUME 37

X-ray Crystallography of Proteins

J. P. GLUSKER

The Institute for Cancer Research, Fox Chase Cancer Center, Philadelphia, Pennsylvania

1. Introduction
2. Methods for Obtaining Suitable Protein Crystals
   - 2.1. Description of the Crystal—The Crystal Lattice
   - 2.2. Crystal Symmetry—Choosing a Unit Cell
   - 2.3. Indexing Crystal Faces—The Meaning of h, k, l
3. Obtaining a Diffraction Pattern
   - 3.1. X-ray and Neutron Scattering by an Atom
   - 3.2. X-ray Scattering by a Molecule and by a Crystal
   - 3.3. The Reciprocal Lattice and Ewald Sphere
4. Combining Scattered Waves in Order to Obtain an Image of the Diffracting Material (Simulating an X-ray Lens)
   - 4.1. Relative Phase Angles
   - 4.2. The Use of Fourier Series
   - 4.3. The Electron-Density Map
   - 4.4. Fourier Transforms (Between Crystal and Diffraction Space)
   - 4.5. Effects of the Atomic Arrangement
   - 4.6. Effects of Atomic Displacements and Vibrations
5. Measuring Diffraction Data $[|F(hkl)|]$
   - 5.1. The Diffraction Equations
   - 5.2. Sources of X rays
   - 5.3. Detectors of X rays
   - 5.4. Preparing a Crystal for Diffraction Study
   - 5.5. Measurements at Low Temperatures
   - 5.6. Laue Diffraction
   - 5.7. Problems with Large Structures
   - 5.8. Putting the Intensity Data on an Absolute Scale—The Wilson Plot
6. Measuring Unit-Cell Dimensions and Space Group
   - 6.1. Unit-Cell Dimensions
   - 6.2. Space-Group and Crystal Symmetry
   - 6.3. Crystal Density and Unit-Cell Contents
   - 6.4. The Experimental Data Set
7. Phase Determination
   - 7.1. Preparation of Heavy-Atom Derivatives of Proteins

Bioanalytical Instrumentation, Volume 37, Edited by Clarence H. Suelter.
ISBN 0-471-58260-3   © 1993 John Wiley & Sons, Inc.

- 7.2. The Patterson Map
- 7.3. Relative Phases by the Method of Isomorphous Replacement
- 7.4. Superposition and Molecular Replacement Methods
- 7.5. Anomalous Dispersion
8. Calculating the Electron-Density Map of the Protein,
    - 8.1. Electron-Density Modification
    - 8.2. Use of Noncrystallographic Symmetry
    - 8.3. Entropy Maximization
9. Interpreting the Electron-Density Map
    - 9.1. The Resolution of an Electron-Density Map
    - 9.2. Macromolecular Model Building
    - 9.3. Refinement of a Protein Crystal Structure
    - 9.4. Difference Maps for Macromolecules
    - 9.5. The Precision of the Derived Protein Structure
10. The Native Protein Structure
    - 10.1. Torsion Angles
    - 10.2. Accessing the Protein Data Bank
    - 10.3. Bond Distances and Angles
    - 10.4. Vibration Parameters
    - 10.5. Motifs in Protein Structures
    - 10.6. Molecular Architecture
11. Protein–Ligand Interaction
    - 11.1. Binding of Drugs to Cytochrome P-450$_{cam}$
    - 11.2. Binding of Drugs to Dihydrofolate Reductase
    - 11.3. Protein-Catalyzed Reactions
    - 11.4. Studies with Crystalline Mutant Enzymes
    - 11.5. Reactions in the Crystal
12. Conclusions

1. INTRODUCTION

The determination of the three-dimensional structure of biological macromolecules by an interpretation of the X-ray diffraction pattern of appropriate crystals (1) will be described here. The result of such a structure analysis is a representation of the electron density, and hence the atomic arrangement, within that crystal. It would have been simpler to use a "supermicroscope" to see this atomic arrangement. We are able to see the details of an object with a microscope provided that these details are separated by at least half the wavelength of the radiation used to view them (2). Visible light has wavelengths of the order of 4 to 8 $\times 10^{-5}$ cm, and therefore, it cannot be used to view atoms that are separated in molecules by distances of the order of 10^{-8} cm. On the other hand, X rays have wavelengths of the order of 10^{-8} cm so that the supermicroscope that we should build to view atoms would have to employ X rays rather than visible light. Unfortunately, no appropriate material has yet been found to construct a lens that can focus X rays. As a result it is not yet possible to build an X-ray "supermicroscope." Some studies with longer-wavelength X rays (3), or with a scanning tunneling microscope (4), have enabled us to obtain a view of atoms directly, but only on the surface of a structure, and sometimes with problems in interpreting the image so obtained.

There is, however, an excellent way to view molecules. It is the method described here, *X-ray* or *neutron diffraction*. The procedure used can be compared with that used in microscopy to magnify an image. The initial stage in the action of an optical microscope is carried out. A beam of X rays strikes a crystal and is scattered by it. The second stage of the action of an optical microscope involves the focusing of light waves by a lens. Since this cannot be done for X rays, the focusing is done by a mathematical summation—the scattered beams are combined (summed) with due attention to their relative phases (which are not measured but are derived in one of a variety of ways). The result of such an analysis is a complete three-dimensional determination of the arrangement of atoms in the crystal with interatomic distances and other features of molecular geometry measured to a known degree of precision (5-8). The method is comprised of several experimental and computational stages as follows:

1. the growth of a diffraction-quality crystal,
2. the measurement of the scattering angles and intensities of each diffracted beam,
3. the determination of the unit-cell dimensions and the molecular contents of the unit cell,
4. the determination of the relative phases of each of the diffracted beams (also called Bragg reflections),
5. the mathematical combination of the diffracted beams to form an image of the diffracting material,
6. the interpretation of the image so obtained, and
7. the introduction of substrate or inhibitor molecules into the crystal and analysis of their binding sites and the implications of these with respect to the mode of action of the enzyme.

2. METHODS FOR OBTAINING SUITABLE PROTEIN CRYSTALS

A *crystal* is a solid that has within it a regularly repeating internal arrangement of atoms. By contrast, the internal arrangement is not regular in amorphous materials such as glasses or gums. Proteins can form beautiful crystals when the conditions are appropriate, as are occasionally found during purification. For example, deep red crystals of hemoglobin were described as early as 1830 (9). The first report of crystals of an enzyme, urease, was published by J.B. Sumner in 1926 (10). In 1930, J. H. Northrop crystallized pepsin, trypsin, and chymotrypsin (11). Even more spectacular was the crystallization of living material, tobacco mosaic virus, by W. Stanley (12). These protein crystals did not, however, diffract X rays in the way that crystals of smaller molecules do. In 1934, J.D. Bernal and D. Crowfoot Hodgkin discovered that good diffraction patterns could be obtained (13), provided the protein crystals were maintained in contact with their mother liquor during the diffraction experiment, rather than being dried in air. It was many years before the extensive diffraction patterns of any proteins could be analyzed correctly so that their molecular structures could be found. The first protein structures so determined were those of myoglobin (14), hemoglobin (15), and lysozyme (16). Now the crystal structures of a large number of protein crystal structures, including several large protein complexes, have been reported (17).

The process of crystallization occurs when molecules separate from solution and settle onto the surface of a growing crystal. These molecules must be laid down in a completely regular manner if a crystal is to form. The extent of regularity of this packing of molecules determines the quality of a crystal for X-ray diffraction studies. The rate of growth of the crystal is affected by the rate of transport of molecules to the surface of the growing crystal and the probability that they will find a good site of attachment. It is surprising that proteins, with so many functional groups on their surfaces can, under appropriate conditions, form such good crystals, rather than the disordered arrangement found with some polyfunctional smaller molecules such as glycerol.

Crystal growth may be roughly divided into three stages. The first stage is *nucleation*, in which a few molecules or ions form a stable aggregate on which more molecules will be laid down. The second stage is *growth*, that is, the orderly addition of further molecules or ions in a regular manner. In the final stage, called *termination*, growth ceases. Protein crystals are characterized by a much higher solvent (water) content than is generally found in crystals of smaller molecules. Side chains on the surface of the molecule may take up different conformations from protein molecule to protein molecule. This, together with the variability of positions of water molecules in the center of areas of water in the crystal, means that protein crystals are usually less ordered than crystals of smaller molecules.

Diffraction-quality protein crystals are generally grown by preparing a solution that is nearly saturated with the protein (18, 19). The protein must be pure and usually needs to be concentrated to the range of 5–100 mg/ml. Experimental conditions such as pH or the amount of added precipitating agent, are then changed very slowly to achieve a slight supersaturation, while the number of separate nuclei that are formed is controlled (for example, by removing most small nuclei by filtration). The solution should then be left undisturbed. Protein solubility is generally at a minimum near its isoelectric point, and therefore, variation in pH may be useful in screening for the best conditions for growing crystals. The degree of supersaturation of the solution should be such as to support continued, ordered growth to give large, single crystals rather than amorphous material or a microcrystalline precipitate. In this way a few crystals of suitable size for diffraction experiments may be obtained. The formation of crystals is usually most effective if it occurs over several days. The aim is to obtain crystals that are several tenths of a millimeter in each dimension. Experiments are now in progress to test the effects of microgravity and the absence of liquid currents (on the space shuttle) on the quality of the crystals grown (20).

In aqueous solution, a protein is covered with water molecules. Added ions, however, will interact with some of these water molecules and will temporarily leave sites on the protein that are free to bind other protein molecules, leading to protein aggregation and, possibly, crystallization. The agents generally used for this are those highly soluble inorganic salts, such as ammonium sulfate or sodium chloride or organic polyethers, such as polyethyleneglycols of a selected molecular weight range, which do not denature the protein. A commonly used precipitating agent is 2-methyl-2,4-pentanediol (MPD). Changes in experimental conditions may produce different types of crystals of the same protein, and these are described as *polymorphs*. They have different unit-cell dimensions and different packings of molecules. A protein crystallographer may report unit-cell dimensions for several polymorphs of a macromolecule and will then select the most suitable one for further study.

A typical method used for growing protein crystals is the hanging-drop vapor-diffusion technique (Fig. 1). A small drop of a concentrated protein solution containing suitable precipitants is suspended from a silicone-treated microscope cover glass placed over a spot plate reservoir containing a more concentrated solution of the precipitants. The cover glass forms a tight seal over this reservoir. In order to achieve vapor equilibrium, some of the liquid in the hanging drop distils into the more concentrated solution in the reservoir. This increases the concentration of the protein in the solution in the hanging drop, so that protein crystals may separate out.

Biological macromolecules are generally only available in very small quantities and are often difficult to crystallize. The strategy, therefore, is to use a combinatorial method in which selected combinations of different pH ranges, different salt concentrations, and different precipitating agents are tested for their ability to achieve the supersaturation necessary for crystallization. Very small amounts of protein are needed. For those interested, a recipe for growing lysozyme crystals is provided on page 95 of reference 18. Such recipes for growing crystals are usually specific for a particular protein, but they can serve as a starting point for crystallizing a similar protein. Membrane-bound proteins are more of a problem to crystallize than are water-soluble proteins, and it is only

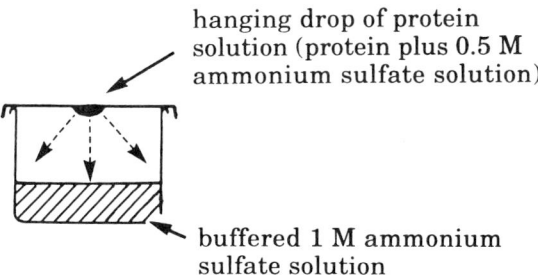

Fig. 1. The vapor diffusion method for growing protein crystals. Water equilibrates from the more dilute protein solution in the drop into the more concentrated solution at the bottom of the beaker. This concentrates the solution slowly in the hanging drop.

recently that some have been successfully crystallized (18). The trick used in this case is to add some detergent (composed of molecules with both polar and nonpolar regions) to increase the solubility of the protein.

Sometimes it is necessary to add extraneous nuclei in order to achieve crystal growth. Appropriate surfaces for crystal growth are provided generally by small freshly grown crystals ("seed crystals") of the same material, or by insoluble crystalline minerals with the appropriate periodicity in their atomic arrangements (21). The arrangement of molecules on the surface of the crystalline mineral promotes an oriented overgrowth of protein molecules, and good crystals have been obtained this way. Several unit cells of the mineral may be needed to match one unit cell of the protein. For example, crystals of lysozyme (large unit cell dimensions) grow well on the mineral apophyllite (small unit cell dimensions) (21-23). If good crystals cannot be obtained, it may be necessary to further purify the protein or to obtain protein from a different source and try again.

2.1. Description of the Crystal—The Crystal Lattice

The regularity of the internal atomic arrangement in crystal is described in terms of a *unit cell* with specific (and measurable) dimensions. By convention the size of this repeat unit is described by edge lengths a, b, and c with angles α, β and γ between them (α between b and c, β between a, and c and γ between a and b). Values for each of these six parameters, with their precision, are listed in each report of a crystal structure determination. Also, by convention, the unit of measurement of unit cell dimensions and interatomic distances is an Ångström unit (Å), named for a Swedish spectroscopist (1 Å = 10^{-8} cm, 10 nm); bond distances are of the order of 1-2 Å.

The regularity of stacking of unit cells can be represented by a "crystal lattice," which expresses only the periodicity of the crystal contents, not the actual molecule that is repeated. As shown in Fig. 2, the crystal structure is a combination (*convolution* is the mathematical term) of a motif (the macromolecule or a group of macromolecules within a unit cell) and the crystal lattice. It is only necessary to find the atomic arrangement in one unit cell, and then this atomic arrangement is repeated in a manner defined by the crystal lattice to give the entire crystal structure.

2.2. Crystal Symmetry—Choosing a Unit Cell

There are seven crystal systems, listed in Table 1, that result from the possible symmetry of the crystal lattice (24). For example, if the crystal lattice describes a cubic unit cell, rotations of 90° or 120° or 180° about appropriate directions will give a lattice indistinguishable from the original; this can be verified by examination of a cube. The unit cell conditions ($a = b = c$, $\alpha=\beta=\gamma=90°$) follow from the lattice symmetry. If, however, $a \neq b \neq c$ and neither α nor β nor $\gamma=90°$, then the symmetry of the crystal lattice is low (triclinic).

The choice of a unit cell is arbitrary as it merely involves the selection of a repeating unit and this can often be done in several ways, as shown in Fig. 3. If,

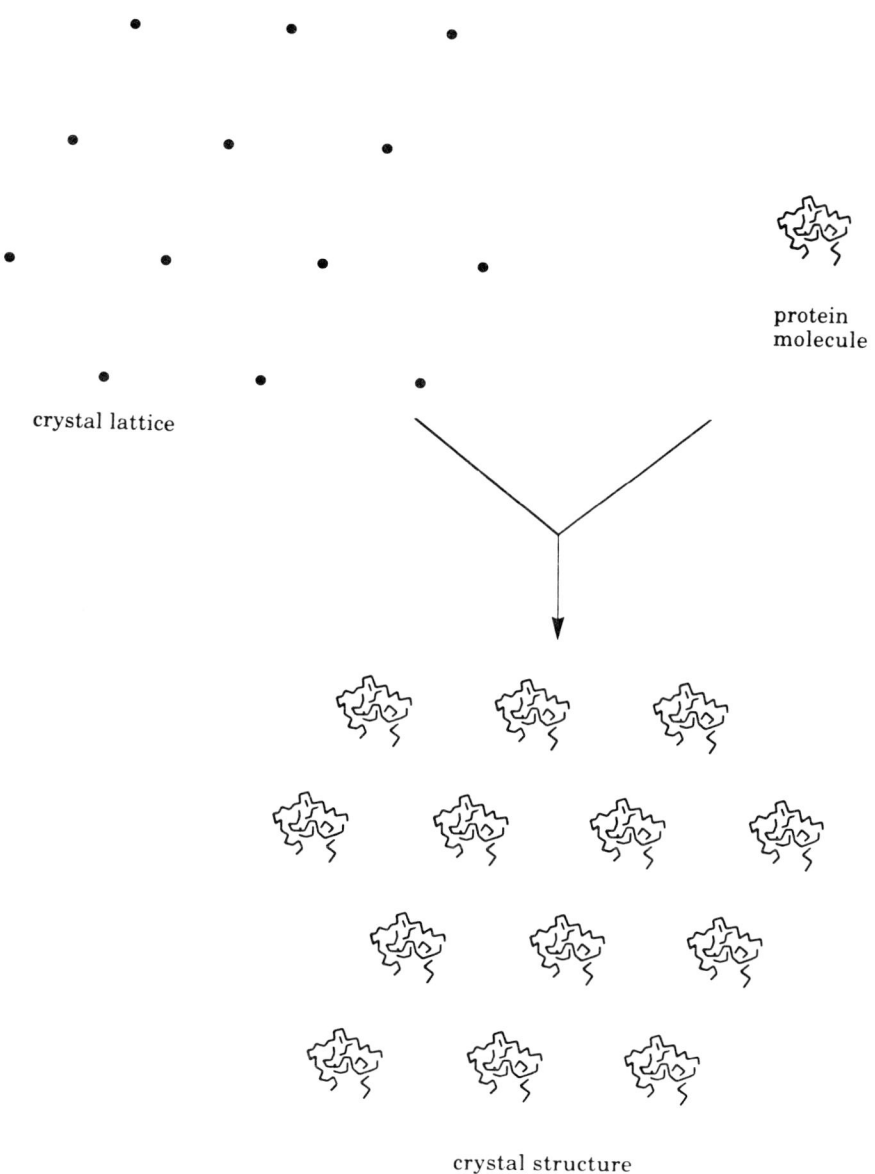

Fig. 2. The crystal lattice is a series of points. When combined (convoluted) with a motif such as protein molecule, a crystal structure is obtained.

TABLE 1

The Seven Crystal Systems

Crystal System (7)	Characteristic Symmetry	Lattice (Laue) Symmetry	Axial and Angular Constraints from Symmetry*
1. Triclinic	Identity or inversion (onefold rotation or rotatory-inversion axis) in any direction	$\bar{1}$	$a \neq b \neq c$ $\alpha \neq \beta \neq \gamma$
2. Monoclinic	A single twofold rotation or rotatory-inversion axis along **b**	2/m	$a \neq b \neq c$ $\alpha = \gamma = 90°, \beta \neq 90°$
3. Orthorhombic	Three mutually perpendicular twofold rotation or rotatory-inversion axes along **a**, **b**, and **c**	mmm	$a \neq b \neq c$ $\alpha = \beta = \gamma = 90°$
4. Tetragonal	A single fourfold rotation or rotatory-inversion axis along **c**	4/mmm	$a = b \neq c$ $\alpha = \beta = \gamma = 90°$
5. Cubic	Four threefold axes along **a+b+c**, **−a+b+c**, **a−b+c**, **−a−b+c**	m3m	$a = b = c$ $\alpha = \beta = \gamma = 90°$
6. Trigonal	A single threefold rotation or rotatory-inversion axis along **a+b+c**	$\bar{3}$m	$a = b = c$ $\alpha = \beta = \gamma \neq 90°$ $\gamma < 120°$
7. Hexagonal	A single sixfold rotation or rotatory-inversion axis (along **c**)	6/mmm	$a = b \neq c$ $\alpha = \beta = 90°$ $\gamma = 120°$

*These follow from the definition given in the characteristic symmetry column, rather than being the direct definition of the crystal system. Note that the symbol $\neq$ means that values are not equal for symmetry reasons; they may, however, accidentally be equal.

however, a unit cell with interaxial angles of 90° and as many dimensions equal to each other as possible can be chosen, that is the best selection. The unit cell with highest symmetry may, however, require that there be more than one repeat unit in the unit cell. When this is the case, there are 14 rather than seven types of crystal lattices, known as Bravais lattices and shown in Fig. 4. The additional seven lattices have an extra lattice point in the center of one face (A, B, C), each face (F), or in the center of the unit cell (I). The chosen unit cell of a protein should correspond to one of these 14 Bravais lattices (Fig. 4).

The symmetry of the atomic arrangement within the crystal can be described by *space group theory*, that is, the theory of the various ways of arranging objects in three dimensions such that a continuation of the symmetry operations gives the next unit cell and so forth (24). For protein molecules, which are by nature asymmetric, the important symmetry operations are rotation axes and screw axes. An object is said to have an n-fold rotation axis if, when an object is rotated $(360/n)°$, it appears like the original. For isolated objects, by point group theory, n may have any value. On the other hand, if the object is in a crystal (with its regular

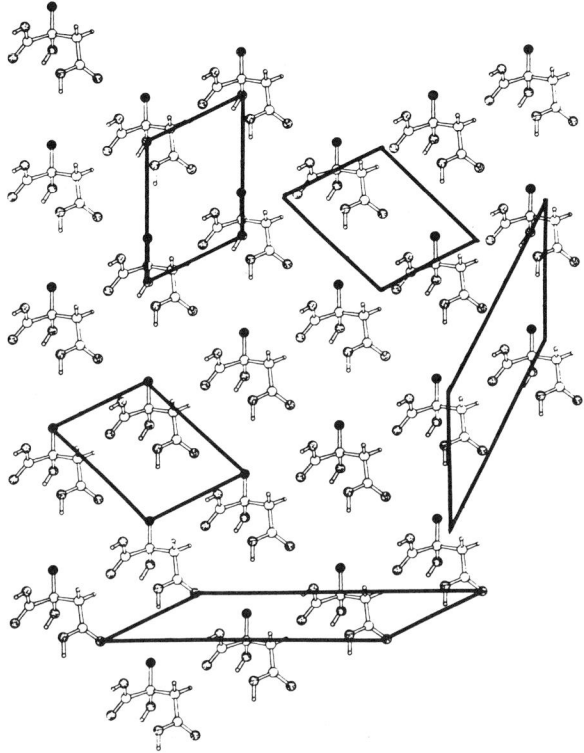

Fig. 3. Unit-cell choices. In this crystal structure the various unit-cell choices each enclose the equivalent of one molecule, and therefore each are appropriate choices. It is, however, more convenient to choose a unit cell with as many angles near 90° as possible.

internal periodicity) then n may only be 1, 2, 3, 4, or 6. If the object in a crystal is rotated about the n-fold axis and then moved (translated) some fraction of the unit cell length (r/n), the crystal is said to have an n_r, screw axis. For example, a twofold screw axis 2_1 involves a rotation of 180° ($= 360°/2$) and a translation of ½ the unit cell length. A twofold axis along the c axis would convert an object at x, y, z to one at $-x, -y, z$. The corresponding twofold screw axis along c would give an object at $-x, -y, ½+z$. For more information the reader is referred to *International Tables*, Volumes 1 or A (24).

2.3. Indexing Crystal Faces—The Meaning of h, k, l

The integers that characterize crystal faces are called *Miller indices* h, k, and l, and it is rare to find h, k, or l larger than 6, even in crystals with complicated shapes. The geometrical construction used to index a crystal face is as follows: from a point inside the crystal, three noncolinear axes x, y, and z are chosen and assigned, by con-

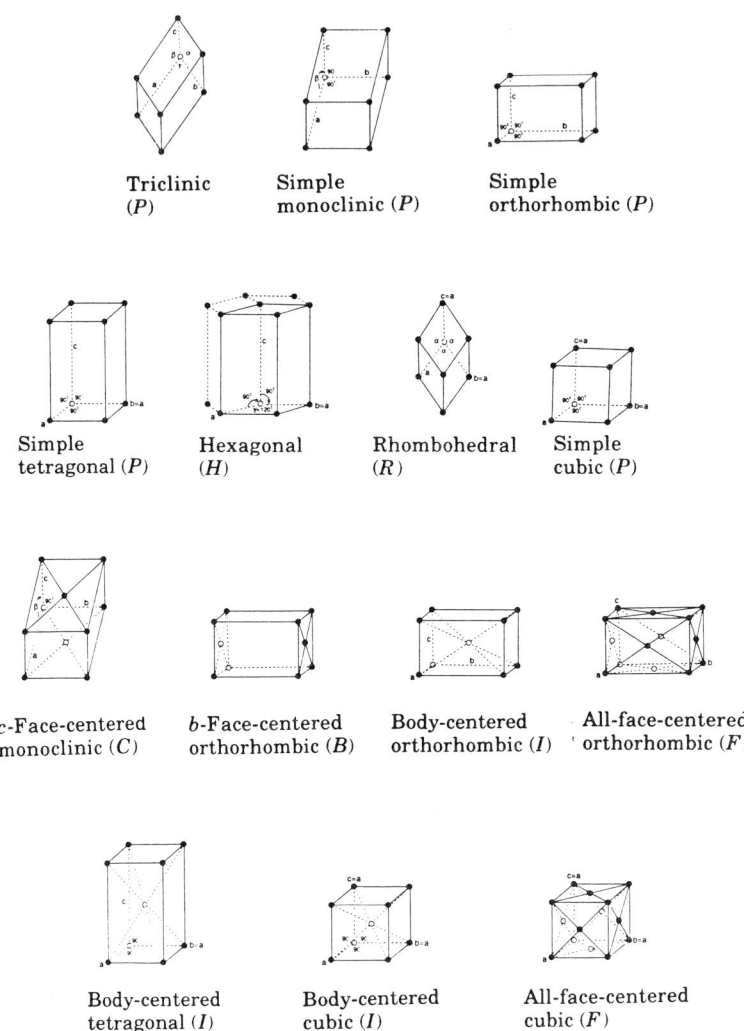

Fig. 4. The 14 Bravais lattices.

vention, in a right-handed system. If we assume that one corner of the unit cell lies at $x = a, y = b$, and $z = c$ along these three axes and another at $x = 0, y = 0, z = 0$, then a crystal face, designated hkl, will make intercepts on the three axes at $x = a/h, y = b/k$, and $z = c/l$, as shown in Fig. 5. When the indices are negative, they are designated $-h, -k, -l$. If the values of h, k, and l are small for all observed crystal faces, then a reasonable unit cell has probably been chosen. For reasons of symmetry, planes in hexagonal crystals are described conveniently by four axes, three in a plane at 120° to each other. This leads to four indices, $hkil$, where $i = -(h + k)$, and equivalent crystal faces will have similar indices.

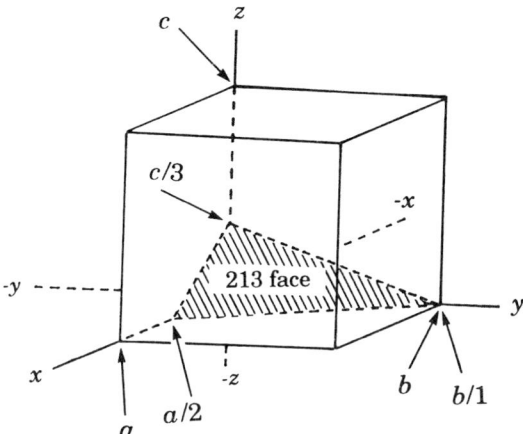

Fig. 5. Indexing the faces of a crystal. If the unit cell is cut at a/h, b/k, and c/l, the indices of the face are hkl. This is shown for the 213 face.

In summary, we choose the unit cell so that the faces can be described by small values for h, k, and l. Not only do h, k and l describe crystal faces, they also describe planes parallel to these faces passing through appropriate crystal lattice points, as shown in Fig. 5.

3. OBTAINING A DIFFRACTION PATTERN

A diffraction pattern can be seen if one looks through a silk umbrella at a distant street light at night (Fig. 6). The street light will appear to have additional spots of light (diffracted rays) regularly disposed around it. The arrangement of these diffracted rays, each with a measurable position and intensity, is referred to as the *diffraction pattern* (25). The spacing between these diffracted rays is inversely related to the spacing between the threads of the material. It is also a function of the distance that the fabric is from one's eyes. All objects may individually diffract radiation of the appropriate wavelengths, but the diffraction effect is reinforced when there is a regular periodicity of structure (as in fabric or a crystal). This periodicity greatly increases the intensity of the diffraction and makes it more readily observable.

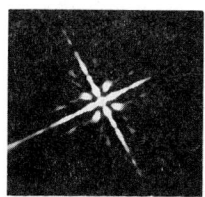

Fig. 6. The diffraction pattern obtained by shining a fine beam of visible light through a fine mesh.

There is a reciprocal relationship between the periodicity of the fabric and its diffraction pattern; the larger the periodicity of the fabric the smaller the periodicity of the diffracted beams in the diffraction pattern and vice versa. A coarser fabric (larger distance between the threads) gives a finer diffraction pattern (smaller distance between spots at the detecting system), while a finer weave of fabric will give a wider spacing between spots in the diffraction pattern. If the scale of this diffraction experiment is decreased several orders of magnitude, so that the fabric (with its regular weave) is replaced by a crystal (with its periodic internal structure), and the visible light is replaced by X rays (with shorter wavelengths), the same effects are observed.

3.1. X-ray and Neutron Scattering by an Atom

X rays are scattered by the electrons around atoms in the crystal structure, and the amplitude of the scattered wave is proportional to the number of electrons at the point of scattering; at atomic positions this corresponds to the atomic number Z_j of the atom (j). Since most of the electron density in a crystal is near an atomic center, the extent to which an individual atom scatters X rays is expressed by an atomic scattering factor f_j. This is defined as the ratio of the scattering of X rays by an individual atom to the scattering by a single electron under the same conditions. Atoms with high atomic numbers ("heavy atoms") scatter X rays more than those with low atomic numbers. At high scattering angles there is a decrease in scattering power that is a result of the size of the atom; this size causes interference between waves scattered by the various regions of the atoms. The wider the electron cloud around an atom, the greater the fall-off in diffracted X-ray intensity at high scattering angles as a function of the scattering angle (described, by convention, as 2θ) (Fig. 7). A list of X-ray scattering factors for individual atoms as a function of 2θ may be found in *International Tables for X-ray Crystallography*, Vol. III (26).

Neutrons, on the other hand, are scattered by the nuclei of atoms. The nucleus is miniscule, of the order of $10-12^{-12}$ cm in diameter, compared with the larger size of electron clouds that are several orders of magnitude larger (10^{-8} cm). Therefore, scattering by different parts of the nucleus is not a consideration, and there is almost no fall-off in the scattering of neutrons as a function of scattering angle. Wavelengths of the order of 1 Å are commonly used for neutrons for crystal diffraction studies. Unlike the case for X rays, neutron scattering factors for atoms may be negative, as for hydrogen atoms. In addition, it is found that scattering amplitudes for neutrons bear no resemblance to the values for X-ray scattering. This makes neutron diffraction useful for distinguishing isotopes (such as those of hydrogen) or atoms with similar atomic numbers (copper and zinc, for example).

3.2. X-ray Scattering by a Molecule and by a Crystal

When the path length differences for X rays diffracted by atoms separated by one unit-cell translation are an integral number of wavelengths, there is reinforce-

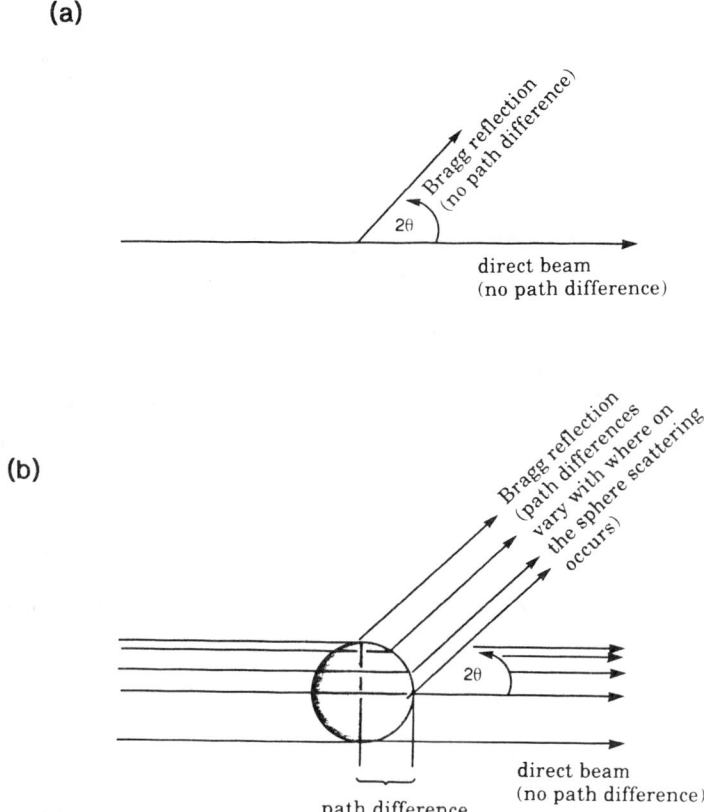

Fig. 7. A wider atom shows a greater fall-off in intensity because of interference between X rays scattered by different parts of the atom. (a) A very small atom and (b) a large atom.

ment of the diffraction effect, and this causes the diffracted beam to have sufficient intensity to be observable. The interference between X rays scattered by different atoms in the same unit cell causes the intensities of the different diffracted beams to have different values (some weak, some intense), giving what is called a *diffraction pattern*. The directions at which reinforcement occurs (2θ values) and a diffracted beam is observed give a measure of the dimensions of the unit cell of the crystal.

When there is only one atom j in a unit cell, the amplitude of the scattered beam $|F(hkl)|$ is proportional to f_j. When there are several atoms in the unit cell, the manner by which X rays (or neutrons) are diffracted by each atom and how these diffracted beams interfere with each other (constructively or destructively) determines the value of $|F(hkl)|$, the structure factor amplitude. The intensities of the individual diffracted beams give a measure of how much X rays, scattered from various atoms in the unit cell, interfere with each other. Thus, $|F(hkl)|$ is a measure

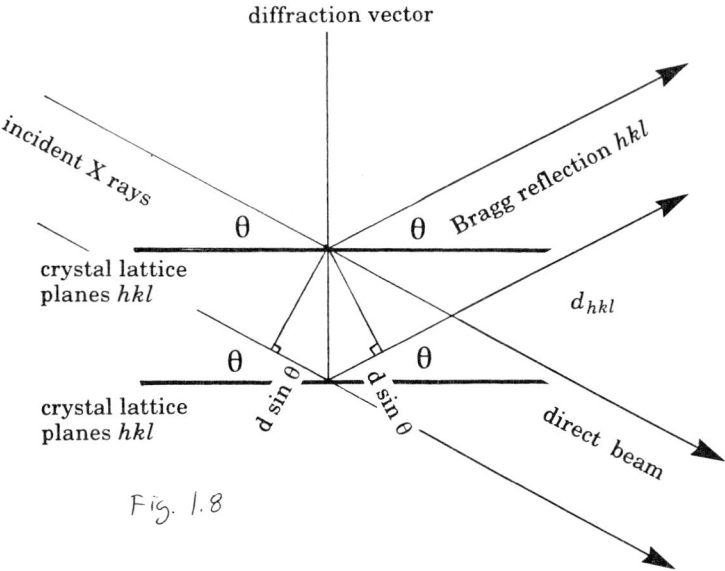

Fig. 8. Bragg's law $n\lambda = 2d_{hkl}\sin\theta$, where hkl defines both the crystal lattice planes and the Bragg reflection.

of the scattering power and position of each atom in the unit cell; *it contains information on the positions of all atoms in the crystal structure.*

W. L. Bragg showed that the angular distribution of scattered radiation from a crystal behaves as if the diffracted beam (hkl) were "reflected" from a plane passing through a crystal lattice point (hkl). Each diffracted beam is considered as a "Bragg reflection" from a lattice plane, so that the angle of incidence of radiation, $(90°-\theta_{hkl})$, equals the angle of reflection (27). As shown in Fig. 8, if the angle between a Bragg reflection consisting of X rays of wavelength λ and the perpendicular to a set of crystal lattice planes is $(90° - \theta_{hkl})$, and if the perpendicular spacing of the lattice planes is d_{hkl}, then:

$$\lambda = 2\, d_{hkl} \sin\theta_{hkl} \tag{1}$$

This is known as *Bragg's Law* and describes the fact that the path differences of the X rays scattered from parallel lattice planes hkl are an integral number of wavelengths. If λ and θ_{hkl} are known, values of d_{hkl} may be determined. When an X-ray beam strikes a crystal, diffraction will occur when, and only when, Bragg's Law is satisfied. The spacing between lattice planes d_{hkl} is a function of the unit cell dimensions and the indices h,k,l of those crystal planes, so that if $2\theta_{hkl}$ is measured for several different Bragg reflections (with different hkl values), the unit-cell dimensions can be found.

3.3. The Reciprocal Lattice and Ewald Sphere

The X-ray or neutron diffraction pattern of a crystal consists of isolated spots, not a continuum (Fig. 9). There is, however, an evident intensity variation across an X-ray diffraction photograph. The spots on the photograph are positioned at the "sampling regions" within the "envelope profile" (which is the diffraction pattern of all the atoms in one unit cell). These sampling regions are arranged on a lattice that is reciprocal to the crystal lattice, and hence, is called the *reciprocal lattice*. The X-ray diffraction pattern of a crystal is the sampling of the X-ray diffraction pattern of the contents of a single unit cell at the reciprocal lattice points.

To construct the reciprocal lattice, one starts with the crystal lattice and, after selecting a crystal lattice point as the origin, looks for series of equally spaced parallel planes through the crystal lattice points, for example, the 10, 01, or 11 series of planes. If the distance between these crystal lattice planes is d_{hkl} (in Å^{-1}), then the distance of the reciprocal lattice point from its origin in a direction perpendicular to the planes is $\lambda/d_{hkl} = d^*_{hkl}$ (see Fig. 10). This operation is repeated for each possible set of planes. The result is a new lattice called the reciprocal lattice. Each set of crystal lattice planes in a crystal structure, specified by the Miller indices h, k, and l, is represented by a single reciprocal lattice point, designated hkl.

Before we can measure the intensity of a Bragg reflection, we need to determine where and from what direction to orient the X-ray detector. A geometrical description of diffraction, the *Ewald sphere*, allows us to calculate which Bragg reflections will be formed if we know the orientation of the crystal with respect to the incident X-ray beam. In the Ewald construction (shown in two dimensions in Fig. 11), a sphere of radius $1/\lambda$ is drawn with the crystal at its center and the reciprocal lattice on its surface. A Bragg reflection is produced when a reciprocal lattice point touches the surface of the Ewald sphere. As the orientation of the crystal is changed, so is the orientation of its reciprocal lattice.

Generally the Bragg's Law is useful in describing the directions, $2\theta_{hkl}$, of diffracted beams, but it provides no information that will help us to understand the

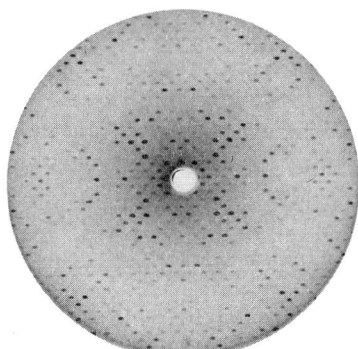

Fig. 9. Diffraction pattern of D-xylose isomerase (Courtesy of H. L. Carrell).

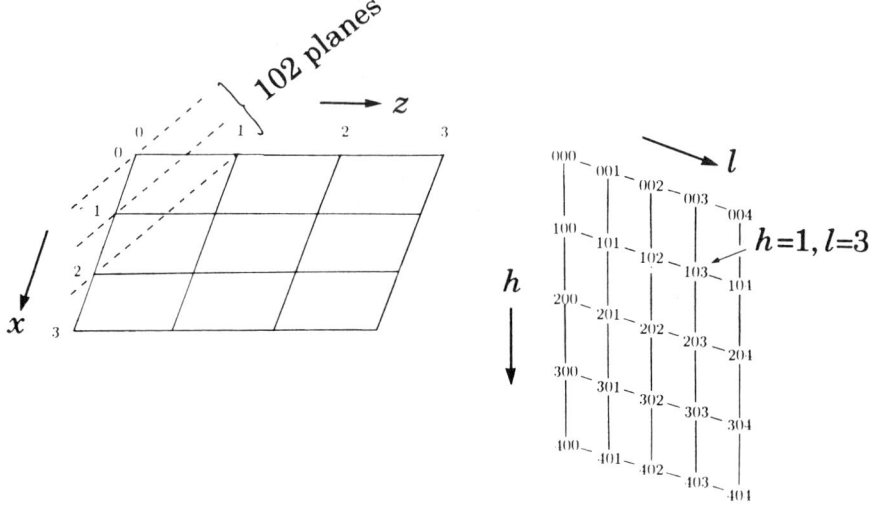

Fig. 10. Constructing a reciprocal lattice. The reciprocal lattice point *hkl* represents the reciprocal of the distance between the crystal lattice planes *hkl*, is perpendicular to these planes, and has distances $d^*(hkl) = \lambda/d(hkl)$.

magnitudes of the intensities (which depend on the arrangement of atoms in the crystal). To this end, the concept of sampling the transform of the contents of one unit cell is much more informative.

4. COMBINING SCATTERED WAVES IN ORDER TO OBTAIN AN IMAGE OF THE DIFFRACTING MATERIAL (SIMULATING AN X-RAY LENS)

In the X-ray diffraction experiment the number of scattered X-ray beams (Bragg reflections) that must be recombined (summed) is large. These would have been focused by an X-ray lens if such a lens could have been devised. This summation is done mathematically, and there are several algebraic representations of waves that are convenient for this. Values for the amplitudes $|F(hkl)|$ of the waves necessary are obtained from the intensities of the Bragg reflections. Values for the relative phases of these waves, however, are not obtained experimentally. Unfortunately, in the summation of a series of waves, the contributions of the relative phase angles are as important as, and generally more important than, the contributions of the amplitudes of the diffracted beams.

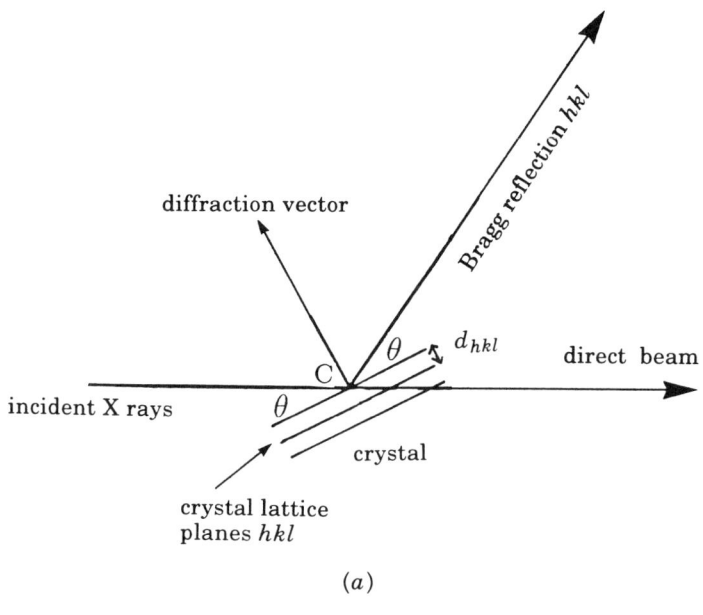

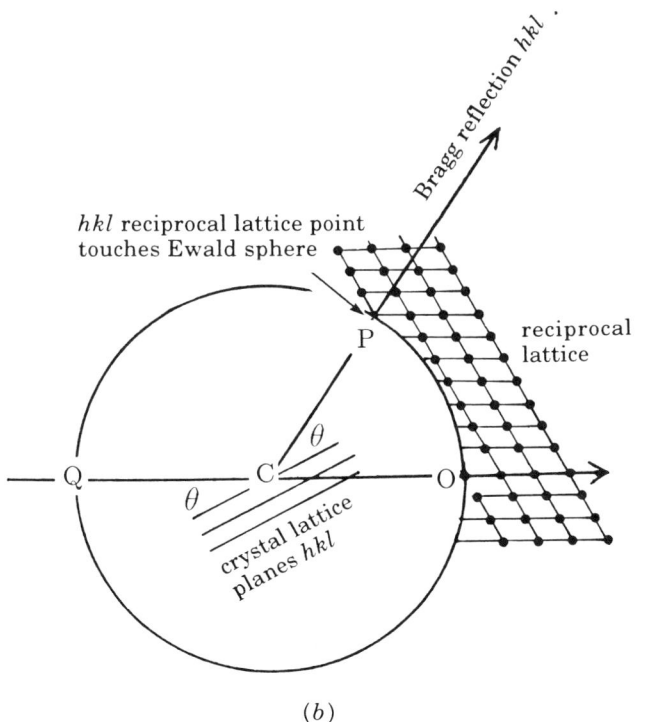

Fig. 11. The Ewald sphere construction, used to determine which Bragg reflections will be obtained for a given orientation of a crystal. (a) The geometry of diffraction. (b) The Ewald sphere and its relationship to the reciprocal lattice.

4.1. Relative Phase Angles

The relative phase of a sinusoidal wave is a measure of a relative position of the crest of the wave measured from some particular vantage point; this may either be the crest of another wave traveling in the same direction, or a chosen origin. In X-ray diffraction, the phase of a Bragg reflection is defined relative to the phase of an imaginary Bragg reflection at the origin of the unit cell, as shown in Fig. 12. The origin of the unit cell of the crystal is merely a convenient geometrical construction that may be chosen in one of several ways. This means that there is no such thing as the "absolute phase" of a Bragg reflection, and the phase we use is one relative to the origin that we have selected. Once the origin of the unit cell is chosen, all atomic coordinates will be given with respect to it, and all phase angles of Bragg reflections will be relative to that chosen origin.

Various methods have been used for the representation of the amplitude and relative phase angle of a Bragg reflection. It is usual to diagram the phase angle on a circle, with 360° as the periodicity (wavelength). By convention this angle is measured in a counterclockwise direction. A line is drawn whose length is proportional to the amplitude and whose orientation gives a measure of the phase angle. If the phase angle is 0°, the line is horizontal; if the phase angle is 90°, the line is vertical (Fig. 13). The line may then be considered a vector $F(hkl)$, with an angle $\alpha(hkl)$ and a length $|F(hkl)|$, the structure factor amplitude. The summation of waves such as these then becomes a geometrical problem, and it can be done vectorially, as shown in Fig. 13.

4.2. The Use of Fourier Series

Any mathematical function that repeats in a regular periodic manner can be represented as the sum of sine and cosine functions of appropriate amplitudes and phases. The periodicities of the terms are the appropriate integral fractions of the

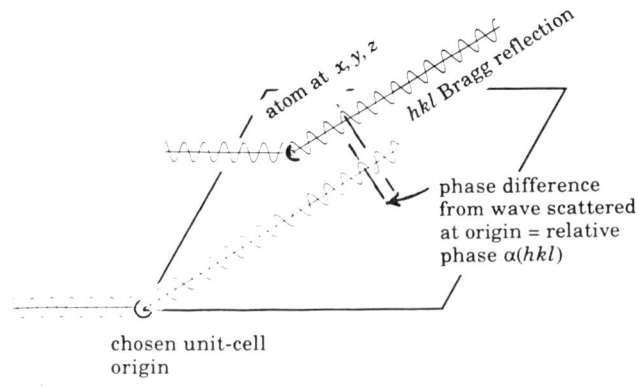

Fig. 12. Definition of a phase angle with respect to an imaginary wave scattered at the unit-cell origin.

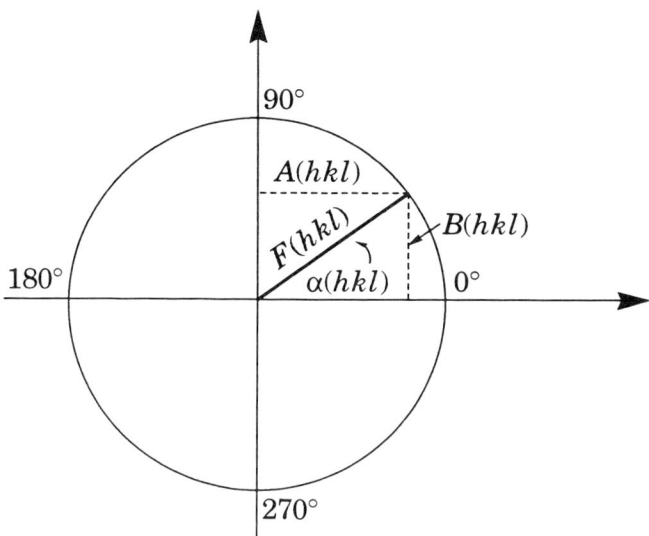

Fig. 13. Representation of a phase angle on a circle. The structure factor $F(hkl)$ can be considered to be composed of a component $A(hkl)$ parallel to the horizontal axis (phase 0°) and a component $B(hkl)$ in a direction perpendicular to it (phase 90°). The ratio of the magnitudes of these two components express the phase angle $\alpha(hkl)$.

repeat unit. These statements constitute the Fourier theorem (28). The electron density in a crystal precisely fits these requirements since an exact repeat of the electron density occurs from unit cell to unit cell. Therefore, a Fourier series is a useful method for representing the electron density in a crystal provided the component terms involved in its calculation are available. The amplitudes and periodicities of the terms are obtained from the intensities of the appropriate Bragg reflections, but the relative phases of these components are not known.

A Fourier analysis involves breaking down a periodic (repetitive) function into its component waves, and then deriving the amplitude, frequency, and phase of each of these component waves. In effect, a Fourier analysis takes place in the diffraction experiment when the scattering of X rays by the electron density in the crystal produces Bragg reflections, each with a different amplitude $|F(hkl)|$ and a relative phase α. Thus, X-ray diffraction gives the components to be summed to give an electron density map (without the necessary relative phase angles).

A Fourier synthesis is a mathematical calculation whereby, in the case of X-ray diffraction, the scattered waves (with correct amplitudes and relative phases) are recombined to give the electron density in the crystal. It is essentially the opposite of a Fourier analysis and is the equivalent of image formation by a lens. It is the stage of the experiment in which the crystallographer and the computer act as the lens of a microscope. Provided the relative phases can be found, an electron density map can be calculated (Fig. 14).

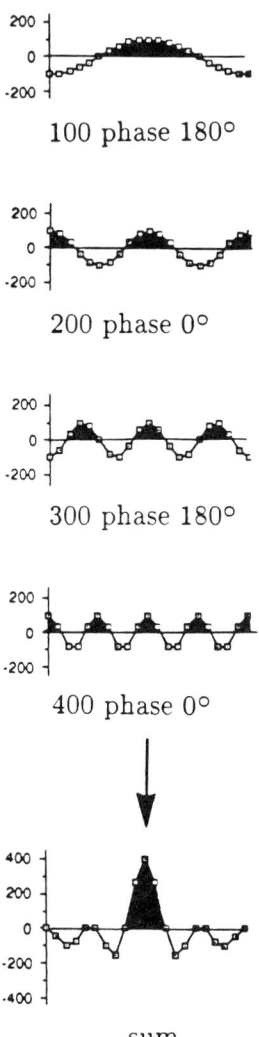

Fig. 14. Combining phases to obtain an electron-density map. Values of h, k, l and the phase angle are given for each term in the summation.

4.3. The Electron-Density Map

The electron density ($\rho(xyz)$) at a point x,y,z in the unit cell is expressed in electrons per cubic Å and is highest near atomic centers. Electron density values can be plotted at constant intervals to give a representation of features of such a map. The equation for the electron density at any point x,y,z in the unit cell then involves $|F(hkl)|$, the structure factor amplitude, and α_{hkl}, the relative phase angle.

$$\rho(xyz) = \frac{1}{V} \sum_h \sum_k \sum_l |F(hkl)| \cos 2\pi(hx + ky + lz - \alpha_{hkl}). \quad (2)$$

If a single point within the unit cell is chosen with fractional coordinates x,y,z (distance xa parallel to a, yb parallel to b, and zc parallel to c), the electron density, $\rho(xyz)$, at that point is given by Equation 2. The right-hand side of this equation involves a summation that includes all measured Bragg reflections. If, for example, the intensities of 6,000 diffracted beams are measured, there will be 6,000 $|F(hkl)|$ values included in the summation in Equation 2 for just one point in the electron-density map. The summation must be repeated for each electron density point, x, y, and z, and it is usual to choose ⅓ to ½ of the minimum d spacing of the measured data as the spacing between points at which the electron density is computed.

If the electron density of a crystal could be accurately described by a single cosine wave that repeats three times in the unit cell dimension, d, that is, has a periodicity of $d/3$, its diffraction pattern would have intensity only in the third order (only one Bragg reflection, 3 0 0). Conversely, if only one order of the diffraction spectrum is observed (the Bragg reflection $h = 3$, for example), then the diffracting density amplitude must correspond to a cosine wave with frequency d/h ($d/3$) (29, 30). This can be considered as an "electron-density wave," one of the components summed to give an electron density map. Each Bragg reflection provides an "electron-density wave" that contributes to Equation 2, the total electron density. In this way the relationship between the order (hkl) of a Bragg reflection and its contribution to the electron density is established.

The Bragg reflections that we are summing are X rays (electromagnetic radiation) that have a constant wavelength and frequency. These are different from the electron-density waves in Equation 2. To avoid confusion, it is best to think of the electron-density waves, the mathematical components of Equation 2, as having periodicity rather than frequency or wavelength. Thus, the contribution of the (400) Bragg reflection to the electron-density map is a wave that has four crests in the length a. Each Bragg reflection contributes an electron-density wave of known amplitude, periodicity, and relative phase to the total electron density (29–31).

4.4. Fourier Transforms (Between Crystal and Diffraction Space)

We have shown that the electron density can be expressed as a Fourier series with the structure factors as coefficients. In an analogous way, the structure factors can be expressed in terms of the electron density. There is a mathematical way of expressing these analogies, and it involves Fourier transforms (Fig. 15). The electron density is the Fourier transform of the structure factor, and the structure factor is the Fourier transform of the electron density. If the electron density can be expressed as the sum of cosine waves, that is, a Fourier series, its Fourier transform gives a function with high values at positions corresponding to intensity in the diffraction pattern. The Fourier transform provides the possibility of using a

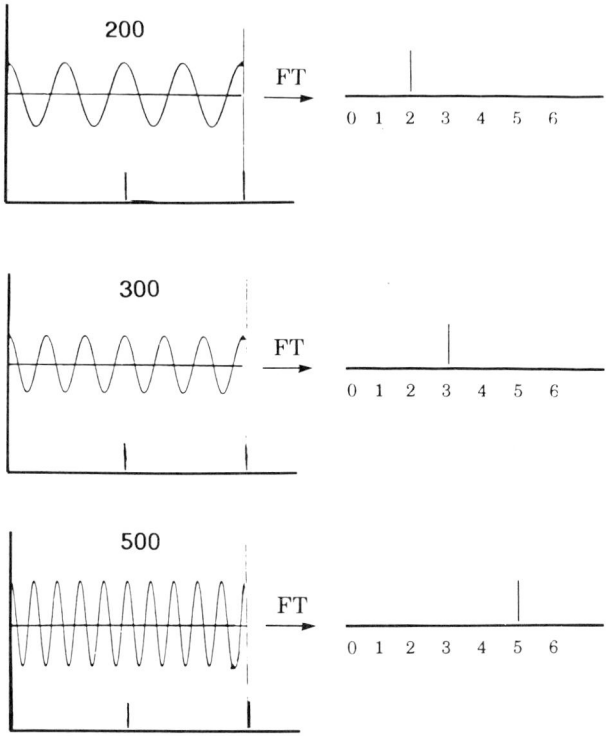

Fig. 15. The Fourier transform. Two unit-cell length are shown on the left and the Fourier transforms of their contents on the right.

mathematical expression to go back and forth between reciprocal space (structure factors) and real space (electron density) (32). Thus, if the electron density is known correctly, then structure factors and their relative phases can be computed by Fourier transform techniques, and vice versa.

4.5. Effects of the Atomic Arrangement

The structure factor amplitude (magnitude) $|F(hkl)|$ is the ratio of the amplitude of the radiation scattered in a particular direction by the contents of one unit cell to that scattered by a single electron at the origin of the unit cell under the same conditions. The structure factor $F(hkl)$ has both a magnitude (amplitude) $|F(hkl)|$ and a phase angle α_{hkl} measured relative to the origin of the unit cell. The structure factor may also be defined as the Fourier transform of the unit cell contents sampled at reciprocal lattice points, hkl; $|F(hkl)|$ is derived from the square root of the intensity of a Bragg reflection. The experimentally measured "observed" structure factor amplitudes are denoted by $|F_o(hkl)|$; those calculated for a proposed crystal structure (a model) are designated $|F_c(hkl)|$.

Once x_j, y_j, and z_j are known for each atom in the unit cell, it is possible to calculate F_{hkl} and its components, $|F_c(hkl)|$ and $(\alpha_{hkl})_c$. This calculation is broken into two parts, one, A_{hkl}, involving a cosine function and the other, B_{hkl}, involving a sine function (Fig. 16). For each atom the scattering factor f_j at the value of $\sin\theta/\lambda$ appropriate for the diffracted beam, hkl, is multiplied by a cosine or sine function that contains h,k,l and x,y,z in it. This is done for each atom in the structure, and the values are all summed to give A_{hkl} and B_{hkl}. This entire computation is then repeated for every Bragg reflection.

4.6. Effects of Atomic Displacements and Vibrations

A precise register of the positions of atoms from unit cell to unit cell is not found in practice, and deviations from this exact register, referred to as "disorder," increase as the temperature is raised because atomic vibrations are thereby made larger. What is obtained in an X-ray diffraction photograph is the equivalent of a snapshot of vibrating molecules, each with atoms in different unit cells displaced randomly to different extents from their average positions, because the frequencies of X rays are four orders of magnitude greater than the frequencies of atomic vibrations.

If the molecule vibrates much, or is disordered in some way or another so that atoms in it lie in slightly different locations from unit cell to unit cell, the overall view of the molecule becomes more fuzzy, and therefore the apparent size of an

For an atom j at x, y, z, with scattering factor f_j at the $\sin\theta/\lambda$ value of the Bragg reflection hkl

the structure factor $|F(hkl)|$ is calculated from:

$$[\,|F(hkl)|\,]^2 = [A(hkl)^2 + B(hkl)^2\,]$$

where:

$$A(hkl) = \Sigma f_j \cos 2\pi(hx_j + ky_j + lz_j)$$

and

$$B(hkl) = \Sigma f_j \sin 2\pi(hx_j + ky_j + lz_j).$$

The phase angle $\alpha(hkl) = \tan^{-1}[B(hkl)/A(hkl)]$

Fig. 16. Formulae for calculating structure factors when atomic positions are known.

individual atom is increased. We stressed earlier that a larger size in real space means a smaller size in reciprocal space. Thus, as the atoms appear to become broader and fuzzier, the diffraction pattern decreases in extent and becomes narrower, and the diffracted beams decrease in intensity (Fig. 17). This fall-off can be approximated by an exponential factor, $\exp[-B_j(\sin^2\theta/\lambda^2)]$, that is, applied to the scattering factors, where B_j, the displacement parameter for an atom, is related to the root-mean-square amplitude of atomic vibration, $\bar{u}^2$:

$$f_j = f_{jo} e^{-B_j(\sin^2\theta/\lambda^2)} \tag{3a}$$

where f_{jo} is the scattering factor for a stationary atom at the same value of $\sin\theta/\lambda$, and

$$B = 8\pi^2 \bar{u}^2 \tag{3b}$$

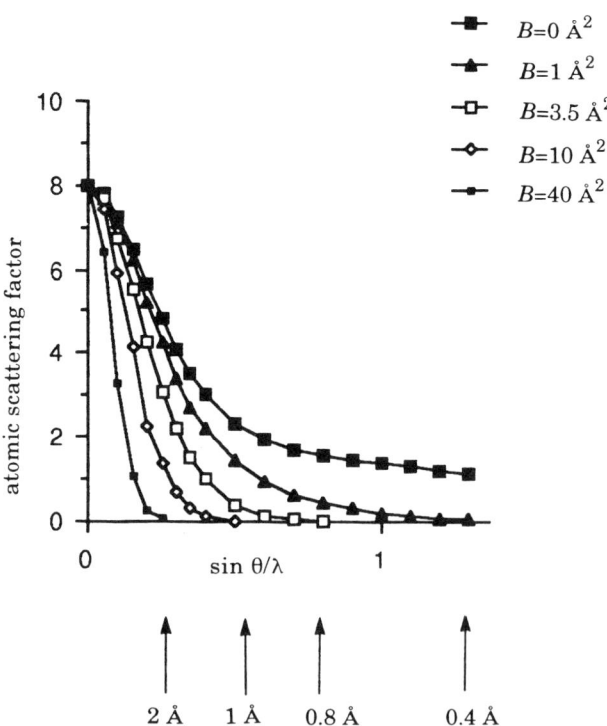

Fig. 17. The effect of atomic vibrations on the scattering factor for an oxygen atom. If the vibration is high enough, data will not be measurable beyond the resolution indicated at the bottom of the figure.

5. MEASURING DIFFRACTION DATA [$|F(hkl)|$]

The apparatus we use today for measuring the intensities of Bragg reflections (Fig. 18) consists of the same three components that were used in the first diffraction experiment (1).

1. A source of X rays (or neutrons),
2. A crystal appropriately placed in a beam of this radiation, and
3. A means of detecting and recording the spatial direction and intensity of each Bragg reflection.

5.1. The Diffraction Equation

Diffraction occurs when a reciprocal lattice point passes through the Ewald sphere and satisfies the Bragg law (Equation 1). The measured intensity, $I(hkl)$, of a diffracted X-ray beam can be calculated for a crystal rotating with a uniform angular velocity, ω, through a reflecting position

$$I(hkl) \text{ is proportional to } \frac{I_o \lambda^3 V_x \cdot L \cdot p \cdot A}{\omega V^2} |F(hkl)|^2. \qquad (4)$$

This equation shows that the intensity, $I(hkl)$, of a diffracted beam is proportional to the square of the structure factor amplitude, $|F(hkl)|$. The intensity of the incoming incident X-ray beam, I_o, and the wavelength λ of the radiation are selected by the experimenter. V is the unit cell volume. As V_x, the volume of the crystal in the incident X-ray beam, increases, so do the intensities of the diffracted beams. The Lorentz factor L takes into account the relative time each reflection is in the diffracting position. It is a geometric correction, calculated as a function of the scattering angle 2θ. The polarization factor p accounts for the polarization of the X-ray beam after diffraction and, like the Lorentz factor, is a function of the scattering angle. The absorption factor A accounts for the absorption of each diffracted beam as it travels through the crystal. The correction for absorption is a function of the wavelength of the radiation, the atomic contents of the unit cell, and the path length of each diffracted beam through the crystal.

5.2. Sources of X rays

When choosing X rays for a diffraction study the longer wavelength radiation will give a better separation of Bragg reflections and less radiation damage to the crystal, while shorter wavelength radiation scatters to a higher resolution and is absorbed less readily. The radiation generally chosen for protein studies is the characteristic CuKα radiation, obtained when high-speed electrons hit a copper anode, ejecting electrons from the metal atoms. As L-shell electrons move to K-shells, characteristic radiation is emitted, wavelength 1.5418 Å. The X-ray beam used for diffraction studies is carefully collimated by a hollow, straight metal tube designed to produce a narrow, nearly parallel, beam of radiation so that all

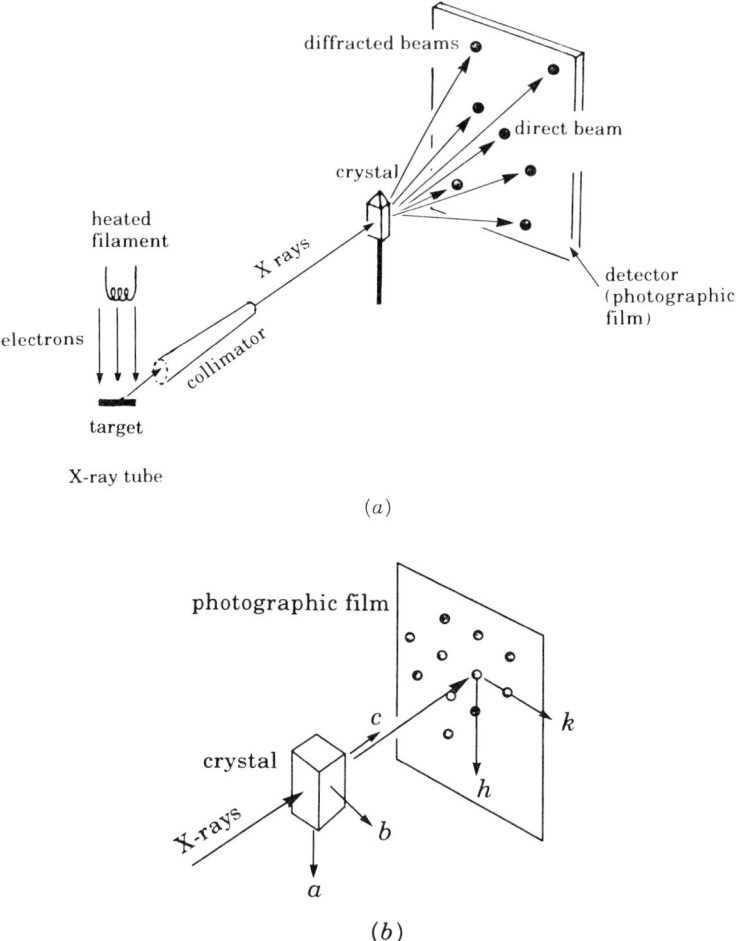

Fig. 18. (a) The general setup of apparatus used to measure X-ray diffraction intensities. In practice, film is often replaced by electronic or other detection devices. (b) The relationship of the chosen unit-cell axes to the indices of Bragg reflections in the diffraction pattern. An orthorhombic crystal is shown for simplicity.

points on the crystal may be completely bathed in X rays of a uniform intensity.

Intense sources of radiation increase I_o in Equation 4 and give better diffraction because the peak-to-background ratio is increased. This intensity is obtained with rotating anode generators. Since so much heat is produced in an X-ray tube, the anode is rotated at high speed, and the electron beam is directed to the outer edge of this rotating target so that the heat can be dissipated more readily. Even more intense radiation is obtained from synchrotron sources. Synchrotron radiation is extremely intense X radiation emitted by very high-energy electrons that travel at nearly the speed of light and have their direction of travel bent in a circular path by powerful magnets. The radiation so obtained consists of pulses (on a nanosecond scale) with a high degree of polarization and has a continuous spectral distribution (is polychromatic) so that it can be "tuned" to a selected wavelength.

There are three accessories used to produce monochromatic (one-wavelength) radiation: metal foil filters, crystal monochromators, and focusing mirrors. An element with atomic number Z can be used as a selective filter for radiation produced by an element of atomic number $Z + 1$. Alternatively, an intense Bragg reflection from a crystal (a monochromator crystal) can be used as the incident beam for X-ray diffraction studies. Third, focusing mirrors, designed to produce a beam that is not only monochromatic but also convergent, may be used. In this case the incident beam is doubly deflected by two perpendicular metal sheets.

5.3. Detectors of X rays

Films, counters, and imaging plates are commonly used to record X-ray diffraction data. They are each highly sensitive to X rays and can provide a precise measure of the intensities of the diffracted beams. X rays, like visible light, interact with the silver halide contained in the emulsion of photographic film. When the film is developed, black metallic silver is deposited at the positions at which the diffracted rays have hit it. The extent of the blackening by silver is directly proportional to the intensity of the diffracted beam that hit it. This blackening is measured by use of a photometer, which enables the optical density to be measured and recorded at regularly spaced grid points on the film, giving numerical values for the extent of blackening (and hence the incident X-ray intensity) at each measured point.

Scintillation counters, Geiger counters, and proportional counters directly measure the intensity of a diffracted X-ray beam, as opposed to a photometric device that measures the intensity of the effect of that beam on photographic film. Counters, however, are involved in sequential measurement, one Bragg reflection after another, so that the crystal needs to be reasonably stable in the X-ray beam for the data to be precise over the long periods of time required for data measurement. During such measurements, it is necessary to relate each Bragg reflection to its reciprocal lattice index (h,k,l). This requires information on the orientation of the crystal with respect to a set of fixed orthogonal axes that are related to the geometry of the detection device (33). Each reciprocal lattice point hkl of the crystal can be related to the camera system by Equation 5:

$$\text{(orientation matrix of crystal)} \times \text{(orthogonalization matrix to camera geometry)} \times (hkl). \quad (5)$$

This equation makes it possible to calculate the relative orientation of crystal and detector device for each Bragg reflection, hkl, so that the detector can be oriented and then the intensity of the diffracted beam can be measured.

The area detector is an electronic device for measuring many diffracted intensities at one time. It is a two-dimensional, position-sensitive detector that records the intensity of a Bragg reflection (diffracted beam) and its precise direction (as a location on the detector); it acts like an electronic substitute for film. This detection device is now used extensively for crystals of biological macromolecules. Such a detector may involve a multiwire proportional counter coupled to an electronic device or a television imaging system; both devices permit a recording of the data in a computer-readable form. Alternatively, imaging plates may be used. These have phosphorescent material layered on them and store information on the extent of X-ray exposure until scanned by a laser, when the intensity and location of the light then emitted is recorded.

There are two sources of error that have to be taken into account when assessing the experimental error in a measured intensity. Errors can arise from the random fluctuations in the source of radiation and in the detection system; these errors follow a Poisson distribution and are proportional to the square root of the measured value (the count, hence the term *counting statistics*). The second type of error to be considered is the instability of the instrument and the crystal. This is monitored by the measurement of a few chosen Bragg reflections at regular intervals during data collection. Any fall-off in intensity as a function of time is considered to indicate crystal decay. This problem is usually obviated for proteins by measuring large numbers of Bragg reflections at the same time, as in the Laue method (described later); in this case all Bragg reflections were measured with the same state of the crystal and radiation source.

5.4. Preparing a Crystal for Diffraction Study

The results of an X-ray diffraction experiment are only as good as the quality of the crystal and the intensity data that result from it. Time and care invested in obtaining a good crystal and setting up a careful experiment for measuring diffraction data result in more precise results in terms of a molecular model. Crystals may be examined under a polarizing microscope for imperfections such as cracks or voids, and to make sure that they are single. If the crystal is not single, the diffraction pattern will be a superposition of two or more patterns, in differing orientations, and may be very difficult to interpret.

A good diffraction pattern from protein crystals generally requires mounting the crystal in contact with its mother liquor inside a thin-walled glass capillary tube along with a drop or two of the mother liquor so that an equilibrium atmosphere is maintained (Fig. 19). The capillary containing the crystal is then mounted on the goniometer head. Alternatively, a flow cell, in which liquid with a controlled composition is allowed to flow over a crystal, can be used to maintain the required crystal environment.

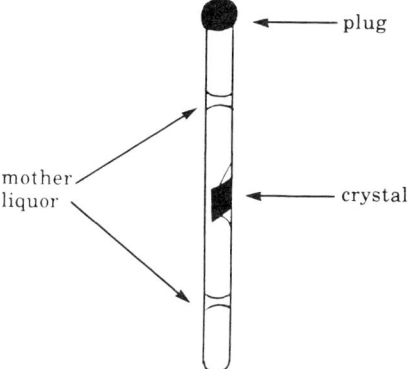

Fig. 19. The mounting of a protein crystal in a capillary tube.

5.5. Measurements at Low Temperatures

Lower temperatures reduce the rate of decay of protein crystals. In addition, thermal motion is reduced at lower temperatures, so that Bragg reflection data at higher scattering angles are enhanced in intensity. This increases the effective resolution of the data. Macromolecules may be studied at low temperatures, preferably by mounting them on a fiber rather than in a capillary tube. It is important that the temperature be kept within a small range ($\pm$ K) during the data collection, and that steps be taken to prevent the crystal from receiving a coat of ice from, moisture in the environment.

5.6 Laue Diffraction

A Laue photograph is produced by irradiating a stationary crystal with a beam of X rays that has a wide range of wavelengths ("white" radiation). It differs from all of the other methods for collecting diffraction data in that the crystal is stationary throughout the experiment. Diffraction is therefore dependent on the multiwavelength feature of the incident beam. With the current accessibility of synchrotron radiation, Laue methods are often used for macromolecular studies, and it is possible to collect large numbers of Bragg reflections on a single photograph in very short times (34, 35).

5.7. Problems with Large Structures

Large macromolecular structures have specific experimental requirements that result from the greater size of the unit cell, which, in turn, leads to a high number of diffracted Bragg reflections. If the unit cell is very large, reciprocal lattice points may be so close together that Bragg reflections will overlap, making it very difficult to obtain precise intensity measurements. In addition, for large molecules there are fewer unit cells per crystal leading to weaker intensities. The use of rotating anodes or synchrotron radiation sources, the use of longer wavelength radiation, increasing the crystal-to-film distance, and sophisticated analyses of the shapes of the measured Bragg reflections (peak profiles) when these are scanned instrumentally, are typical ways used for overcoming these experimental problems.

5.8. Putting the Intensity Data on an Absolute Scale—The Wilson Plot

For each Bragg reflection, the raw data normally consist of the Miller indices (h,k,l), the integrated intensity $I(hkl)$ and its standard deviation $[\sigma(I)]$, and the direction in which the diffracted beam traveled. The conversion of $I(hkl)$ to $|F(hkl)|$ involves the application of corrections for X-ray background intensity, Lorentz and polarization factors, absorption effects, and radiation damage (see Equation 4). The absorption correction is complicated by the glass from the capillary and the liquid from mother liquor that the X-ray beam encounters. Therefore an experimental measure of the variations in this correction is usually made (36). This process of obtaining values of $|F(hkl)|$ is known as *data reduction*.

Structure amplitudes $|F(hkl)|$ are on an absolute scale when they are expressed relative to the amplitude of scattering by a single electron under the same conditions. In order to obtain the necessary scale factor, the average intensity from a crystal, as a function of scattering angle, is compared with the theoretical values to be expected for a completely random arrangement of the same atoms in the same unit cell:

$$K<I(hkl)> = \sum f_j^2 \{\exp[-2B \sin^2\theta/\lambda^2]\} \tag{6a}$$

$$\ln(<I(hkl)>/\sum f_j^2) = -\ln K - 2B \sin^2\theta/\lambda^2, \tag{6b}$$

where $<I(hkl)>$ is the mean value of $I(hkl)$ in the chosen range of $\sin\theta$ and $\sum f_j^2$ is the sum of the squares of the scattering factors of the protein atoms. This graph, known as a *Wilson plot* (37), provides both the scale factor K and an average temperature factor B for the crystal under study. Since the disposition of atoms in proteins is not entirely random, there is appreciable scatter of points in their Wilson plots.

6. MEASURING UNIT-CELL DIMENSIONS AND SPACE GROUP

6.1. Unit Cell Dimensions

Unit cell dimensions are obtained from measurements of 2θ values of several Bragg reflections for which the indices h, k, and l are known. Values of $2\theta_{hkl}$ are measured as accurately as possible and, since the wavelength λ, of the radiation used is known, a value of d_{hkl} may be found by Bragg's equation. The value of d_{hkl}

is related to the unit cell dimensions and, if $2\theta_{hkl}$ values are measured for several reflections, values of the unit cell dimensions may be derived. The selected group of reflections chosen to do these calculations should contain a distribution of Miller indices (h,k,l), and they should have relatively high $2\theta_{hkl}$ values.

6.2. Space-Group and Crystal Symmetry

The internal symmetry of the crystal is revealed in the symmetry of the Bragg reflection intensities and this symmetry will give information on the crystal system and space group. Friedel noted that the intensity distribution in the diffraction pattern is centrosymmetric (38).

$$\text{Friedel's Law: } I(hkl) = I(\overline{hkl}). \tag{7}$$

As a result, there are Bragg reflections with different (but related) indices that have identical intensities. The only exception to Friedel's Law is found if atoms scatter radiation anomalously, as discussed later.

There may also be more symmetry in the intensities in the diffraction pattern than that implied by Friedel's Law. This additional symmetry in the diffraction pattern is called *Laue symmetry* (because it can be displayed on Laue X-ray diffraction photographs of an appropriately aligned crystal). For example, if a crystal is monoclinic, then the intensities $I(hkl)$ and $I(h\overline{k}l)$ are the same, although $I(hkl)$ does not equal $I(\overline{h}kl)$. Orthorhombic crystals, with three mutually perpendicular twofold rotation axes, have more symmetry than monoclinic crystals and their intensities $I(hkl) = I(\overline{h}kl) = I(h\overline{k}l) = I(hk\overline{l})$. If, by chance, the crystal is monoclinic with the β angle equal to 90°, the Laue symmetry in the X-ray diffraction pattern will show that $I(hkl)$ does not equal $I(\overline{h}kl)$. It is the symmetry of the diffraction pattern that tells us that a crystal lattice is orthorhombic, not the fact that $\alpha = \beta = \gamma = 90°$.

Systematic absences in the diffraction pattern show that there are translational symmetry elements relating components in the unit cell (Fig. 20). The intensity of every member of a particular group of Bragg reflections may be zero, for example, $hk0$ when $h + k$ is odd. A twofold screw axis, parallel to the unit cell translation, a, will convert an atom at x to one at $x + ½$. This repetition every $a/2$ implies apparent halving of the spacing between x values for atoms (y and z do not contribute to this argument). Therefore, in this case, a Bragg reflection $h00$ will be absent if h is odd, provided all atoms in the structure are repeated at $a/2$. The spacing in reciprocal space has been doubled, $2h$, because the repeat unit in real space has been halved, $a/2$.

A careful tabulation of systematic absences in the X-ray diffraction pattern allows the space group of the crystal to be assigned. This, after the determination of unit cell dimensions, is the next piece of information that the X-ray crystallographer obtains. Space group determination is done systematically considering each class of Bragg reflection ($h00$, $0k0$, $00l$, $hk0$, $h0l$, $0kl$, hkl) and checking that examples of reflections in these classes with appreciable intensities for even and odd values of each index either are observed or are consistently

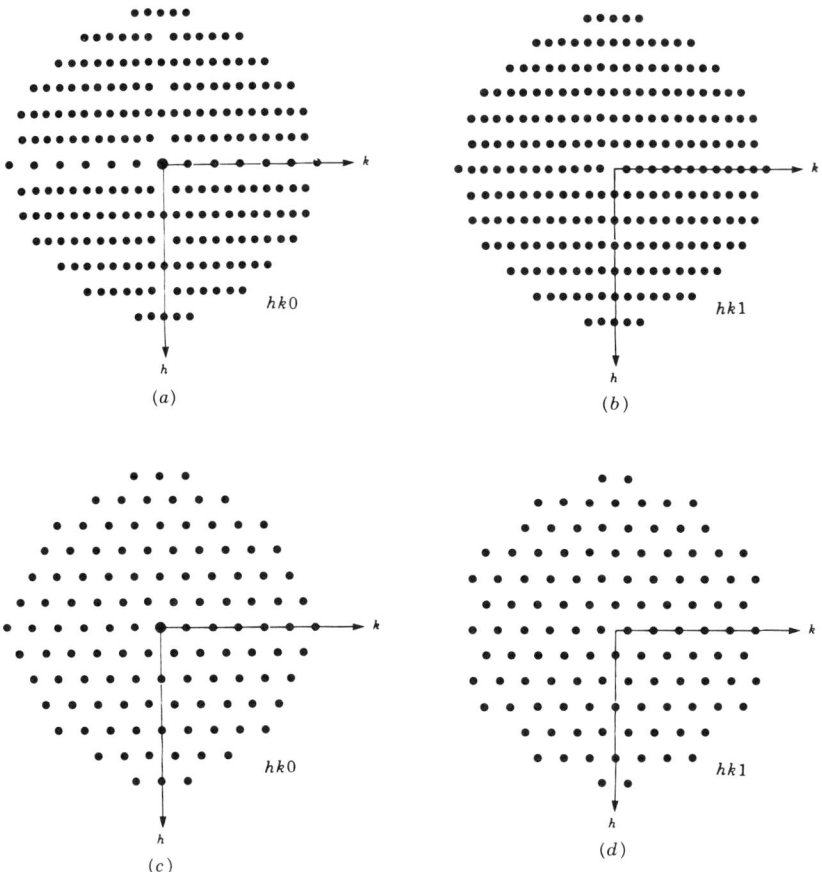

Fig. 20. The use of sysematic absences in X-ray diffraction patterns to determine the space group of the crystal. (a) $P2_12_12_1$ $hk0$, (b) $hk1$, (c) $I222$ or $I2_12_12_1$, $hk0$, and (d) $hk1$. In the second case there is a space-group ambiguity.

absent. There are 230 possible space groups for crystals, but if the molecule has asymmetry, as for proteins, this number is reduced to 65 (Table 2). The crystallographer can determine the space group by consulting the appropriate page from the *International Tables* (24). Sometimes two or more space groups fit the conditions imposed by crystal symmetry on the Bragg reflections, giving rise to a space group ambiguities.

A knowledge of the space group symmetry is essential for the structure determination because only the contents of the asymmetric unit, combined with the relationships between positions of atoms listed for the space group, are needed

TABLE 2

Space Groups of Unsymmetrical (Chiral) Objects

	Noncentrosymmetric, Chiral Molecules		
1	Triclinic	C_1	$P1$
3–5	Monoclinic	C_2	$P2, P2_1, C2$
16–24	Orthorhombic	D_2	$P222, P222_1, P2_12_12, P2_12_12_1, C222_1, C222, F222,$ $I222, I2_12_12_1$
75–80	Tetragonal	C_4	$P4, P4_1, P4_2, P4_3, I4, I4_1$
89–98		D_4	$P422, P42_12, P4_122, P4_12_12, P4_222, P4_22_12, P4_322,$ $P4_32_12, I422, I4_122$
143–146	Trigonal	C_3	$P3, P3_1, P3_2, R3$
149–155		D_3	$P312, P321, P3_112, P3_121, P3_212, P3_221, R32$
168–173	Hexagonal	C_6	$P6, P6_1, P6_5, P6_2, P6_4, P6_3$
177–182		D_6	$P622, P6_122, P6_522, P6_222, P6_422, P6_322$
195–199	Cubic	T	$P23, F23, I23, P2_13, I2_13$
207–214		O	$P432, P4_232, F432, F4_132, I432, P4_332, P4_132, I4_132$

to calculate $F(hkl)$. The space group $P2_12_12_1$ has only one asymmetric unit that needs to be studied. The symmetry operations: x,y,z; ½$-x,-y$,½$+z$; ½$+x$, ½$-y,-z$; $-x$, ½$+y$,½$-z$, give the rest of the contents of the unit cell and of the crystal structure.

6.3. Crystal Density and Unit-Cell Contents

The weight of all the atoms or ions in the unit cell can be calculated if the unit cell dimensions and the density of the crystal are known. This means that crystal density measurements are important in the early stages of a crystal structure analysis. A vertical column with a density gradient, calibrated with crystals of known density, is used to measure protein densities. These are determined by where along the column the crystal settles and so that, by interpolation, its density can be estimated (39). This method must be done with care to ensure that the measurement is good. The density of a crystal is given by

$$D_x = \frac{M_c \times n}{V \times N_{Avog}} \qquad (8)$$

where M_c is the sum of the atomic weights of all atoms in the asymmetric unit, V is the volume of the unit cell in Å^3, n is the number of asymmetric units per unit cell and N_{Avog} is Avogadro's number (6.022×10^{23}).

Most macromolecules crystallize with about 50% or more solvent of crystallization. If an approximate value for the molecular weight is known, then the number of subunits per cubic Å may be found. Matthews (40) introduced a useful measure V_m, the ratio of the unit cell volume to its molecular weight content. The value of V_m varies between 1.7 and 3.5 Å^3 per dalton and gives a measure of the molecular weight of protein in the unit cell. If the value of V_m, obtained from the measured density lies outside this range, it needs adjustment by an integral factor n that indicates how many protein subunits there are in the asymmetric unit.

6.4. The Experimental Data Set

The data set is now available, generally as a file in a computer-accessible storage device. The relevant data for each Bragg reflection are h, k, and l, the observed structure factor magnitude $|F|$ and its estimated standard deviation $\sigma(F)$. Additional information for each reflection in the Bragg reflection data file includes whether the reflection is above or below a threshold value (observed or unobserved), $\sin\theta/\lambda$ value of the Bragg reflection, and atomic scattering factors, f_j, for each atomic type j in the crystal. Other information in the computer data file include unit cell dimensions with their estimated standard deviations and the space-group symmetry.

7. PHASE DETERMINATION

The principal method used to determine the relative phases of a biological macromolecule is the method of isomorphous replacement. Phases can be estimated by comparing intensities of isomorphous (isostructural) crystals that differ only in the identity of one atom, and otherwise contain identical atomic arrangements. The isomorphism is generally between the crystalline macromolecule and its heavy-atom derivative obtained by replacing some of the solvent in the crystal by a compound containing a heavy atom.

Two other methods for deriving relative phases are still in the developmental stage. One method involves experimental measurements of scan profiles of "multiple Bragg reflections" (41). These Bragg reflections occur when the primary diffracted beam is rediffracted in the crystal, denoted by two reciprocal lattice points lying simultaneously on the Ewald sphere. Intensity changes occur, which may yield some phase information. The instrumentation is complicated, and the method requires much experimental care. A second method requires diffraction data at or near atomic resolution, a situation rarely possible for proteins. Relative phase angles can be estimated by statistical methods that are based on the concept that the electron density is never negative and that it consists of isolated, sharp peaks at atomic positions (42). These statistical methods are called *direct methods*. This is the present method of choice for small molecules, but its use for macromolecules is only just being investigated.

7.1. Preparation of Heavy-Atom Derivatives of Proteins

The method of isomorphous replacement is the primary method used to determine phases for protein structures. Isomorphism is the similarity of crystal shape, unit cell dimensions, and structure between substances of nearly, but not completely, identical chemical composition. The arrangements of atoms in the isomorphous crystals are the same, but the identity of one or more atoms in this arrangement has been changed. Ideally, isomorphous compounds are so closely similar in composition that they can form a continuous series of solid solutions. The best known examples are provided by the alums $(M_1)_2(SO_4) \cdot (M_3)_2(SO_4)_3 \cdot 24H_2O$, where M_1 is a monovalent cation such as potassium or ammonium and M_3 is a trivalent cation such as aluminum or chromium (43).

Isomorphous replacement is now employed in the determination of the structures of biological macromolecules (44, 45). Proteins crystallize with 50% or more of the crystal volume filled with solvent. Compounds containing heavy atoms can be diffused into the crystals through the solvent channels when they are added to the crystallizing mother liquor. Hopefully, the heavy (metal) atom will settle on preferred sites on protein molecules in the crystal. If the crystal still has the same unit-cell dimensions, it is considered isomorphous to the unsubstituted ("native") crystal. The intensities in the diffraction patterns of the native crystal and the heavy-atom derivative are then compared. Some commonly used heavy-atom compounds include uranyl nitrate, p-chloromercuribenzoic acid, and platinum complexes such as $Pt(ethylenediamine)Cl_2$. Attention must be paid to pH, since insoluble hydroxides form at high pH, and protein side chains are protonated at low pH. In addition, concentrations of heavy atom derivatives should not be too high, generally in the 0.001–0.1 M range (8). The method of isomorphous replacement, when applied to a protein, requires the preparation of several heavy-atom derivatives, each with metal attached to different sites on the protein.

7.2. The Patterson Map

All methods of deduction of the relative phases for Bragg reflections from a protein crystal depend, at least to some extent, on a Patterson map, commonly designated $P(uvw)$ (46, 47). This map can be used to determine the location of heavy atoms and to compare orientations of structural domains in proteins if there are more than one per asymmetric unit. The Patterson map indicates all the possible relationships (vectors) between atoms in a crystal structure. It is a Fourier synthesis that uses the indices h,k,l, and the square of the structure factor amplitude $|F(hkl)|$ of each diffracted beam. This map exists in vector space and is described with respect to axes u, v, and w, rather than x,y,z as for electron-density maps.

$$P(uvw) = \frac{1}{V} \sum_h \sum_k \sum_l |F(hkl)|^2 \cos 2\pi(hu + kv + lw). \tag{9}$$

This equation has the same form as Equation 2 for electron density, but note that there is no phase angle in the expression!

A peak at location u,v,w in the Patterson map represents a vector from the origin of the Patterson function 0,0,0 to the point u,v,w. A Patterson map is centrosymmetric, irrespective of whether or not the space group of the crystal studied is centrosymmetric or noncentrosymmetric. Any pair of atoms yield two vectors, one at u,v,w, and one at $-u,-v,-w$; atom 1 to atom 2 and atom 2 to atom 1. If any two atoms in the unit cell are separated by a vector u,v,w, then there will be a corresponding peak in the Patterson map at u,v,w (Fig. 21). This peak represents the vector between two atoms, one at x_1,y_1,z_1 and one at x_1+u,y_1+v,z_1+w; it tells us what uvw are, but not what x_1,y_1,z_1 are. All other interatomic vectors appear as peaks in the Patterson map, each vector having one end at the origin of the map; thus for N atoms there are N^2-N peaks in the Patterson map, and there result too many overlapping peaks to be useful for interpretation. The heights of the peaks are approximately proportional to the products of the atomic numbers $Z_i\,Z_j$ of the atoms at the two ends of the vector. The peak at the origin of the map, representing the vector between each atom and itself, has a value proportional to $\sum Z_i^2$.

A Patterson map has been likened to the impressions of 100 strangers at a cocktail party (48). While there are only one hundred invitations, there are 5000 introductions to be made. The concept of the interatomic vectors and their directionality is likened to the guests having their shoes nailed to the ground and having to twist in different directions and extend their arms by differing lengths, with differing strengths of grip, in order to greet everyone. These handshakes are equivalent to the vectors between atoms in the Patterson map.

High peaks in the Patterson map represent either vectors between heavy atoms or the overlap of several interatomic vectors, and can be analyzed to give the fractional atomic coordinates of the heavy atoms. If there is only one heavy atom in the asymmetric unit, the interpretation of the Patterson map is simplified since, to a first approximation, the map is dominated by heavy-atom–heavy-atom vectors. For example, in the space group $P2_1$, the general atomic positions are:

$$x, y, z \quad \text{and} \quad -x, \tfrac{1}{2}+y, -z.$$

The vectors between symmetry-related atoms can be found by subtracting the coordinates of one atom from that of another atom at a symmetry-related site. In this particular case the vectors between like atoms in different asymmetric units lie at:

$$u = 2x, \quad v = \tfrac{1}{2}, \quad w = 2z.$$

Apart from the high peak at the origin (the sum of the vectors from each atom to itself), there should be one high peak in the Patterson map at $v = \tfrac{1}{2}$, and the position of this peak will give values for x and z of the heavy atom in the unit cell (with the screw axes at $x = 0, z = 0$). The y coordinate of one atom is arbitrary in this particular space group. This is evident in the general positions of the space group (listed above) since if there is an atom at y there is also one at $\tfrac{1}{2}+y$, but no infor-

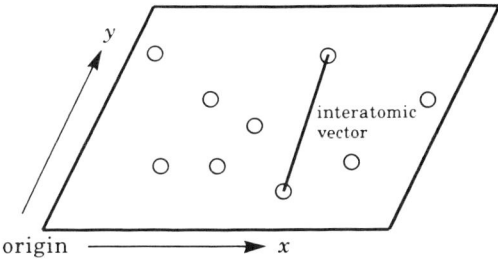

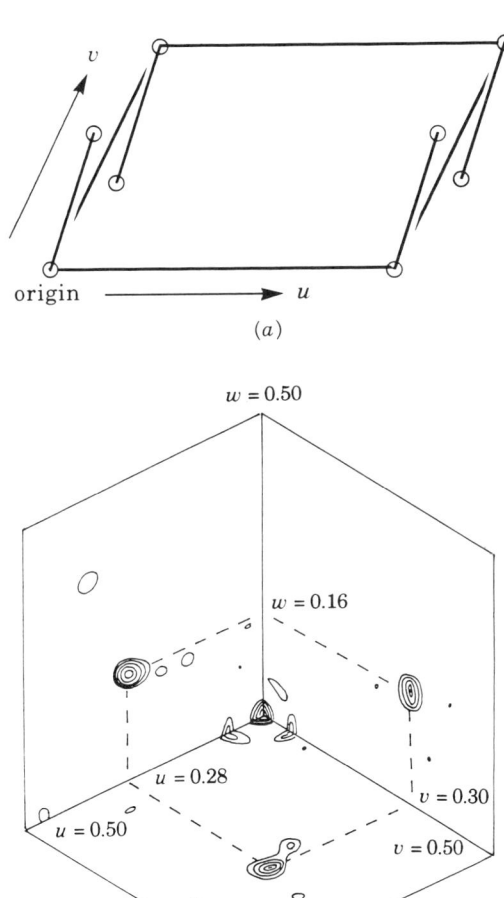

Fig. 21. Vectors in a Patterson map. (a) A peak in a Patterson map indicates that the vector defined between the origin of the Patterson map and the peak in it must be found between atoms in the crystal structure. (b) Harker sections of the Patterson map for a heavy-atom derivative of D-xylose isomerase (Courtesy of H. L. Carrell).

mation on where y is. The y value of a second atom is necessarily not arbitrary, but is relative to that of the first atom. Specific peaks, first described by David Harker (49), are associated with the vectors between atoms related by these symmetry operators. These peaks are found along "lines" or in "sections" (such as $v = \frac{1}{2}$ for the space group $P2_1$).

The unit cell of a protein is assigned with respect to a right-handed system of axes. Once a heavy atom has been located, its phase angle may be $+\alpha$ or $-\alpha$ for F_H, since it is not known whether the interpreted peak in the Patterson map is from atom 1 to atom 2 or atom 2 to atom 1. Several methods have been developed to remove this ambiguity of which the most decisive are those that involve the preparation of a derivative containing both heavy atoms and/or anomalous dispersion measurements (5).

7.3. Relative Phases by the Method of Isomorphous Replacement

The relative phases of the native protein data set can be deduced from the changes in intensity of Bragg reflections between the native protein and its heavy-atom derivative. These are small intensity differences between two large numbers. The value of $|F(hkl)|^2$ should be replaced by the difference between the square of the structure amplitude for the heavy-atom derivative and that of the native protein $|F_{PH}^2 - F_P^2|$. In practice the coefficient $(F_{PH} - F_P)^2$ is used to give an isomorphous difference-Patterson map. This makes it possible to locate the heavy atom and determine its three-dimensional coordinates with respect to the unit-cell dimensions. These heavy-atom positions are generally refined by least-squares methods.

When a heavy atom has been located in the unit cell, its structure factor F_H can be used to estimate the relative phase angle F_P of the native protein. The value of F_P depends on the positions and scattering powers of each atom in the unit cell. If a new atom is added, and it does not cause any disruption in the crystal structure, values of the modified structure F_{PH} will be obtained by a vectorial sum as follows:

$$F_{PH} = F_P + F_H, \tag{10}$$

where P represents the original protein and H represents the atom with a changed identity. Each of these structure factors has a phase and an amplitude. If the magnitudes of $|F_p|$ and $|F_{PH}|$ have been measured, and if it has been possible to locate the variable atom or the heavy atom, then the structure factor of the heavy-atom alone F_H (which includes a phase angle) can be calculated. Several different heavy-atom derivatives of the protein PH_1, PH_2, etc., are studied. The construction used to derive native protein phase angles is illustrated in Fig. 22. If there is only one heavy-atom derivative, there is an ambiguity so that at least two heavy-atom derivatives are needed in order to find the phase angle α. This is the reason that multiple isomorphous replacement (MIR) must be used. An absolute requirement of this method is that the positions of the heavy atoms be known relative to the same unit-cell origin. The phases so derived may be refined

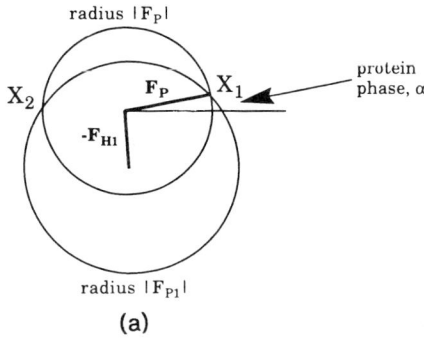

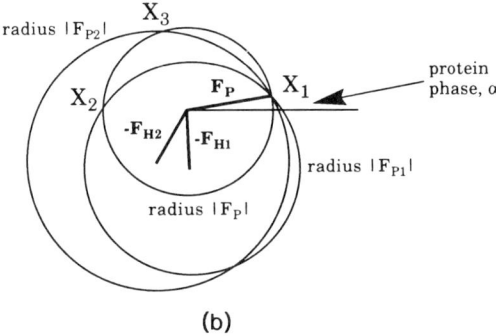

Fig. 22. The construction for determining protein phases from heavy-atom derivatives. (a) with a single heavy-atom derivative two possibilities (X_1 and X_2) are obtained. If there are two heavy-atom derivatives, the phase ambiguity is resolved.

mathematically in order to obtain the best possible set. Once the relative phases are found, an electron density map can be calculated.

This method of isomorphous replacement, together with anomalous dispersion data collection is, to date, the principal method that has been successful for phase determination of macromolecules. It was first used successfully for proteins, myoglobin by John Kendrew and coworkers (14), and hemoglobin by Max Perutz and co-workers (15). Unfortunately, it is common to find that, although a heavy-atom solution has been soaked into a protein crystal, no regular (ordered) substitution has occurred, and solutions of other heavy-atom compounds must be tried.

There are many imprecisions that occur with this method of deriving phases. The bulky heavy-atom compound generally disturbs the positioning of many of the atoms in the area in which it has bound. This may cause local displacements plus changes in unit cell dimensions. Even at 2 Å resolution, changes of 2% in unit cell dimensions can cause problems in phase estimations. In addition, the site occupancy of the heavy atom may vary from crystal to crystal and also on X-ray exposure. Therefore, if data from several crystals are needed for the data set, this must be taken into account. Secondly, the heavy atom position must be correctly determined and should be relative to the same origin for each derivative. Errors in determining the locations of heavy atoms in the unit cell have caused incorrect structures to be reported.

7.4. Superposition and Molecular Replacement Methods

The molecular replacement method used for protein structure determination (50,51) involves determining the orientation and the position in the unit cell of a known structure such as that of a homologous protein that has previously been determined or the same protein in a different unit cell (a polymorph). For the rotation function the Patterson map is systematically laid down upon itself in all possible orientations (Fig. 23). Six parameters that define the position and orientation of the protein in the unit cell are found from maxima in a function that describes the extent of overlap between the two placements of the Patterson function. This function will reveal the relative orientations of protein molecules in the unit cell. The rotation function is thus a computational tool used to assess the agreement or degree of coincidence of two Patterson functions, one from a model and the other from the diffraction pattern.

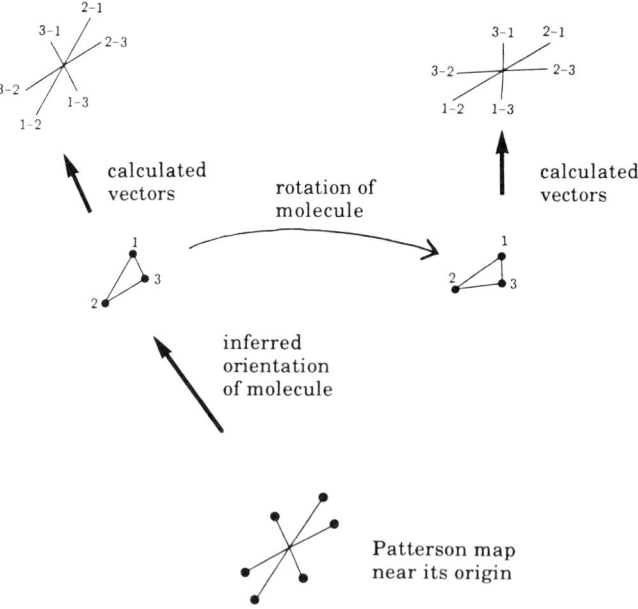

Fig. 23. A rotation function.

Once the orientation of the molecule is determined, it is necessary to position the fragment with respect to the crystallographic axes. A translation function was developed that essentially takes this molecule (whose structure and orientation are now known) and calculates the degree of overlap or coincidence of the two maps (the experimental map and a copy), when one map is translated various amounts with respect to the other (52). These comparisons are made at each position in the unit cell. Computer program systems such as MERLOT (53) exist that do rotation-translation searches for macromolecules, although the translation problem still seems to be more difficult to solve than the orientation problem.

7.5. Anomalous Dispersion

If the wavelength of the X rays used in the diffraction experiment is near the absorption edge of an atom in the crystal structure, then anomalous effects are observed on the diffraction pattern, and these effects can aid in phase determination. Normally, there is a 180° phase shift of a wave on diffraction. This is not so if anomalous dispersion occurs. Therefore, Friedel's Law (Equation 7) no longer holds, and $(hkl) \neq (\overline{hkl})$. Measurements of these two types of Bragg reflections hkl and $\overline{hkl}$ will give information on the absolute configuration of the structure (54–57). For example, an anomalous difference Patterson map with coefficients $[F(hkl)-F(\overline{hkl})]^2$ will give peaks corresponding to vectors between anomalous scatterers. Not only that, it will also give the correct enantiomorph.

Anomalous dispersion has been used in several ways in protein structure determination. In the case of crambin (58) the anomalous scattering of sulfur atoms could be used to solve this structure for which diffraction data were obtained to an unusually high resolution (that of a small molecule). The effect is also used to aid in phase determination (Fig. 24) by helping to resolve any phase ambiguity. It may also help assure that no origin change occurred in deriving heavy-atom position. In addition, enzymes may be engineered with seleno-

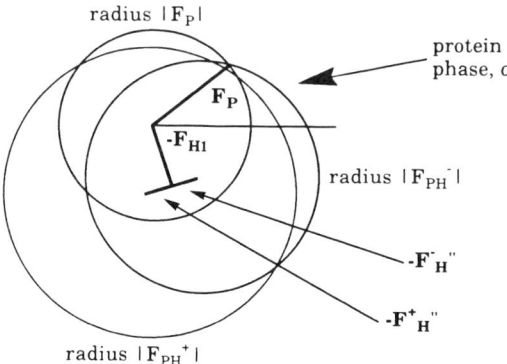

Fig. 24. Anomalous scattering as an aid in protein phasing. $\mathbf{F}_{PH}{}^+$ implies the hkl structure factor, and $\mathbf{F}_{PH}{}^-$ implies the $\overline{hkl}$ structure factor.

methionine replacing methionine (59). Bacterial strains deficient in the methionine metabolic system will not grow unless methionine is added to the growth medium. If selenomethionine is added instead, the bacteria grow fairly well, and each methionine is now replaced by selenomethionine in each enzyme it produces. By this method a heavy atom has been introduced into the protein, and it may be located from a Patterson map. If the diffraction data are measured at various wavelengths (by tuning synchrotron radiation), intensity data are obtained with and without anomalous scattering, so that it may be possible to solve the phase problem.

8. CALCULATING THE ELECTRON-DENSITY MAP OF THE PROTEIN

The question at this stage is "How does one derive the atomic arrangement in a crystal from the intensities in their respective diffraction patterns?" The answer is that X-ray beams, with amplitudes represented by (the square roots of) the measured intensities of the diffracted beams, must be recombined by a Fourier synthesis in a manner similar to that by a lens in the optical microscope. The relative phases in macromolecular crystallography necessary for the Fourier synthesis are derived by the method of isomorphous replacement.

8.1. Electron-Density Modification

The electron density in a crystal structure is positive or zero at all points, and this information has been used in attempts to improve electron-density maps. If phase information is poor, it may still be possible to calculate the electron-density map with these less than optimal phases, modify the electron density in some way to make it more satisfactory (perhaps by use of some chemical information), and then carry out a Fourier transformation to give new relative phases. The assumption is that these new phases are better than the original ones and that the new electron density map will therefore be easier to interpret (60). This process may be repeated. The aim is to suppress any unwanted noise from the maps so that they may be interpreted more precisely. Two assumptions are made for this method. The first is that the electron density in solvent areas, in which there is much motion of atoms, is time-averaged to a fairly constant value, and the second is that there is an absolute value below which the electron density may not fall. These conditions lead to the method of density modification (60–62), which was originally introduced to help "correct" the phases of partially known structures but has proved invaluable for X-ray diffraction analyses of large structures. An example is provided by "solvent flattening," used for macromolecular structures when the unit cell contains a high proportion of solvent, which may be 50% or more for macromolecular crystals. An "envelope" defining the approximate boundary of the molecule is determined from the electron-density map, and all electron density outside this envelope is set to the average value for water background. A new set of phases is then determined by Fourier transformation of this solvent-flattened map.

8.2. Use of Noncrystallographic Symmetry

Many protein crystals exist with more than one molecule per asymmetric unit. These molecules are sometimes related by noncrystallographic symmetry (pseudosymmetry), that is, additional symmetry (such as a twofold rotation axis) that is not part of the symmetry defined by the space group. This feature can be very useful in finding the molecular structure using rotation functions. Additionally, it is also possible for a compound to crystallize in different forms with different packing (polymorphism). If the molecular transforms of the components of the crystal are known, it is possible, by the methods described above, to determine their positions and orientations in the respective unit cells.

If there is one protein molecule per asymmetric unit, but it has some local symmetry such as a twofold axis of rotation relating two identical subunits, then use can be made of this information in order to solve the structure. For example, the orientation of the twofold rotation axis can probably be determined from the rotation function. In a similar way, the translation function can lead to information on the positions of the two related subunits. If the twofold axis can be located, then the electron density for the two symmetry related portions of the molecule can be averaged. An envelope is drawn in the electron-density map that essentially defines the edge of each molecule. The electron density in the two independent molecules is then averaged and the solvent region is flattened. By Fourier transformation a better set of phases is obtained for a new electron-density map, which may be more readily interpreted (63).

8.3. Entropy Maximization

There are many cases in crystallography where not all the necessary Bragg reflection data are available. The reason may be because the resolution is limited (because of poor crystal quality or wavelength restrictions) or because of problems in obtaining phases. What is generally done with data that cannot be measured is to ignore them in any calculations, that is equate them all to zero. This may be satisfactory in early stages but it causes problems when refining structures. Entropy maximization methods can be used in an attempt to obtain more unbiased information from measured data (64–66). Lack of pattern, which is what is aimed at for unmeasured data, can be interpreted as "chaos" or "maximum entropy." The unmeasured data are constrained to provide the least regular (most chaotic) perturbations to the known data. The procedures involved are highly complex, but the result is a less biased map than that obtained by current methods.

9. INTERPRETING THE ELECTRON-DENSITY MAP

There are two general ways in which electron-density maps are used in structure determination. Each will be considered in turn.

a. Electron-density maps may be used to locate atoms and determine the arrangement of the atoms in the crystal structure.
b. Difference electron-density maps may be used to refine a trial structure, to find a part of the structure that may not yet have been identified or located, to identify errors in a postulated structure, or to refine the positional and displacement parameters of a postulated structure. This type of map is useful for small molecules and is also very useful in studies of the structures of crystalline macromolecules, since it can be used to find the location of substrate or inhibitor molecules when the crystal structure of the macromolecule is known.

Since the electron density is calculated only at grid points, it is necessary to contour the values of the electron density at each of these grid points so that the locations and shapes of peaks in this map can be evaluated (Fig. 25). Contouring is the drawing of lines connecting points of equal value (generally electron density) and is done for successive plane sections, usually, but not always, parallel to planes defined by two of the three crystallographic axes. This is analogous to the contouring for altitude on geographic maps found in atlases. Contouring can be done manually, but more frequently it is now done by computer. Contoured sections are then superimposed at appropriate distances apart to obtain a three-dimensional image of the contents of the map.

The availability of high-performance computer graphics systems has made it possible for contoured electron-density maps to be displayed on a graphics screen (67). The maps so calculated may have peaks that look like chicken wire. A stick model of the molecular fragment can be fit, again by computer graphics technology, into this electron density. The whole map with the molecule posi-

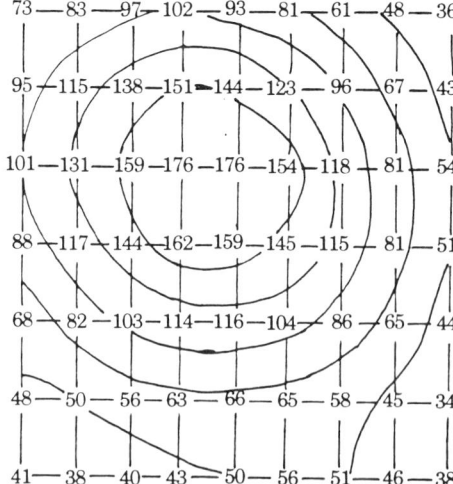

Fig. 25. Contouring an electron-density map that has numerical values at grid points.

tioned in it, can then be rotated and viewed from all directions in order to check the correctness of the fit. The coordinates of the atomic positions in the trial structure with respect to the axes of the electron-density map are then automatically determined and recorded.

9.1. The Resolution of an Electron-Density Map

Resolution is the ability to distinguish two close objects as separate entities rather than as a single, blurred object. In X-ray crystallography, resolution is defined as $\lambda/2\sin\theta_{max}$, the maximum value of $\lambda/2\sin\theta$ in the measured data set. This measure is used by consideration of Bragg's Law $\lambda = 2d\sin\theta$, so that $\lambda/2\sin\theta$ gives d, the interplanar spacing; often the resolution is alternatively described in terms of d spacings. As shown in Fig. 26, the values of θ_{max} are determined by how far out from the center of an X-ray diffraction photograph the diffraction data are visible. Most X-ray structures of small molecules are determined with Cu$K\alpha$ radiation, $\lambda = 1.54$ Å, and the maximum "resolution" that can be obtained (because 2θ cannot be greater than $180°$) is $\lambda/2\sin\theta = 0.77$ Å. A limit to how high a resolution may be obtained experimentally is also imposed by the scattering power of the atoms present and the quality of the crystal (since disorder or high thermal motion will reduce θ_{max}). It is essential, when inspecting an electron density map, to know resolution of the map. Many crystals, however, do not scatter to high enough 2θ for the maximum possible resolution to be achieved.

The resolution is higher (better), if diffraction data have been obtained to high scattering angles (higher $\sin\theta$), but the value of the resolution is represented by a lower number. At a resolution of 0.8 Å, each atom is distinct. For protein crystal structures, on the other hand, the resolution is seldom better than 1.5 Å and is generally poorer. An example is given in Fig. 26 as projections of electron density calculated (with precise phase angles) for a small molecule. At high resolution, the atomic positions in the molecular line diagram provide an excellent fit to the electron-density map. At low resolution, however, the electron density is more diffuse, and, while the line diagram fits reasonably well, it might have been more difficult to find the exact atomic arrangement if it were not already known.

9.2. Macromolecular Model Building

High-performance computer graphics systems and their associated software have greatly improved methods of model building. Several computer programs have been developed over the years that display the electron density and allow one to construct a trial structure (model). Although the features of model-building programs vary, the basic ideas behind them are the same. The program FRODO (67) provides an extensively used example. The electron-density map can be viewed from different directions, and coordinates of peptide or side chain fragments can be docked into it by simple manipulations with dials or a light pen.

Initially the polypeptide backbone of the protein ((-NH-CHR-CO)n) is sought in the map, and this procedure is called *tracing the chain*. The model used

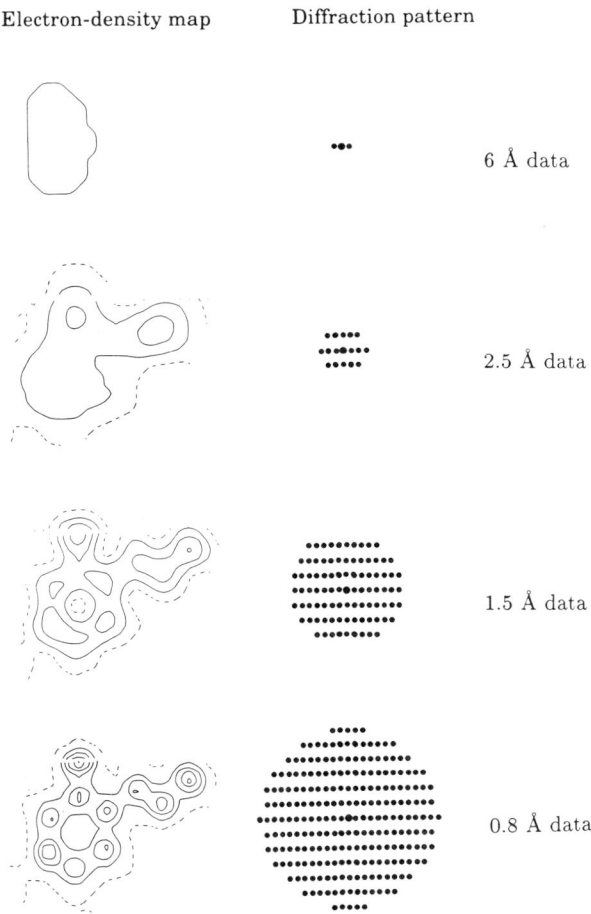

Fig. 26. The resolution of a crystal structure shown on the left is the appearance of a small molecule at various resolutions and on the right a diagram of the relative number of Bragg reflections included in the calculation.

for this is usually polyalanine (R = CH_3). It is generally possible to see the electron density associated with the carbonyl oxygen of the main chain, and that will fix the orientation and position of the peptide unit. Later the electron-density map is searched for evidence of the locations of the side chains (R), provided the amino acid sequence of the protein is known. The number of atoms in such a model is so large that only a small region of the electron-density map is used for selecting atomic positions. The individual atoms are generally not resolved, but models of side chains or backbone are best positioned to fit the electron density. Sometimes in the refinement the investigator will omit part of the structure, usually in a specific area in the unit cell. Then an electron density map should show an image of the omitted portion at half the expected peak height. In this way some measure of the precision of the overall model can be obtained.

It is usual to start with a low resolution map, to about 5 Å resolution. At this resolution only the overall shape can be found, while sometimes cylindrical rods, characteristic of α helices can be seen. At about 3 Å resolution the map will show the polypeptide backbone of the molecule clearly enough for it to be traced. Finally, electron-density maps are calculated at high resolution (1.5–2 Å resolution), at which point the side chain conformations can be defined more clearly.

9.3. Refinement of a Protein Crystal Structure

Most protein structures are refined by methods that involve reciprocal space rather than real space. The procedure involves improving the model obtained for the crystal structure while keeping the measured structure amplitudes $|F(hkl)|$ fixed. If the data have been obtained to very high resolution, the atomic coordinates x, y, and z and a temperature factor are adjusted (68, 69) to give the best agreement between the observed structure factors and those calculated for the molecule model; thus

$$\sum w \, |F_o - F_c|^2$$

is minimized (where w is a weight, often the inverse of a measure of the reliability of the value of F_o). This method can only be used if there are many Bragg reflections measured for each parameter to be refined. At 2 Å resolution, the number of positional parameters (three per atom) to be determined is approximately equal to the number of Bragg reflections measured. At lower resolution, there are fewer observations than parameters, so that it is not possible to locate individual atoms.

This problem is addressed by fixing the geometry of various groups (such as the peptide group, which is planar), thereby reducing the number of parameters by introducing constraints on the way the refinement may go. Restraints may alternatively be used in which the parameters may vary about defined values. The refinement involves many cycles because of approximations used in setting up the equations to be fitted by the experimental data. The refinement is considered to be complete when individual changes in parameters are no longer significant, that is, some small fraction of the estimated standard deviation (e.s.d.) of the parameter. The e.s.d. is a quantity determined during the refinement and gives a measure of the precision of the analysis. The final results from a least-squares refinement are used to calculate a difference electron-density map (which, ideally, should be near zero at all grid points if the model is correct). Any high peaks or deep troughs in this map should be investigated as they may indicate a problem with the model.

The trouble with this method of least squares is that the result of the refinement may be a local rather than global minimum. Therefore it is common to refine a protein crystal structure by use of molecular dynamics. This means that each atom is assigned a velocity, and equations of motion are solved under the restraints of X-ray information. The assigned velocities, while initially corresponding to room temperature (~300 K), are increased to correspond to velocities to 2000–4000 K, and then are reduced. This method, called *simulated annealing*, allows that the conformation of the model to change over appreciable energy barriers and should give, on refinement, a better approximation to the structure at the global energy minimum. The "temperature" aspects of the refinement are purely mathematical and input by the assigned atomic velocities. The computer program XPLOR (70) is used for this.

The refinement procedure should involve a final check of the electron-density map to insure that no major part of the structure lies in an area devoid of much electron-density. This has to be done judiciously. In all cases a knowledge of the amino acid sequence is needed. Those areas of the protein that are in high motion or disordered will be hard to find in the electron-density map because the electron density is smeared over such a large volume. Therefore it may be found that some of the ends of the polypeptide chain cannot be located.

9.4. Difference Maps for Macromolecules

Difference electron-density maps are of great use in protein crystallography, particularly for locating the binding of small molecules such as substrates and inhibitors (Fig. 27) and for examining the effects of changes in pH. Since the protein crystal is approximately half water (generally in the 27–65% range), there is plenty of space for such molecules to be soaked into the crystal. If, however, some conformational change takes place as the small molecule binds, the crystal may crack or the molecular contents may become disordered so that no diffraction is observed. Conditions must then be found that avoid this problem. An entire set of Bragg reflections to as high a resolution as possible must be measured for a crystalline protein-ligand complex. The phase problem is considered initially solved by use of the phases for the native (unsubstituted) protein. Any change that occurs when substrate, inhibitor, or drug binds or when a mutation occurs can then be found in electron-density maps.

When a substrate or inhibitor binds to a protein, it displaces water. As a result, electron density for ordered water is replaced by electron density for part of the ligand molecule. This means that there will be no appreciable peak in the difference map. In addition, because of somewhat incorrect phases, the substrate or inhibitor will appear in a difference map with reduced electron density, usually about half that of a well-phased map. Consequently, the practice has sometimes been to enhance coefficients in the Fourier expression. Often a map that combines the features of a difference map, enhanced by a factor of 2, and of the native protein map, is used. The coefficients of the Fourier synthesis are then:

$$[(n + 1)|F_o| - n|F_c|],$$

where n is usually 1, but may be as large as 5. By this method the macromolecular difference electron-density map is enhanced to emphasize

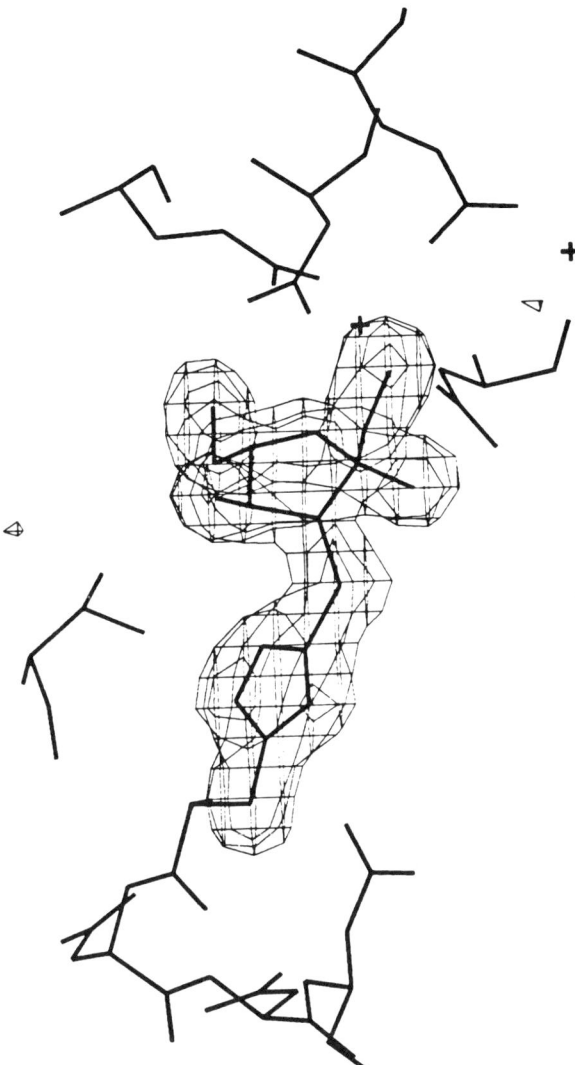

Fig. 27. A difference electron-density map showing inhibitor binding (Courtesy of H. L. Carrell).

$$2|F_o| - |F_c| = |F_o| + (|F_o| - |F_c|).$$

Thus, $2(F_{PH} - F_P) + F_P$ is used rather than $F_{PH} - F_P$. Note that it is assumed that the phases of the protein-ligand crystal structure are the same as those of the native structure (protein only).

At this point it must be clear to the reader that the *electron-density maps or other*

Fourier maps obtained are only as good as the relative phase angles used in their calculation. The structure amplitudes are important in this calculation, but if the phase angles are not approximately correct, the crystallographer may be in the situation of refining an incorrect structure, often to a quite reasonable result, but to a local minimum in deviations between observed and calculated structure amplitudes. If there are unusual features in a crystal structure, such as nonbonded atoms that approach each other too closely, or unusual features of folding, it is best to return to the determination of the phase angles and check that there was not a better set of phase angles that could have been used.

9.5 The Precision of the Derived Protein Structure

There are many possible sources of errors in a protein structure determination. None of the intensities measured are totally precise, but contain the experimental errors normally encountered for the methods used. In addition, the heavy-atom positions and relative phase angles derived from them may contain errors due to disturbances to atomic positions caused by introduction of the heavy atom.

Some protein crystal structures reported in the scientific literature initially have some regions of their polypeptide backbones erroneously interpreted. The problems have mainly been in loop regions, where the protein molecule is often decidedly floppy. The reasons for this are that (for example) it is not possible to see individual atoms at 2.8 Å resolution. In essence the problem is analogous to that found when looking at an object through a low-power microscope; sometimes it is difficult to tell what is joined to what. At higher power this problem diminishes, as does the problem of locating the backbone and side-chain atoms in protein when higher-resolution X-ray diffraction data are obtainable.

The similarities in folding of analogous proteins in different crystal structures and the catalytic activities of crystals indicate that conformations of proteins in crystals are similar (if not identical) to those in solution. Local pH in the crystal structure may, however, differ from that of the solution from which the protein crystal was grown (71), so that the protonation state of various side chains may not be clear.

10. THE NATIVE PROTEIN STRUCTURE

Once the arrangement of atoms in the macromolecule has been determined, one can proceed to analyze the results. At this stage it is essential to take into consideration the resolution of the structure determination, the extent to which refinement of the structure has been possible, the constraints and restraints used in the refinement, and the overall R value. At 6 Å resolution a protein molecule looks like a folded solid tube, and no atomic detail can be found. At 3.5 Å resolution more information is obtained and bulky side chains might even be discerned. Many protein structure determinations can only be carried out to 2.2–2.5 Å resolution and the overall atomic arrangement can be deduced by use of geometric building blocks (planar peptide groups, most likely side chain conformations) and refined to check this. At 1.6–1.8 Å resolution the overall atomic arrangement will be clearer, and at 0.8 Å resolution (as is the case for crambin), complete atomic detail results. This is the resolution to which most small-molecule crystal structures are determined. There are several tasks for the crystallographer when the atomic model has been determined. These include a calculation of torsion angles and a Ramachandran plot, a calculation of hydrogen bonding, and the identification of α-helix- and β-sheet components and a calculation of bond distances and angles. These will be described in turn.

The polypeptide chain of a protein folds systematically into a specific three-dimensional structure (72). There are various levels of structure in such a folded protein. *Primary structure* refers to the amino acid sequence, *secondary structure* is any regular local structure of a segment of a polypeptide chain, while *tertiary structure* is the overall topology of the folded polypeptide chain. *Quaternary structure* describes the aggregation of folded polypeptides with each other by means of specific interactions, such as the aggregation of subunits to give a complete protein molecule.

The polypeptide backbone of a protein consists of a repeated sequence of three atoms—the amide nitrogen (N_i), the α-carbon (C_α), and the carbonyl carbon (C_i), where i is the number of the residue starting from the amino end (remember the order of words in "amino acids"). The repeat distance between peptide units in an extended *trans* conformation is approximately 3.8 Å. The peptide group has a permanent dipole moment with the negative charge on the carbonyl oxygen atom, and is, as noted first by Pauling in 1933 (73), nearly planar and can exist in either the *cis* or the *trans* form; in practice the *trans* form is that most commonly found.

10.1. Torsion Angles

The torsion angle was introduced originally in order to facilitate a description of steric relationships across single bonds. If one looks directly along a bond, the two atoms of the bond become apparently superposed as a point. Groups attached to the two atoms of the bond stick out radially. A graphical description of this is familiar to organic chemists as a Newman projection (74); the atom nearer the viewer is designated by radii spaced at 120° angles while the atom further from the viewer is designated by a circle with equally spaced radial extensions. The bonds to the nearer atom are drawn so that they penetrate the circle, while bonds to the further atom do not. X-ray diffraction results can give values for the angles between these radii.

A torsion angle is determined by how much a bond has to be twisted to cause two substituents on the atoms it connects to be eclipsed (Fig. 28). By definition, a torsion angle is positive if a clockwise twist about the central of three connected bonds that is needed to make the near (upper) bond ($A-B$ in a series of four-bonded atoms $A-B-C-D$) eclipse the far (lower) bond ($C-D$); a negative torsion angle is a counterclockwise twist. The direction of view along the bond [that is, which is nearer to you ($A-B$) or ($D-C$)] is immaterial; reversing the sequence of

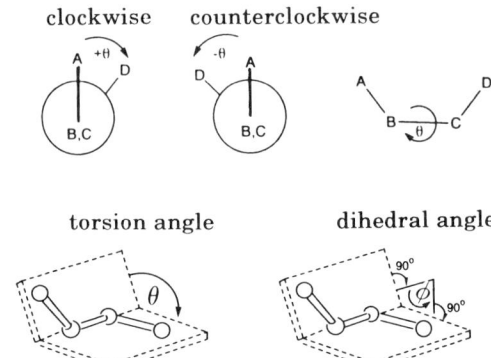

Fig. 28. A torsion angle.

view from *A-B-C-D* to *D-C-B-A* does not change the direction of the twist necessary to make *A-B* and *C-D* lie on top of each other. There is no overall change in the shape of the molecule if you turn it over to look at it from the other end.

The conformational analysis of protein structures involves three torsion angles, φ (phi), ψ (psi), and ω (omega) for the main chain (Fig. 29). The angle ω describes rotation about the peptide bond −CO−NH−, and this is generally 180°, sometimes 0°. Each peptide group essentially lies on a flat rectangle, and it is necessary only to describe how each of these rectangles is related to the next, so that only two angles are needed to describe how the peptide planes of adjacent residues lie with respect to each other. The torsion angle φ is the torsion angle about the −NH−CHR− bond, and ψ is the torsion angle about the −CHR−CO− bond. Torsion angles, in addition, may be used to designate the conformation of the side chains. These are denoted by χ(χ^1 to χ^n chain away from C$_α$). The steric interactions within the side chains in the *trans* form of the peptide bond (ω = 180°) are much more favorable than those in the *cis* form (ω = 0°), where there may be steric interference with side chains from residues $i + 2$. If the residue $i + 1$ is proline, however, the *cis* and *trans* forms (Fig. 30) have similar energies. Proline is the only amino acid taking part in a *cis* peptide that is normally encountered in proteins.

When a protein structure is obtained by fitting a model to the electron-density map, the φ and ψ torsion angles are calculated for each amino acid residue. These angles give a good indication of the types of folding throughout the back-

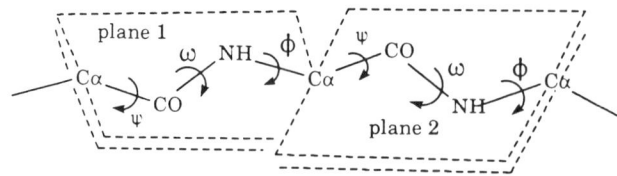

Fig. 29. The three torsion angles of a peptide unit in a protein.

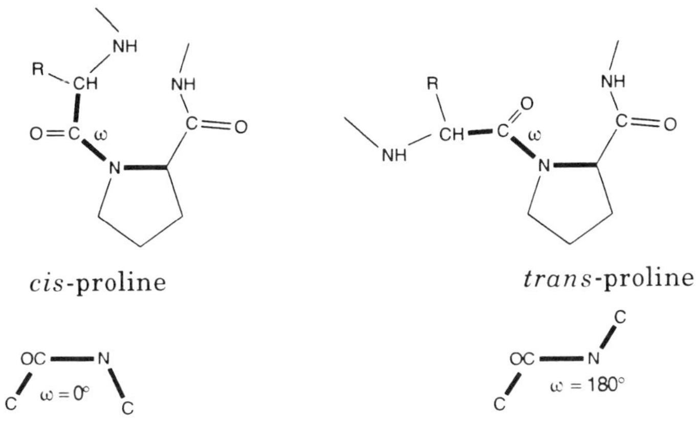

Fig. 30. *cis* and *trans* proline.

bone of the protein. Geometrical restraints on φ and ψ occur for steric reasons. For example, φ and ψ cannot both be zero, otherwise the carbonyl oxygen and the hydrogen atom on the nitrogen would overlap each other. Glycine has only a hydrogen atom as the side chain and therefore has more conformational flexibility than the other amino acids. It is usual to construct a conformational map, obtained by plotting φ versus ψ for each amino acid to give a Ramachandran plot (Fig. 31), named after G. N. Ramachandran (75) who was the first to perform

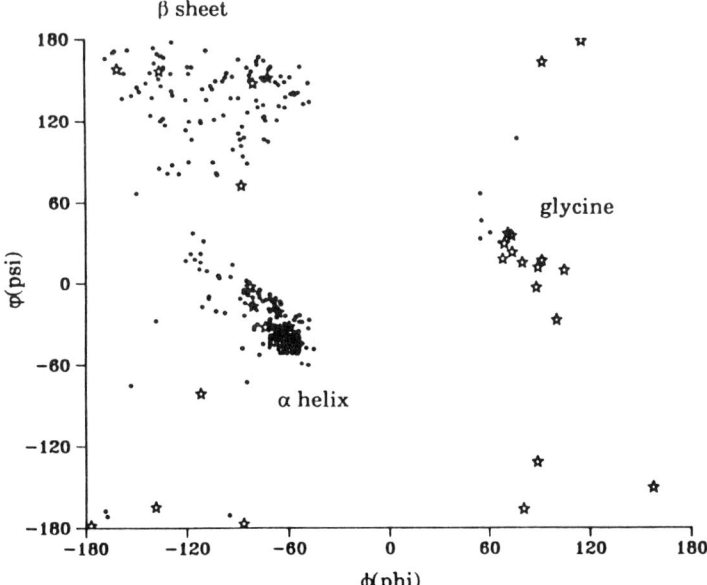

Fig. 31. The Ramachandran plot of a β-barrel structure (D-xylose isomerase) (Courtesy of H. L. Carrell).

such an analysis. This plot shows that points are clustered in specific regions in the map; these regions represent α helix, β sheet, and other ordered features of secondary structure. If too many points lie outside these areas, there may be something wrong with the interpretation of the electron-density map.

10.2 Accessing the Protein Data Bank

When three-dimensional structural data on a protein are required, the place to seek them is the Protein Data Bank (PDB) (76). This data bank is a computerized archive for the three-dimensional structures of biological macromolecules determined by X-ray and neutron diffraction. This database is organized and maintained at Brookhaven National Laboratory. The files are available on request from personnel at the PDB and are best accessed by computer. At present, the structural data (atomic coordinates, journal references, and in many cases, structure factors for protein crystal structures) for about 1200 biological macromolecules, over 550 with atomic coordinates, are available. Also available from the same source is a list of proteins that have been crystallized and the experimental conditions used. Each protein structure in the Protein Data Bank report has an identifying code (ID-CODE), a header record containing useful information on the protein such as the name and source of the protein, a series of references to published articles on the protein, and the resolution of the data. Data on the refinement, such as the programs used, the R value, the number of Bragg reflections, the root-mean-square deviation from ideality, and the number of water molecules that have been located, are also provided. Then follows a description of the protein, its amino acid sequence, including an analysis of which parts are helix, sheet, and turns. The main entry is a list of atomic coordinates (ATOM) and information on coordinates of metals, substrates, and inhibitors bound to the protein (HETATM). Coordinates of a protein or nucleic acid can readily be extracted in a suitable form for use with a computer-graphics system so that the three-dimensional structure can be viewed and analyzed.

10.3. Bond Distances and Angles

Three atomic coordinates, x, y, and z, describe the position of an atom in the asymmetric unit of the crystal structure. This is the way that atomic positions are listed in the Protein Data Bank. They are each measured as a fraction of the distance along the appropriate unit cell edge and therefore are also called fractional coordinates. The value listed for x, for example, is a fractional coordinate relative to the length of the a dimension and is measured parallel to this unit cell edge. One unit cell repeat, whatever its length, corresponds to 1.000 in fractional coordinates, so that if the value of x is 0.6725, the atom lies 67.25% along the a spacing, measured parallel to the a axis. The y and z coordinates are defined in an analogous manner (Fig. 32). Because atomic arrangements in crystal structures are periodic from unit cell to unit cell, part of a molecule may be in one unit cell and another part in an adjacent unit cell. Values that are reported usually correspond to a distinct molecule so that, for convenience, some atoms may have negative values for their coordinates (x, y, and z), and coordinates of other atoms have

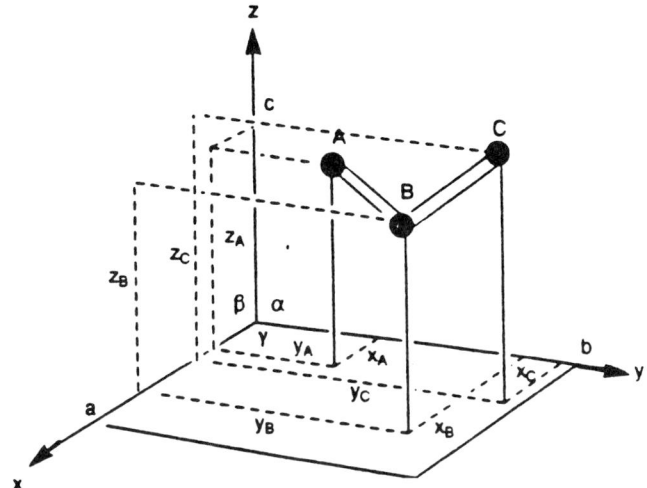

Fig. 32. The meaning of x, y, and z coordinates for three atoms.

values greater than 1.0000. If there is an atom at x, there is another at $1 + x$, and another at $n + x$, n being any integer. This information is used to determine how the molecules pack together.

The notation may be simplified by setting $\Delta x = x_1 - x_2$ so that it equals the difference in x values for the two atoms at the ends of the bond for which the distance if being calculated. This gives

$$r^2 = (a\Delta x)^2 + (b\Delta y)^2 + (c\Delta z)^2 + 2ab\cos\gamma\Delta x\Delta y \quad (11)$$
$$+ 2ac\cos\beta\Delta x\Delta z + 2bc\cos\alpha\Delta y\Delta z.$$

The bond angle, A-B-C, is the angle between the bonds A-B and B-C formed by three atoms A, B, and C, connected in that order. If the length of A-B = ι_1, B-C = ι_2, $A\cdots C$ = ι_3, then the angle A-B-C = δ may be calculated from:

$$\cos\delta = \frac{\iota_1^2 + \iota_2^2 - \iota_3^2}{2\iota_1\iota_2} \quad (12)$$

While values of x, y, and z are listed in the Protein Data Bank (76), these are generally derived from a model involving fragments of specified geometry that have best fit the electron density map. The coordinates listed are good for viewing on a graphics screen, but generally do not indicate individually determined atomic positions.

10.4. Vibration Parameters

The temperature factors for a protein, output with x, y, and z from the least squares refinement, while generally only isotropic (the same in all directions), can give some useful information on the general mobilities of various portions of the protein. These temperature factors contain information on static disorder, caused for example by a conformational variability of the side chains, and on the dynamic motion as a function of temperature. These two situations can generally be distinguished by studies at various temperatures. An overview of molecular flexibility can be obtained by examining a backbone model of the protein molecule with temperature factors listed for each peptide group. In the case of lysozyme this indicated that the two lobes of the structure could move with respect to each other (77); this may be important in the catalytic mechanism. The side chains are more mobile than the backbone. Much more sophisticated analyses of vibration parameters are now underway.

10.5. Motifs in Protein Structures

During the folding of a water-soluble globular protein, hydrophobic amino acid residues are mostly folded into the interior of the protein, while the hydrophilic amino acid residues tend to lie on the surface of the protein. The result is a protein molecule with a hydrophobic core and a hydrophilic surface. These remarks apply to the side chains of the protein. The main chain, however, contains polar C=O and N–H groups, and their polarities must be neutralized when the main chain is in the hydrophobic core. Two types of secondary structure, α helices and β sheets, satisfy this requirement by forming hydrogen bonds between the C=O and N–H groups. It is necessary to avoid hydrophobic patches on the surface of a protein during folding because, as shown by the mutation of hemoglobin to sickle-cell hemoglobin, such hydrophobic areas tend to favor aggregation of the protein, and therefore sickle-cell hemoglobin forms fibers.

The polypeptide chain in an α helix folds in a right-handed spiral manner with 3.6 residues per turn and a translation per residue along the helical axis of 1.5 Å (3.6 × 1.5 = 5.41 Å for one complete turn). This folding was first described by Pauling (78). Hydrogen bonds are formed between the carbonyl oxygen atom of residue i and the amide nitrogen of residue $i + 4$. These hydrogen bonds all point in the same direction, nearly parallel to the helix axis, and therefore the α helix has a dipole, with the N terminus positive and the C terminus negative. The α-helix is a very stable folding pattern. The backbone atoms of an α helix are packed together tightly inside the helix and are in van der Waals contact so that there is no empty space in the middle of the helix. Side-chain residues point out from the helix and do not interfere with the helix formation. For proline residues, the amide group is part of a five-membered ring, and rotation about the C–N bond is not possible. Therefore, proline is considered a "helix breaker," and if found in an α helix, a bend or kink in the helix is usually observed at that point.

The periodicity of 3.6 residues per α-helical turn means that if the helix is viewed down its axis, the side chains stick out approximately every 100°. Therefore, if the helix is on the outside of the protein, the nature of the side chains must change from hydrophobic for the part that faces the interior of the protein to hydrophilic for the part of the protein that can interact with solvent. This can be represented by a "helical wheel." It is often found that one side of the wheel contains hydrophobic residues that pack into the interior of the protein, while the other contains hydiophilic residues that point toward the surface of the protein. Such a helix, with side chains changing from hydrophobic to hydrophilic and vice versa every three to four residues, is described as an amphipathic helix.

A second type of hydrogen bonding between different portions of a polypeptide chain, also first postulated by Pauling and found later in many structures, is called the β sheet (79). Individual polypeptide chains lie side by side and form hydrogen bonds between the carbonyl group oxygen atom of one chain and the amide-hydrogen atom of another. The polypeptide chains in this side by side packing may be either parallel or antiparallel. Antiparallel sheets have unevenly spaced hydrogen bonds perpendicular to the strands and the narrowly spaced bond pairs alternate with widely spaced pairs. Parallel sheets, on the other hand, have hydrogen bonds that are more evenly spaced. When examined end on, the β sheet appears to be pleated in that successive C_α atoms and their substituent side chains point alternately above or below the sheet plane. Parallel strands tend to be buried in the interior of the protein. Often, β sheets have a right-handed twist that varies form 0 to 30°. When the twists of two β sheets are approximately zero, they can pack readily. When, however, they have different degrees of twists, one β sheet needs to be rotated with respect to the other for them to pack. It is found that β sheets are connected to each other by means of an α helix, a coil, or another strand, generally with a right-handed topology. The α helices follow the curvature of the β sheet.

After a protein structure is determined it is usual to calculate interatomic distances, including hydrogen bonding. This calculation will give a good overall picture of the protein folding. The C=O group of the backbone will form a hydrogen bond to the NH group of the backbone three amino acid residues away for a 3_{10} helix and four residues away for an α helix. Other types of hydrogen bonding, involving two groups of amino acid residues that are widely separated, probably indicate β-sheet formation and turns. It is usual to try to simplify reported crystal structures of proteins, and this is done by representing α helices by cylinders and β strands by flat arrows, the arrow pointing to the C terminus.

10.6. Molecular Architecture

Proteins are constructed of modular systems or domains. These are portions of the polypeptide chain that can fold independently into a stable structure. A protein may be just one domain, or may be comprised of many domains. A typical domain may be roughly 25 Å in diameter and consist of 100–150 amino acid residues. For example, two α helices, joined by a loop (the helix-loop-helix motif) can give a calcium-binding motif or a DNA-binding motif. The *so called* Greek key motif consists of four antiparallel β strands arranged in a pattern reminiscent of one found in ancient Greek friezes (80). The βαβ motif consists of two β strands that are parallel but not necessarily adjacent connected by an α helix, which shields the strands from solvent. Some examples of these types of motifs are shown in Fig. 33.

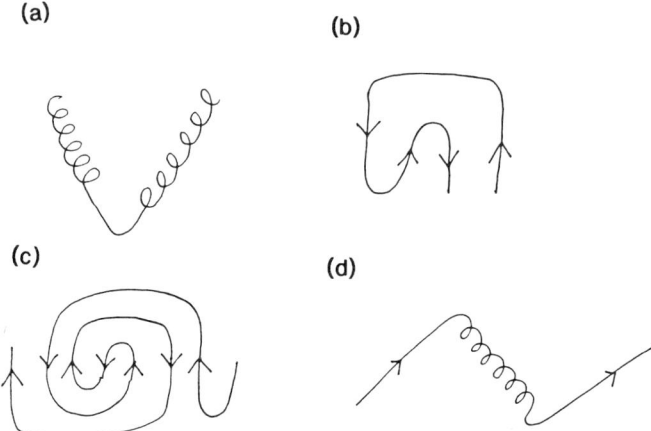

Fig. 33. Some motifs in protein structures. The α helix is represented as a coil and β sheet as a line with an arrow. (a) helix-turn-helix, (b) Greek key, (c) Jelly roll, and (d) βαβ structures.

M. Levitt and C. Chothia (81) constructed taxonomic classifications of protein structure and noted that α helix and β sheet structures within protein domains are organized in a limited number of ways. The four main subcategories are named α for α-helical proteins, β for proteins that are primarily composed of β sheets, α + β for proteins with both α helices and β sheets, and α/β for those proteins that have alternating α-helix- and β-sheet arrangements.

When more than 60% of the amino acid residues in a protein adopt a helical conformation, the protein is usually classified as α. The α helices are usually in contact with one another. The simplest antiparallel α structure, a four-helix bundle, consists of four α helices, aligned to form a hydrophobic core. The structure is cylindrical and has a left-handed twist of approximately 15° between the helical axes. Hemerythrin provides an example of this topology. The "globin fold," so-named because it was found in myoglobin and hemoglobin, consists of eight α helices arranged to enfold the active site heme group. The angles between the α-helical axes are approximately 50°.

The packing of α helices can be considered in terms of knobs (the side chains) and holes (or grooves and ridges). Three classes of α-helix- to α-helix interactions are found, and these have different angles between the main axes of two α helices. When the interhelical angles are near 90°, the contact residues are small, but if the α helices are more parallel to each other, it is found that the residues at the contact point are larger. For example, if the contact point on each helix involves glycine, the helices pack with an angle of 80° between helix axes. The second class includes the side chains alanine, valine, isoleucine, serine, or threonine, and the angle between helices is 60°. In the third class, the interhelical angle is small, approximately 20°, and the side chains are generally leucine or threonine (hence the "leucine zipper"). Polarity also affects the nature of the contact. Large residues will not allow close penetration of the α helices, and polar residues will resist being buried in a nonpolar environment.

Proteins that are composed mainly of β sheets are folded either as barrels or as layered sheets. Barrel structures are made with either six or eight strands of β structure wrapped into a cylinder, forming the surface of the barrel. Barrels are also formed by folding of a slightly more complicated variation of the Greek key motif and the jelly-roll motif (which has an extra swirl in the Greek key pattern and a slightly different topology of connections).

Another class, the α + β class, has approximately equal amounts of α-helical and β-sheet structure. Often there is a cluster of helices at one or both ends of a single β sheet. One type of α + β structure is the open-face sandwich β structure. The subtilisin inhibitor has a five-membered β-sheet structure with two α helices protecting its open face.

Proteins that are classified as α/β generally have one major β sheet per domain. This is usually parallel sheet. The α helices occur in the loops connecting the strands and pack against the β sheet. The simplest arrangement is the singly wound parallel barrel consisting of eight parallel β strands forming a central β barrel with eight α helices surrounding the strands. This motif was first identified in triose phosphate isomerase (82) and is found in many proteins (Fig. 34). Since βαβ units generally have a right-handed connection, the direction of the strands

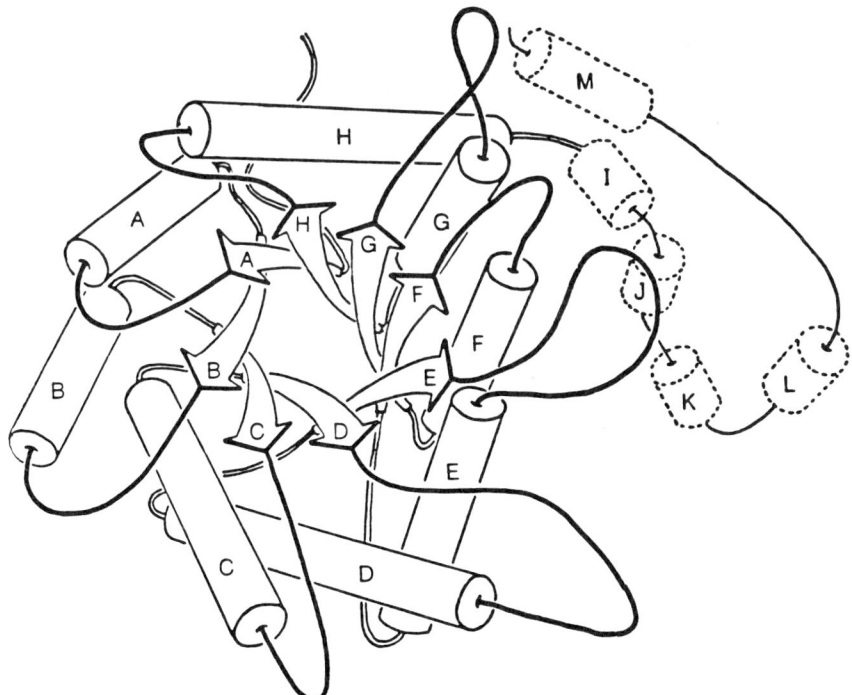

Fig. 34. The (βα)$_8$ barrel structure of D-xylose isomerase.

around the barrel is the same in all such structures. The active site of the enzyme is generally found at the C-terminal end of the β strands and involves residues that are part of the β sheet or very near it.

Another metal-binding motif found in proteins, such as troponin C, calmodulin, and parvalbumin, is the "helix-loop-helix motif," also known as an *EF* hand, which binds calcium ions in an octahedral manner (83, 84). This is named by analogy to a right hand in which the index finger is the *E* helix, a curled second finger is the loop, and the thumb is the *F* helix (Fig. 35). The vertices of the octahedron are designated X, Y, Z, $-X$, $-Y$, $-Z$ in a right-handed system. The position $-Y$ is filled by a conserved backbone carbonyl group and $-Z$ by the shared glutamic acid group. Generally in proteins with this calcium-binding site X is aspartic acid; $-X$ is a water molecule. The sequence by which amino acids bind to the calcium ion is the order $X(i)$, $Y(i+1)$, $Z(i+2)$, $Y(i+3)$, $-X(i+4)$, $-Z(i+5)$.

The viruses with icosahedral symmetry are impressive structures resulting from protein assembly. Sixty copies of a fundamental unit are arranged on the surface of a shell. These structures have 532 symmetry (twofold, threefold, and fivefold symmetry axes) (85). Icosahedral symmetry is the most efficient arrangement for a closed shell, since it has the lowest surface-to-volume ratio, most nearly approximating a sphere. A regular three-dimensional icosahedron can be generated from an extended planar hexagonal net of equilateral triangles. It is possible to put 60 equivalent objects on the surface of an icosahedron in such a way that each is identically situated and related to the others by a rotational operation, yet many virus structures have a capsid with more than 60 units. To account for the additional subunits, it is necessary to relax the requirement that each subunit have the same environment.

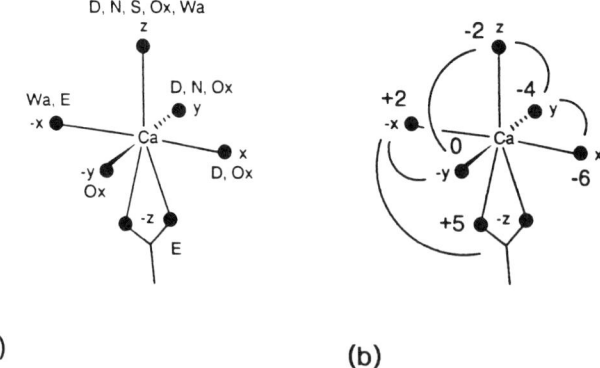

Fig. 35. The EF hand. (a) Residues found in several calcium-binding proteins are shown. Wa = water, ox = main chain carbonyl, D,N,S,E, as amino acids. (b) The order in which the protein winds around the calcium starting at amino acid residue $i(-6)$ then binding $i+2(-4)$, $i+4(-2)$, $i+6(0)$, $i+8(+2)$, and $i+11(+5)$.

11. PROTEIN–LIGAND INTERACTIONS

Just as compounds containing heavy atoms can be soaked into a protein crystal, so substrates and inhibitors can be soaked in and may thereby give some information on the location of the active site. If such a binding of substrate causes very little change in the overall shape of the protein, the native protein and the ligand-protein complex may be isomorphous and crystallize in the same space group with the same unit-cell dimensions and a similar atomic arrangement. The relative phases used to calculate the electron-density map of the protein-ligand complex are, by necessity, assumed to be approximately the same as those found for the native protein. In order to enhance the image of the ligand bound to the protein, a difference map is calculated (Fig. 36). The difference map contains peaks where there is more electron density in the ligand-protein map than in the native protein map and troughs where there is less electron density. Therefore, besides having peaks in the area of ligand binding, this difference map will also indicate which protein side chains have moved. Binding of a ligand, even if it is a substrate, at a given site does not always assure one that the active site has been located. Care must be taken in interpreting such maps. Information on both productive and nonproductive binding sites may be obtained.

The binding of drugs and other small molecules to biological macromolecules involves a fit between the two that depends not only on shape, but also on the distribution of charge on the surfaces of both so that a complementary fit may be obtained. Molecules with similar but not identical formulae may bind to the same receptor in the same or different ways, depending on the nature of the functional groups in each. Some actual measurements of the way that small molecules bind to proteins will now be described. The data so obtained are of great use to the pharmacologist in modeling, for example, the possible interactions of a drug with a receptor of unknown structure.

11.1. Binding of Drugs to Cytochrome P-450$_{cam}$

The cytochrome P-450 enzyme system that has been characterized structurally is the form cytochrome P-450$_{cam}$, which catalyzes the oxidation of the monoter-

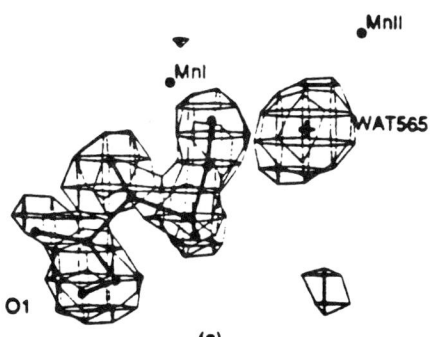

Fig. 36. A difference map showing water and substrate binding (Courtesy of H. L. Carrell).

pene camphor (2-bornanone). This reaction is the first step in a catabolic process that allows the organism to use camphor as a sole carbon source. The enzyme places a hydroxyl group on the 5-*exo* position of camphor. The crystal structure of this P-450$_{cam}$ has been determined to 2.2 Å resolution in the absence of bound substrate and to 1.6 Å resolution in the presence of camphor (86). The enzyme contains 414 amino acids of known sequence. X-ray studies on the camphor-P-450$_{cam}$ complex show a pentacoordinated iron atom with only one axial ligand, the sulfur atom of Cys-357 (Fig. 37). The camphor was located above the protoporphyrin IX. Cys-357, the axial thiolate ligand to the iron of the protoporphyrin, is protected in a pocket of Phe-350, Leu-358, and Gln-360.

The active site of the enzyme, which is lined with the hydrophobic residues Phe-87, Tyr-96, Leu-244, Val-247 and Val-295, lies deep inside the protein, above the heme group (on its distal side) and next to the dioxygen-binding site. Camphor, the substrate, is held in place in this active site by a hydrogen bond with the hydroxyl group of Tyr-96, which binds to the carbonyl oxygen atom of camphor. Other contacts holding the camphor in place, such as Phe-87, Val-247, and Val-295, are made with the methyl groups of the camphor and are hydrophobic.

The camphor lies 4 Å above the pyrrole A ring of the protoporphyrin IX. The orientation of the camphor implies that a heme-bound "activated" oxygen atom can add only from the direction that gives 5-*exo*-hydroxycamphor, the product that is actually found. Thus the stereospecific nature of the oxidation is demonstrated in the crystal structure. Abstraction of hydrogen can occur from either the *exo* or *endo* positions C5 of camphor, but oxygen adds stereospecifically only to the *re* face to give only the *exo* compound.

Crystal structures have been reported for the enzyme with several bound drugs (87). Metyrapone- and 1-, 2- and 4-phenylimidazole-inhibited complexes of cytochrome P-450$_{cam}$ have been refined to 2.1 Å resolution. Except in the 2-phenylimidazole complex, each complex forms an N–Fe bond to the heme iron atom. In the 2-phenylimidazole complex, water or hydroxide coordinates the heme iron atom and the inhibitor binds to the camphor pocket.

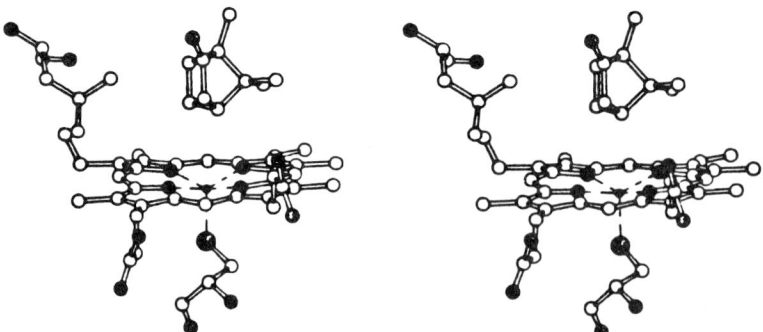

Fig. 37. Stereoview of part of the active site of cytochrome P-450$_{cam}$ showing the porphyrin ring, the sulfur ligand to the ion (below), and the bound camphor (above).

The three-dimensional information so obtained is now being used to model the shapes of other less soluble cytochromes P-450 and the manner by which they catalyze reactions with the appropriate substrates.

11.2. Binding of Drugs to Dihydrofolate Reductase

Another enzyme for which X-ray diffraction studies have aided in an analysis of the mode of action is the enzyme dihydrofolate reductase. This catalyzes the reduction of 7,8-dihydrofolate to 5,6,7,8-tetrahydrofolate, an essential coenzyme used in the synthesis of thymidylate, inosinate, and methionine. The antitumor agent methotrexate is a powerful inhibitor of dihydrofolate reductase, causing, on binding, a cellular deficiency of thymidylate (the cause of its antitumor activity). The crystal structures of the enzyme from two bacterial sources—*Escherichia coli* and *Lactobacillus casei*—and from chicken liver have been studied (88–90). Both the *E. coli* and *L. casei* enzymes have been studied as complexes with methotrexate bound at the active site, and, in the case of the *L. casei* enzyme, the cofactor, NADPH, was also present.

The methotrexate molecule is found to lie in a cavity, 15 Å deep, in the *E. coli* enzyme. The pteridine ring of methotrexate lies nearly perpendicular to the aromatic ring of the *p*-aminobenzoyl glutamate group. The entire inhibitor molecule is bound by at least 13 amino acid residues of the enzyme by a system of hydrogen-bond- and hydrophobic interactions. A similar binding is found for the *L. casei* enzyme complex.

The absolute configuration of tetrahydrofolate (in the Cahn–Ingold notation, 91) is *S* at C6 (92, 93), but if dihydrofolate is bound in the same orientation as methotrexate, the configuration should be *R*. When dihydrofolate (or tetrahydrofolate) is bound in the enzyme in the same way as methotrexate, the hydrogen atom that is transferred to NADPH points in the wrong direction. Thus, dihydrofolate and methotrexate must bind to the enzyme with different orientations as established by X-ray crystallographic studies. The binding of folate and methotrexate are shown in Fig. 38.

11.3 Protein-Catalyzed Reactions

The first crystal structure of an enzyme, that of lysozyme, was determined by D. C. Phillips in 1965 (16). The most striking feature in the three-dimensional structure of this enzyme is a prominent cleft that traverses one face of the molecule. The X-ray structure of lysozyme complexed with a three-residue oligosaccharide showed that this cleft was, indeed, the substrate binding site. The crystal structure of this complex provided the first three-dimensional model for how enzymes work. In its catalytic action, the cleft binds a penta- or hexasaccharide (*ABCDEF*) and the enzyme cuts this between *D* and *E* (Fig. 39). Phillips, however, had only X-ray diffraction data on a bound trisaccharide (ABC) (16). By model building he proposed that glutamic acid 35 transfers its proton to the bridging oxygen atom between sugar rings *D* and *E*. The C–O bond then breaks, leaving a positive charge at C1 that is stabilized by the ionized carboxylate aspartate 52. The *EF* portion of the polysaccharide then leaves, and the carbonium ion

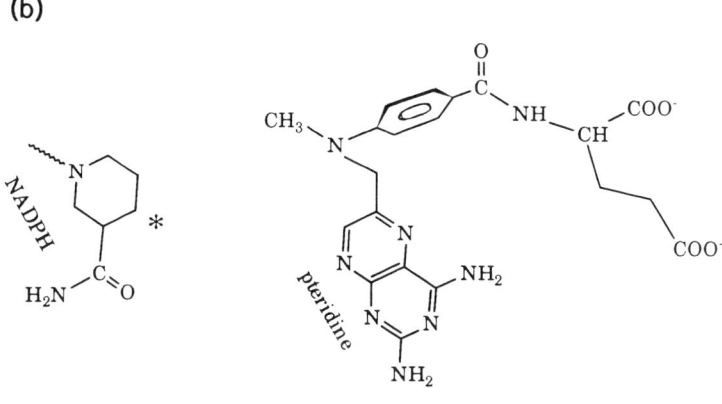

Fig. 38. Binding to dihydrofolate reductase.

interacts with a hydroxyl group, regenerating the active enzyme. Phillips considered that the D-ring is distorted to a half-chair conformation, more like the transition state. Others have suggested that loss of water on substrate binding increases the ability of aspartate 52 to stabilize the transition state. Details of this mechanism are still under investigation.

Considerable structural data for serine proteases are also available. These particular enzymes, so-named because they have a common mechanism, catalyze the breaking of peptide bonds by hydrolysis. The crystal structures of several serine proteases show that the mechanism of hydrolysis is apparently similar for each. The specificity of each enzyme is, however, different and is dictated by the nature of the side chains flanking the scissile peptide bond (the bond that is broken in the catalytic mechanism).

X-ray crystallographic data on chymotrypsin provide a plausible mechanism

(c)

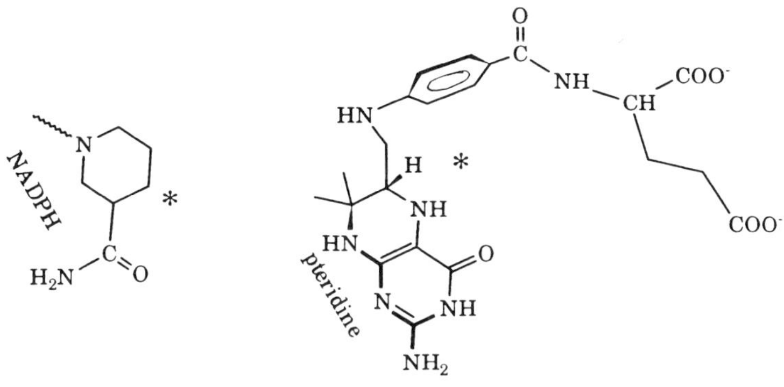

tetrahydrofolate in the same
orientation as methotrexate

(d)

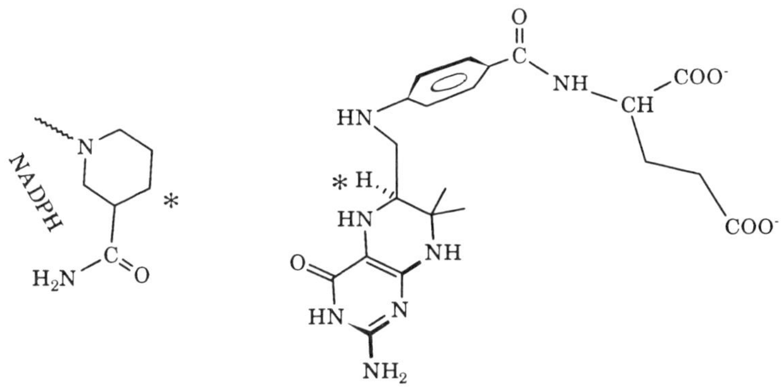

tetrahydrofolate rotated so that
hydrogen can add from the NADPH side

Fig. 38. (*Continued*)

of action of the enzyme (94). In the first step of any catalyzed reaction, the enzyme and substrate form a complex, ES, commonly called the Michælis complex. The hydrolysis of the peptide bond involves three amino acid residues, histidine 57, serine 195, and aspartic acid 102. These form a hydrogen-bonded grouping called a *catalytic triad*. The shape of the active site of chymotrypsin is complementary to the aromatic side chain of its substrates. Serine 195 is then in an ideal position to carry out the required nucleophilic attack on the peptide carbonyl carbon atom. A tetrahedral intermediate is formed. The imidazole ring of histidine 57 accepts the proton thereby liberated. The peptide bond is then cleaved and an acyl-enzyme intermediate is formed (bound through serine 195). The new N terminus of the substrate (now product) is released from the enzyme

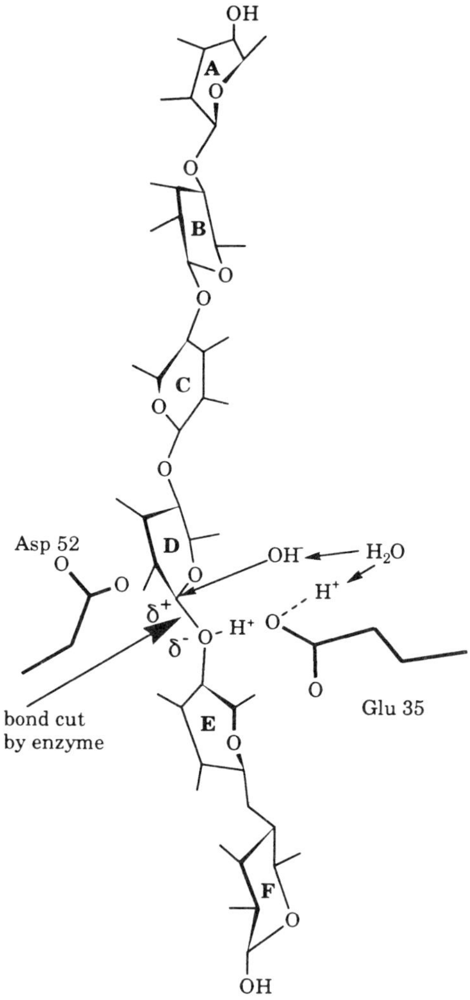

Fig. 39. The mechanism of action of lysozyme, proposed by David Phillips.

and replaced by a water molecule. The acyl-enzyme is then converted back into its native state by the second half of the mechanism, which is a deacylation. A proton is abstracted from a water molecule, and the resulting hydroxide ion attacks the carbonyl group of the acyl group that is covalently attached to serine 195. Histidine 57 donates a proton to the oxygen, the new C terminus of the product diffuses away, and the enzyme is restored to its original active state. The importance of enzyme-bound water to this mechanism has been pointed out.

11.4. Studies with Crystalline Mutant Enzymes

Structural and enzymatic studies of proteins containing a mutation in one amino acid residue, particularly one thought to be catalytically significant in some way, can provide complementary information that aids in the elucidation of the mechanism of action of an enzyme. It is important not to assume that a mutant enzyme necessarily acts by exactly the same mechanism as does the wild-type enzyme. If the mutant enzyme is isomorphous with native enzyme, the phases of the native enzyme can be used to calculate electron-density maps and difference electron-density maps. These will indicate the structural result of the change in the identity of the one amino acid. For example, serine proteases have three residues that are essential for catalysis. The catalytic role of serine 195 and histidine 57 is well established, but the function of aspartic acid 102 is less clear (94). This aspartic acid may either stabilize the conformation of the histidine, maintain the correct tautomer, or stabilize the positive charge that forms during the reaction. A crystallographic analysis was carried out on a mutant enzyme with aspartic acid at position 102 replaced by asparagine (95). This study demonstrated the role of aspartic acid in the native structure. The activity of the mutant enzyme toward a variety of substrates is reduced several orders of magnitude relative to the activity of the native enzyme. In the crystal structure of the mutant enzyme, the histidine 57 does not act as the base that accepts a proton from serine 195. This implies that in the native protein aspartic acid 102 plays a critical role by providing hydrogen-bond stabilization of the functional tautomer and serves to keep the catalytic site (histidine 57 and serine 195) correctly oriented for action.

11.5. Reactions in the Crystal

The elucidation of catalytic mechanisms of enzymes by diffraction methods is very difficult, even when the three-dimensional structure has been determined to a reasonably high resolution. With conventional X-ray sources, diffraction data collection for macromolecules usually takes days, while chemical reactions that are catalyzed by enzymes may occur in fractions of a second. Nonetheless, it has been possible to study catalysis by altering the experimental conditions (e.g., pH, ionic strength, or temperature) such that substrate is much more slowly converted to product. Alternatively, the chemical nature of the substrate and/or

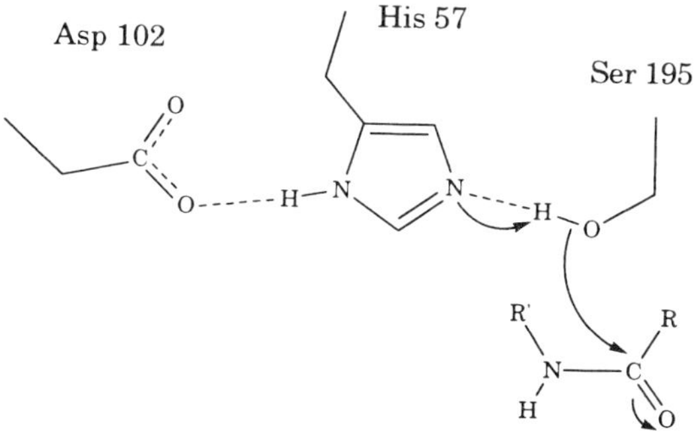

Fig. 40. The catalytic triad in chymotrypsin and a possible mode of action.

enzyme may be altered. As a result, the enzyme-substrate complex can be viewed crystallographically.

The very high-intensity X-ray sources provided by synchrotron radiation give a dramatic reduction in the time needed for diffraction data collection. Transient events that occur on the multisecond time scale can now be investigated. With fast data collection methods, it is possible to obtain a set of crystal structures that represent various stages along the enzyme-catalyzed reaction coordinate. Methods are being developed that will enable the investigator to obtain a direct visualization of the conversion of native enzyme-substrate complex to enzyme-product complex as a function of time (96, 97). This is done by use of Laue methods, in which large numbers of Bragg reflection intensity data are recorded simultaneously.

An example of this method is provided by a study of glycogen phosphorylase b, an enzyme that catalyzes the phospholytic breakdown of glycogen to glucose-1-phosphate. Crystals are catalytically active, but the reaction is too fast to study directly by diffraction methods. A glycosylic substrate analogue, heptenitol, is, however, converted more slowly to heptulose-2- phosphate, presumably by the same mechanism. This substrate analogue provides a useful model system for following the enzyme-catalyzed reaction via crystallographic methods. The rate-limiting step in this reaction is the conversion to product of the Michælis complex, so that the enzyme-substrate complex accumulates transiently in the crystals. Several X-ray diffraction data sets were recorded as a function of time. Each set then provided a snapshot of the reaction

$$\text{heptenitol} + P_i \rightarrow \text{heptulose} \cdot 2 \cdot \text{phosphate}, \tag{13}$$

depicting the formation and transformation of the enzyme-substrate complex into product.

The crystal was mounted in a flow cell (98), substrate solution was flowed over the crystal for about 10 minutes, and Laue photographs were taken with a synchrotron source of white radiation. It was possible to measure over 100,000

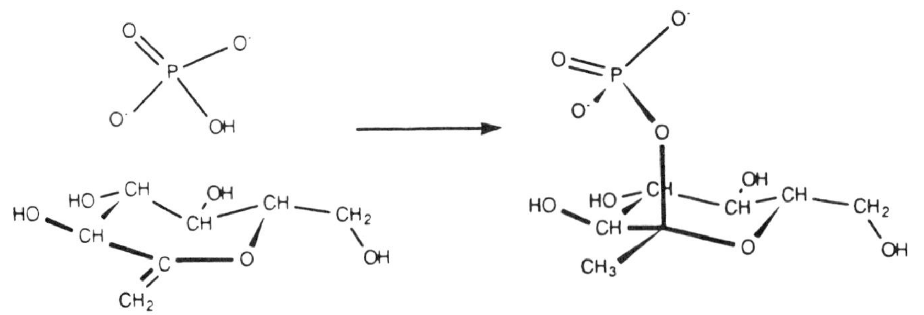

Fig. 41. The reaction of glycogen phosphorylase b studied by synchrotron Laue diffraction.

reflections per second. Data sets, each of one second duration, were taken before, during, and after the initiation of the reaction. The site of binding had already been established by structural work with monochromatic radiation. The course of the reaction was followed by this study (Fig. 41). Unfortunately, if the lifetime of the intermediate is very short (less than 3 seconds), other methods must be used.

12. CONCLUSIONS

There are several pitfalls in the determination of crystal structure of proteins, and some have appeared in published papers, necessitating later retraction. First, the resolution of the diffraction data should always be kept in mind when a crystal structure determination is reported. This limits how well the molecule can be located in the electron-density map. Second, the investigator must assign the correct space group. Third, the reader should check how many heavy-atom derivatives were used and whether or not anomalous scattering was also used in phase determination. It is also necessary to know to what resolution the isomorphism pertained. The more Bragg reflection data, the better the phases. The interpretation of the electron-density map has, in certain cases, caused problems, generally because the electron-density maps were not calculated to a sufficiently high resolution. Finally, any unusual characteristics need further investigation.

ACKNOWLEDGMENT

I thank Drs. H. L. Carrell, G. D. Markham, and H. Roder for helpful discussions. This work was supported by grants CA-10925 and CA-06927 from the National Institutes of Health and by an appropriation from the Commonwealth of Pennsylvania. I thank Pat Bateman for her patient typing of the manuscript.

REFERENCES

1. W. Friedrich, P. Knipping, and M. Laue, *Sitzungsberichte der mathematischphysikalischen Klasse der Königlichen Bayerischen Akademie der Wissenschaften zu München*, 1912, pp. 303–322. English translation: J. J. Stezowski, in *Structural Crystallography in Chemistry and Biology* (ed J. P. Glusker), Hutchinson Ross: Stroudsburg, PA, Woods Hole, MA, 1981, pp. 23–39.
2. E. Abbé, *Archiv für Mikroskopische Anatomie* **9**, 413 (1873).
3. D. Sayre, J. Kirz, R. Feder, B. Kim, and E. Spiller, *Science* **196**, 1339 (1977).
4. G. Binnig, H. Rohrer, C. Gerber, and E. Weibel, *J. Appl. Phys.* **40**, 178 (1982).
5. T. L. Blundell and L. N. Johnson, *Protein Crystallography*, Academic Press: New York, London, San Francisco, 1976.
6. G. H. Stout and L. H. Jensen, *X-ray Structure Determination: A Practical Guide*, Macmillan: London, 1968, 2nd ed., John Wiley: New York, Chichester, Brisbane, Toronto, Singapore, 1989.

7. J. P. Glusker and K. N. Trueblood, *Crystal Structure Analysis: A Primer*, 2nd ed., Oxford University Press: New York, Oxford, 1985.
8. G. Zanotti, in *Fundamentals of Crystallography* (ed. C. Giacovazzo), Oxford University Press: Oxford, 1992, Ch. 8, pp. 535–597.
9. K. H. Baumgärtner, *Beobachtungen über die Nerven und das Blut*. Freiburg, 1830.
10. J. B. Sumner, *J. Biol. Chem.* **69**, 435 (1926).
11. J. H. Northrop, *J. Gen. Physiol.* **13**, 739 (1930).
12. W. M. Stanley, *Science* **81**, 644 (1935).
13. J. D. Bernal and D. Crowfoot, *Nature* **133**, 794 (1934).
14. J. C. Kendrew, R. E. Dickerson, B. E. Strandberg, R. G. Hart, and D. Davies, *Nature* **135**, 422 (1960).
15. M. F. Perutz, H. Muirhead, J. M. Cox, L. C. G. Goaman, F. S. Mathews, E. L. McGandy, and L. E. Webb, *Nature* **219**, 29 (1968).
16. D. C. Phillips, *Scientific American* **215**(5), 78 (1966).
17. J. Deisenhofer, O. Epp, K. Miki, R. Huber, and H. Michel, *Nature (London)* **318**, 618 (1985).
18. A. Ducruix and R. Giegé, *Crystallization of Nucleic Acids and Proteins: A Practical Approach*. IRL Press: Oxford, 1992.
19. H. W. Wyckoff, C. H. W. Hirs, and S. N. Timasheff (eds.), *Methods in Enzymology*, Vols. 114 and 115, Academic Press: London, 1985.
20. L. J. DeLucas, F. L. Suddath, R. Snyder, R. Naumann, M. B. Broom, M. Pusey, V. Yost, B. Herren, D. Carter, B. Nelson, E. J. Meehan, A. McPherson, and C. E. Bugg, *J. Crystal Growth* **76**, 681 (1986).
21. A. McPherson and P. J. Shlichta, *J. Crystal Growth* **85**, 206 (1987).
22. L. K. Steinrauf, *Acta Cryst.* **12**, 77 (1959).
23. W. H. Taylor and S. Náray-Szabó, *Z. Krist.* **77**, 146 (1931).
24. *International Tables for X-ray Crystallography, Volume 1, Symmetry Groups* (eds. N. F. M. Henry and K. Lonsdale), Kynoch Press: Birmingham, 1952 and *Volume A: Space-Group Symmetry* (ed. T. Hahn), 2nd ed., rev., D. Reidel: Dordrecht, Boston, Lancaster, Tokyo, 1987. *International Tables for Crystallography — Brief Teaching Edition of Volume A: Space-Group Symmetry* (ed. T. Hahn), International Union for Crystallography, D. Reidel: Dordrecht, Boston, Lancaster, 1985.
25. E. Hecht, *Optics*, Addison-Wesley: Reading, 1987.
26. *International Tables for X-ray Crystallography, Vol. III. Physical and Chemical Tables* (eds. C. H. MacGillavry and G. D. Rieck), Kynoch Press: Birmingham, New York, 1962, pp. 201–245.
27. W. L. Bragg, *Proc. Camb. Phil. Soc.* **17**, 43 (1913).
28. J. B. J. Fourier, *Théorie Analytique de la Chaleur*, Firmin Didot: Paris, 1822.
29. A. B. Porter, *Phil. Mag.* **11**, 154 (1906).
30. J. M. Bijvoet, N. H. Kolkmayer, and C. H. McGillavry, *X-ray Analyses of Crystals*, Interscience: New York, 1951.
31. J. Waser, *J. Chem. Educ.* **45**, 446 (1968).
32. R. N. Bracewell, *Sci. Amer.* **260**(6), 86–95 (1989).
33. U. W. Arndt and A. J. Wonacott, *The Rotation Method in Crystallography*, North-Holland: Amsterdam, 1977.
34. J. R. Helliwell, *Macromolecular Crystallography with Synchrotron Radiation*, Cambridge University Press: Cambridge, U.K., 1992).
35. H. D. Bartunik, H. H. Bartsch, and H. Qichen, *Acta Cryst.* **A48**, 180 (1992).
36. A. C. T. North, D. C. Phillips, and F. S. Mathews, *Acta Cryst.* **A24**, 351 (1968).
37. A. J. C. Wilson, *Nature* **150**, 152 (1942).
38. G. Friedel, *Comptes Rendus, Acad. Sci. (Paris)* **157**, 1533 (1913).

39. B. W. Low and F. M. Richards, *J. Amer. Chem. Soc.* **74**, 1660 (1952).
40. B. W. Matthews, *J. Mol. Biol.* **33**, 491 (1968).
41. S. L. Chang, *Multiple Diffraction of X-rays in Crystals*, Springer-Verlag: Berlin, Heidelberg, New York, Tokyo, 1984.
42. H. Hauptman and F. Han, *Acta Cryst.* **D49**, 3-8 (1993).
43. J. M. Cork, *Phil. Mag.* **4**, 688 (1927).
44. C. Bokhoven, J. C. Schoone, and J. M. Bijvoet, *Acta Cryst.* **4**, 275 (1951).
45. D. W. Green, V. M. Ingram, and M. F. Perutz, *Proc. Roy. Soc.* **A225**, 287 (1954).
46. A. L. Patterson, *Phys. Rev.* **46**, 372 (1934).
47. A. L. Patterson, *Z. Krist.* **A90**, 517 (1935).
48. H. F. Judson, *The Eighth Day of Creation*, Simon and Schuster: New York, 1979.
49. D. Harker, *J. Chem. Phys.* **4**, 381-390 (1936).
50. M. G. Rossmann and D.M. Blow, *Acta Cryst.* **15**, 24 (1962).
51. M. G. Rossmann (ed.), *The Molecular Replacement Method*, Gordon and Breach: New York, 1972.
52. R. A. Crowther and D. M. Blow, *Acta Cryst.* **23**, 544 (1967).
53. P. M. D. Fitzgerald, *J. Appl. Cryst.* **21**, 273 (1988).
54. D. Coster, K. K. Knol, and J. A. Prins, *Z. Physik* **63**, 345 (1930). English translation: J. J. Stezowski, in *Structural Crystallography in Chemistry and Biology* (ed. J. P. Glusker), Hutchinson Ross: Stroudsburg, PA, Woods Hole, MA, 1981, pp. 158-160.
55. J. M. Bijvoet, A. F. Peerdeman, and A. J. van Bommel, *Nature* **168**, 271 (1951).
56. C. Bokhoven, J. C. Schoone, and J. M. Bijvoet, *Proc. K. ned. Akad. Wet.* **52**, 120 (1949).
57. J. R. Herriott, L. C. Sieker, L. H. Jensen, and W. Lovenberg, *J. Mol. Biol.* **50**, 391 (1970).
58. W. A. Hendrickson and M. Teeter, *Nature* **290**, 107 (1981).
59. W. A. Hendrickson, A. Pähler, J. L. Smith, Y. Satow, E. A. Merrit, and R. P. Phizackerley, *Proc. Natl. Acad. Sci. USA* **86**, 2190 (1989).
60. W. Hoppe and J. Gassmann, *Acta Cryst.* **B24**, 97 (1968).
61. B. C. Wang, *Methods in Enzymology* **115**, 90 (1985).
62. A. G. Leslie, *Acta Cryst.* **A43**, 134 (1987).
63. G. Bricogne, *Acta Cryst.* **A30**, 395 (1974).
64. E. T. Jaynes, *Phys. Rev.* **106**, 620 (1957).
65. D. M. Collins, *Nature* **298**, 49 (1982).
66. G. Bricogne, *Acta Cryst.* **A44**, 517 (1988).
67. T. A. Jones, *J. Appl. Cryst.* **11**, 268 (1978).
68. A. Jack and M. Levitt, *Acta Cryst.* **A34**, 931 (1978).
69. J. L. Sussman, in *Proceedings of the Daresbury Study Weekend—Refinement of Protein Structures* (eds. P. A. Machin, J. W. Campbell, and M. Elder), SRC Daresbury Laboratory: Warrington, U.K., 1981, pp. 13-23.
70. A. T. Brünger, J. Kuriyan, and M. Karplus, *Science* **235**, 458 (1987).
71. S. O. Smith, S. Farr-Jones, R. G. Griffin, and W. W. Bachovchin, *Science* **244**, 961 (1989).
72. C. Branden and J. Tooze, Introduction to Protein Structures, Garland: New York, 1991.
73. L. Pauling and J. Sherman, *J. Chem. Phys.* **1**, 606-617 (1933).
74. M. S. Newman, *J. Chem. Educ.* **32**, 344-347 (1955).
75. G. N. Ramachandran and V. Sasisekharan, *J. Mol. Biol.* **7**, 95-99 (1963).
76. F. C. Bernstein, T. F. Koetzle, G. J. B. Williams, E. F. Meyer, Jr., M. D. Brice, J. R. Rodgers, O. Kennard, T. Shimanouchi, and M. Tasumi, *J. Mol. Biol.* **112**, 535-542 (1977).
77. M. J. E. Sternberg, D. E. P. Grace, and D. C. Phillips, *J. Mol. Biol.* **130**, 231 (1979).

78. L. Pauling, R. B. Corey, and H. R. Branson, *Proc. Natl. Acad. Sci. (USA)* **37**, 205 (1951).
79. L. Pauling and R. B. Corey, *Proc. Natl. Acad. Sci. (USA)* **37**, 251 (1951).
80. J. S. Richardson, *Adv. Protein Chem.* **34**, 167 (1981).
81. M. Levitt and C. Chothia, *Nature* **261**, 552 (1976).
82. T. Alber, D. W. Banner, A. C. Bloomer, G. A. Petsko, D. C. Phillips, P. C. Rivers, and I. A. Wilson, *Phil. Trans. Roy. Soc.* **B293**, 159–171 (1981).
83. R. H. Kretsinger, and C. E. Nockolds, *J. Biol. Chem.* **248**, 3313(1973).
84. N. C. J. Strynadka and M. N. G. James, *Annu. Rev. Biochem.* **58**, 951 (1989).
85. K. Namba, R. Pattanayek, and G. Stubbs, *J. Mol. Biol.* **208**, 307–325 (1989).
86. T. L. Poulos, B. C. Finzel, and A. J. Howard, *J. Mol. Biol.* **195**, 687–700 (1987).
87. T. L. Poulos and A. J. Howard, *Biochemistry* **26**, 8165–8174 (1987).
88. J. T. Bolin, D. J. Filman, D. A. Matthews, R. C. Hamlin, and J. Kraut, *J. Biol. Chem.* **257**, 13650–13662 (1982).
89. D. J. Filman, J. T. Bolin, D. A. Matthews, and J. Kraut, *J. Biol. Chem.* **257**, 13663–13672 (1982).
90. D. A. Matthews, J. T. Bolin, J. M. Burridge, D. J. Filman, K. W. Volz, B. T. Kaufman, C. R. Beddell, J. N. Champness, D. K. Stammers, and J. Kraut. *J. Biol. Chem.* **260**, 381–391 (1985).
91. R. S. Cahn, C. K. Ingold, and V. Prelog, *Experientia* **12**, 81–124 (1956).
92. J. C. Fontecilla-Camps, C. E. Bugg, C. Temple, Jr., J. D. Rose, J. A. Montgomery, and R. L. Kisliuk, *J. Amer. Chem. Soc.* **101**, 6114–6115 (1979).
93. W. L. F. Armarego, P. Waring, and J. W. Williams, *J. Chem. Soc., Chem. Commun.* 334–336 (1980).
94. D. M. Blow, *Acc. Chem. Res.* **9**, 145–152 (1976).
95. S. Sprang, T. Standing, R. J. Fletterick, R. M. Stroud, J. Finer-Moore, N. H. Xuong, R. Hamlin, W. J. Rutter, and C. S. Craik, *Science* **237**, 905–909 (1987).
96. K. Moffat, D. Szebenyi, and D. Bilderback, *Science* **223**, 1423 (1984).
97. J. Hadju, K. R. Acharya, D. I. Stuart, D. Barford, and L. N. Johnson, *Trends Bio. Sci.* **13**, 104 (1988).
98. H. W. Wyckoff, M. Doscher, D. Tsernoglou, T. Inagami, L. N. Johnson, K. D. Hardman, N. N. Allewell, D. M. Kelly, and F. M. Richards, *J. Mol. Biol.* **27**, 563 (1967).

Transmission Electron Microscopy and Scanning Probe Microscopy

KAREN L. KLOMPARENS AND JOHN W. HECKMAN, JR.
Center for Electron Optics, Michigan State University, East Lansing, Michigan

1. Introduction
2. The Transmission Electron Microscope
    2.1. Image Formation
    2.2. Advantages and Disadvantages of TEM
    2.3. Specialized Auxiliary Equipment
        2.3.1. Scanning Transmission Electron Microscope
        2.3.2. Energy Dispersive X-ray Microanalysis
    2.4. Specimen Preparation
        2.4.1. Ultrathin sectioning
        2.4.2. Target-specific Labeling
            2.4.2.1. Pre-embedding General Experimental Procedure
            2.4.2.2. Postembedding General Experimental Procedure
            2.4.2.3. Controls
            2.4.2.4. Quantitation
            2.4.2.5. Double Labeling
            2.4.2.6. Cryosectioning
            2.4.2.7. *In Situ* Hybridization using Gold Probes
        2.4.3. High-resolution Autoradiography
            2.4.3.1. Flat Substrate Method
            2.4.3.2. *In Situ* Hybridization
        2.4.4. Negative Staining
            2.4.4.1. Stain Characteristics
            2.4.4.2. Support Films
            2.4.4.3. Staining Procedures
            2.4.4.4. Beam Damage
            2.4.4.5. Immunosorbent Electron Microscopy
        2.4.5. Shadowing and Replica Techniques
            2.4.5.1. General Preparation Techniques
            2.4.5.2. Molecular Spreading
            2.4.5.3. Evaporation Techniques
            2.4.5.4. Freeze–etch Techniques

Bioanalytical Instrumentation, Volume 37, Edited by Clarence H. Suelter.
ISBN 0-471-58260-3 © 1993 John Wiley & Sons, Inc.

            2.4.5.5. Replica Interpretation
            2.4.5.6. Other Applications
        2.4.6. Energy Dispersive X-ray Microanalysis
3. Scanning Probe Microscopy
    3.1. Image Formation
    3.2. Advantages and Disadvantages of SPM
    3.3. Specimen Preparation
4. Conclusions
4. References

1. INTRODUCTION

In 1986, the Nobel Prize in Physics was awarded jointly to Ernst Ruska, Gerd Binnig, and Heinrich Rohrer. Ruska, of the Fritz Haber Institute in West Berlin, was recognized for his research on electron optics and for the original design of the transmission electron microscope in the 1930s. Binnig and Rohrer, of the IBM Zurich Research Laboratory, shared the award for their design of the scanning tunneling microscope, research conducted in the 1980s. In presenting the Prize for research spanning a half-century, the Nobel Committee noted the significance of electron microscopes in all areas of science, calling them "one of the most important inventions of the century."

When it first became commercially available in the 1955, the *transmission electron microscope* (TEM) provided the highest imaging resolution available to study cellular ultrastructure, as well as to analyze a wide variety of isolated subcellular components. With the advent of molecular biology techniques, the TEM is still the instrument of choice to allow correlation of the now abundant molecular data and probes to their location and function in cells. The TEM should be considered within the full range of image collection devices including *scanning EM*, *light microscopy* (LM), *laser scanning confocal microscopy* (LSM), and *scanning probe microscopy* (SPM) (Table 1).

Often, the most appropriate approach to a research problem in any area of study, is to use one or more of these microscopes to corroborate other data. Images of specimens can provide the necessary visual data to correlate with biochemical or physiological measurements. Correlative research that takes advantage of the capabilities of LM and LSM (1) can also extend the usefulness of TEM data. Although the SEM has unique capabilities and is often complementary to the TEM, its application to biochemical research is limited by its resolution of 3 nm–6 nm and its mostly topographical (surface) images. The SEM will not be discussed here, but the reader is referred to two excellent treatments of this instrument and its uses (2,3). The family of scanning probe microscopes, including scanning tunneling and atomic force microscopes, are just beginning to contribute to atomic resolution imaging of biologically interesting specimens.

Detailed descriptions of the theoretical and practical aspects of TEM operation are covered in numerous texts (4–6). For the purposes of this chapter, we

TABLE 1

Comparison of Selected Characteristics of Microscopes

	Resolving Power	Magnification Range	Special Features
Light microscope	200 nm	5–1500X	Live samples Fluorescence
Laser scanning confocal microscope	100nm	5–64,000X	Live samples Fluorescence Z-sectioning
Transmission electron microscope	0.2nm	500–500,000X	Internal structures
Scanning electron microscope	3–6nm	10–250,000X	Surfaces
Scanning tunneling/atomic force microscopes	≤0.2nm		Surfaces

present a brief overview sufficient to provide the fundamentals necessary to an understanding of the effects of instrument design and operation on the data contained in the final image. Any treatment of the uses of TEM in biochemical research would be remiss without discussion of the hybrid instrument: the scanning transmission electron microscope and the auxiliary analytical device called the energy dispersive X-ray microanalyser. These are introduced in the context of their use in high-resolution imaging and analytical data collection.

This chapter was written to provide a brief discussion of instrumentation, present the advantages and disadvantages of specimen preparation techniques for biochemical research, assess the sources of error and artifact, provide selected examples of applications, and outline general experimental procedures, where they can be prescribed appropriately.

2. THE TRANSMISSION ELECTRON MICROSCOPE

The TEM is roughly analogous to a compound light microscope, that is, it can image particulate samples and ultrathin sections or replicas through samples. The TEM produces data on the ultrastructural morphology of samples, size, and shape of macromolecules, and with specialized specimen preparation, can be used to localize antigens, enzymes, gene products, and elements at the subcellular level.

Transmission EM images are of higher resolution than a LM, by a factor of 1000, and a laser scanning LSM by a factor of 500 (Table 1). This increase in resolving power is due largely to the smaller wavelength of electrons as compared to either light or lasers. Wavelength itself is inversely related to both mass and velocity of the energy source. Hence, the wavelength can be decreased by increasing the velocity of the particles. The shorter wavelengths characteristic of electrons and the ability to accelerate them result in increased resolving power over the longer wavelengths of light.

Although some principles of optics may be applied to both LM and TEM, image formation in a TEM is fundamentally different than that of either a LM or LSM. An understanding of the process of image formation is essential for illustrating the specific applications of such microscopy to biochemical research.

2.1 Image Formation

The process of image formation in a TEM can be described by combining several principles of physics and must be interpreted in the context of the physical design of the microscope. To this end, the general design and function of the microscope components will be combined with the necessary theoretical considerations in the brief description that follows. Voluminous literature exists that describes the TEM in greater detail, and the reader is referred to two texts for more comprehensive information (4, 5).

Pared to its most essential components, a standard TEM consists of an electron beam source, a column of lenses to control the beam and render magnified images, a viewing and recording system, and a system to ensure a high vacuum environment for operation. Of course, the microscope is also designed with the requisite electronic and mechanical components to render it operable. The entire microscope column is kept at high-vacuum, 10^{-6} torr (approximately 10^{-3} Pa). The vacuum system in the TEM is crucial to its operation, because it prevents extraneous scattering events between the electron beam and air or water molecules that would interfere with the desired scattering events between the electron beam and the sample that produce the image. The vacuum system prohibits the imaging of any living (wet) samples (7).

The electron source is a filament of tungsten wire or a crystal of lanthanum hexaboride mounted on a tungsten wire. As a voltage is applied to the filament, electrons are emitted from the tip and are attracted toward an anode. The typical range applied to a filament in a TEM is between 20,000 and 200,000 volts. Just below the anode is the first of the series of electromagnetic lenses that function as the basis of the imaging process. As the current is changed in a lens, the beam of electrons is deflected to control illumination, focus, and magnify the image, depending on the specific lens function. The two condenser lenses at the top of the column function to control the intensity of illumination on the specimen for viewing and recording the images. The second condenser lens is fitted with an aperture that defines the diameter of the cone of illumination that will strike the specimen and also shields the specimen from too many stray electrons.

The next four lenses, *objective*, *diffraction*, *intermediate*, and *projector*, are the magnifying lenses. The objective lens is the most important and complex lens in a TEM. The specimen itself is immersed in the lens and is moveable in X, Y, and Z directions, as well as 60° tilt in each direction and rotation through approximately 360°. Since the specimen itself is a major source of contaminants, the objective lens is also fitted with an anticontaminator that functions as a cryopump to remove most local contaminants.

When the electron beam strikes the specimen, a number of interactions occur that ultimately form the final image. Since these interactions are both numerous and complex, and well documented (3–5), only a brief description of those important for imaging biological, generally noncrystalline samples will be included here.

Inelastic and elastic scattering events are the key interactions in TEM image formation of non-crystalline samples. Absorption, which constitutes the major interaction leading to LM image formation, does not contribute appreciably to TEM images. An exception is the absorption of electrons into stain puddles or dirty areas of the sample. In this case, the electrons contribute to localized heating and distortion of the sample. Diffraction, a scattering of the electron beam through a specific angle, occurs in crystalline samples and can contribute to image contrast.

Inelastically scattered electrons are produced when beam electrons strike the electrons in the atoms of the sample. The beam electrons are scattered through small angles and suffer a loss of energy. Elastically scattered electrons are produced when beam electrons strike the nuclei in the atoms of the sample. These electrons are scattered through large angles with little or no energy loss. In a standard TEM bright-field image, most of the inelastically scattered electrons are transmitted to the viewing screen and most of the elastically scattered electrons are not. The extent to which the proportion of inelastically and elastically scattered electrons occur depends on the **mass-thickness** of the sample. Hence, the thickness of the sample and the variations in atomic number of the various atoms making up the sample affect the specific scattering events. This differential scattering from one portion of the sample to another results in the contrast that forms the final image.

Terms used to describe the contrast are *electron transparent* (*lucent*) and *electron dense*. Electron transparent-areas scatter fewer electrons and appear bright in the image, whereas electron dense areas scatter more electrons and appear dark. Discrete areas of electron transparency and electron density, then, are responsible for the contrast required for image formation.

An objective lens aperture placed just below the sample is used to enhance contrast. A smaller aperture will screen out more electrons creating more contrast. The choice of a small aperture size, however, must be balanced by the resulting increase in aperture contamination and subsequent objective lens astigmatism. Decreasing accelerating voltage can also increase contrast, but here again, a compromise exists. Decreasing accelerating voltage results in a dimmer image and decreased resolving power.

The specimen–beam interactions that form the image also produce damage to the specimen. Careful specimen preparation can help minimize damage. The deposition of hydrocarbons onto the sample (contamination) can be reduced greatly by a good vacuum system and use of the objective lens liquid nitrogen anticontaminator. Mass loss due to electron beam radiation is the most serious problem (8, 9), and can be somewhat reduced, but not completely eliminated.

The most effective means to reduce mass loss appears to be with the use of a cold specimen stage. Different samples lose mass at different rates, but most lose some percent and then are quite stable thereafter. Mass loss becomes important in several quantitative procedures, such as mass mapping of macromolecules and elemental analysis.

Electromagnetic lenses have three major aberrations that can reduce the resolution and image quality in a TEM: spherical and chromatic aberration and astigmatism. Spherical and chromatic aberration are constituitive to microscope design, but astigmatism can be corrected by the user. Astigmatism occurs when the lens symmetry is altered by differing magnetic strengths in two directions at 90° to each other, and hence the resulting electron beam is no longer round. The cause is usually contamination on the lens aperture, and the result is a blurred image that cannot be focused. Each lens that requires it is equipped with a stigmator that is used to correct this defect by compensating for the asymmetrical magnetic strength. Objective lens astigmatism is an especially serious problem in the high magnification imaging of macromolecules, viruses, etc. It is crucial to identify objective lens astigmatism and correct it, especially before each high magnification image is photographed.

2.2. Advantages and Disadvantages of TEM

As stated previously, the fundamental advantage of a TEM over the light microscope is the increase in resolving power. This allows imaging of cell ultrastructure and a myriad of subcellular organelles and components. Although the great magnifying capability of the TEM is often cited as the advantage, in reality, without the resolving power, additional magnification provides no additional information.

The TEM can be used to corroborate other data and is a powerful imaging tool when used along with the light microscope, laser scanning confocal microscope, scanning electron microscope, and/or scanning probe microscope. The data produced by the TEM can be complementary to those produced by other imaging devices, as well as those data produced by biochemical, physiological, and molecular means.

Disadvantages of TEM depend on the researcher's point of view. Since the TEM operates under high vacuum conditions, live specimens cannot be imaged. In addition, specimen preparation for TEM is not trivial. In fact, in some cases, it is as much art as technology. In general, proper specimen preparation requires both auxiliary equipment and expertise. The more complex the research questions, the more complex and tedious the specimen preparation. This, however, can be approached - from a more positive stance. The vast capabilities of the TEM can be exploited by the selection of specific specimen preparation protocols that are suited to specific experimental designs.

Generally, images in the TEM are two-dimensional, because the depth-of-focus of the imaging lenses is greater than the thickness of most biological specimens. Tilting the specimen and viewing stereo pairs, especially of specimens prepared as replicas or that are shadowed, can provide three-dimensionality, as can the more equipment- and time-intensive three-dimensional reconstructions. Reconstructions can be done with serial sections in the conventional TEM mode, or can be accomplished using digital imaging in the scanning transmission EM mode.

The cost of acquiring and maintaining a TEM and the expertise required for some forms of specimen preparation are often considered as obstacles to use. Multiuser facilities can be both effective and cost-efficient ways to provide access to state-of-the-art equipment and the requisite expertise. There are also national resource facilities available for some techniques and specialized equipment.

2.3. Specialized Auxiliary Equipment

A specialized instrument that is a hybrid between a TEM and a scanning EM and is used to view transmitted images is the scanning transmission electron microscope. Auxiliary equipment can be attached to a TEM or a STEM to enhance its capabilities, the most common one being the energy dispersive X-ray microanalyser. These two instruments are described in the following two sections.

2.3.1. SCANNING TRANSMISSION ELECTRON MICROSCOPE

The scanning transmission electron microscope (STEM) can be either a "dedicated" or "nondedicated" instrument. A nondedicated STEM is a TEM on which scanning coils and a separate imaging screen are attached. It produces a conventional TEM image or, when the scan coils are activated, it produces a transmitted image, but one that is formed by a very small beam of electrons that is scanned across the sample surface. The STEM can image sections that are thicker than those that can be imaged in a conventional TEM. Thicker sections are an advantage when doing serial reconstructions and for increasing the signal for X-ray microanalysis (10–13). Because the image is produced point-by-point as the beam scans across the sample, digital imaging capabilities are possible. While the nondedicated STEM can provide advantages for imaging and X-ray microanalysis, the dedicated STEM provides better resolution and additional methods of signal collection especially useful for imaging macromolecules (14).

A dedicated STEM is an instrument that has no postspecimen lenses and images only in the scanning transmission mode. It is generally equipped with a field emission electron source, which is very bright and coherent allowing for maximum resolution (10).

Such an electron source requires very high vacuum, so the instrument is designed with this in mind. These instruments are highly specialized and not as ubiquitous as a conventional TEM; however, the Brookhaven National Laboratories in Upton, New York operates a dedicated STEM as a national resource for research requiring such capabilities.

As Engel and Reichelt (13) point out, there are several imaging modes that are used in STEM, including bright-field, dark-field, and a combination of the two. The STEM can be used to study and map periodic structures (14) since the signal can be digitized and processed. Aebi et al. (15) provide an excellent discussion of the uses of STEM in imaging protein structure and protein-protein interactions, and the recommended methods are described in the specimen preparation section.

Leapman and Andrews (16) discuss the advantages and disadvantages of using a dedicated STEM to examine frozen macromolecules and to conduct mass mapping and elemental analysis. Imaging and analysis of macromolecules, or most biological samples, can be limited by electron beam radiation damage (mass loss) to the sample and by alterations in structures that occur as a consequence of the preparation methods (see section below on specimen preparation). Use of the STEM dark-field signal, that is, collecting the elastically scattered electrons, is particularly useful for mass mapping of unstained proteins because the signal is related linearly to the mass-thickness of the portion of the sample directly under the electron beam (17, 18). In fact, mass mapping of macromolecular weights using low electron beam doses is the most recognized application of STEM in biology.

Traditional methods of measuring mass, e.g., centrifugation and chromatography, rely on averaging a population of molecules. Sequencing would constitute the most accurate technique. The STEM can analyze single macromolecules, while distinguishing the size and shape of the molecule at the same time. Since the intensity of the signal reaching the dark-field detector is linearly related to the number of atoms under the beam and the STEM images one spot at a time, it is possible to directly calculate the mass. The background from the specimen substrate is subtracted, and the resulting intensity multiplied by a calibration constant. Tobacco mosaic virus is used as an internal standard for calculating the calibration constant.

Three additional advantages of STEM mass mapping are that very small amounts of specimen are needed, mass per unit area or length can be calculated, and heterogeneous populations can be examined (17, 18). Engel et al. (19) successfully mapped the hexagonally packed intermediate layer from the cell envelope of *Micrococcus radiodurans* using STEM. They achieved 3 nm resolution with morphology that corroborated images obtained with negative staining. The mass mapping technique also allowed them to examine the mass of individual domains of the monolayer.

Johnson and Wall (20) conducted mass mapping of dynein ATPase and produced a unique image of the structure, which they termed a *three-headed bouquet*. The STEM permitted measurement of molecular weight of the asymmetric dynein and its individual domains and provided high-resolution imaging. Mosesson et al. (21) conducted similar studies on human fibrinogen and provide a thorough discussion of the mathematics required to conduct mass mapping and to calculate expected errors.

Mass mapping is not without possible errors in accuracy. Mass loss accounts for inaccuracies, especially in proteins. Wall and Hainfield (17) report that a dose of $10e^-/0.1 nm^2$ is a reasonable compromise between structural preservation and accuracy of mass measurements. Use of a cold specimen stage is requisite. Since proteins, nucleic acids, and lipids suffer mass loss at different rates, a dose-response curve is recommended as a starting point for each specific experiment. Proper specimen preparation, especially a clean background, is essential for accurate mass mapping. The STEM itself may contribute random errors, such as electron beam fluctuations.

Many of the specimen preparation procedures described in the next few sections apply to the STEM. Imaging of frozen-hydrated specimens in combination with image processing, however, appears to allow for the optimum structural preservation (at least the minimum perturbation) and contrast in STEM imaging. Mass mapping requires placing the sample on very clean carbon-coated specimen grids and freeze-drying at a microscope vacuum level for optimum results.

2.3.2. ENERGY DISPERSIVE X-RAY MICROANALYSIS

A second common auxiliary device is an energy dispersive X-ray microanalyser. X-ray microanalysis is based on the collection and segregation of the X rays that are generated via inelastic scattering events from the interaction of the primary electron beam and the electrons of the atoms of a specimen. Such X rays can be characterized either by their wavelength or by their energy, and each is specific to the element that produces it. Sorting of X rays by wavelength is termed *wavelength dispersive spectroscopy* (WDS) and is performed using a microprobe. Since EDS is more common than WDS in biological applications, it will be considered here. The reader is referred to Goldstein et al. (3) and Anderson (22) for further information on WDS.

Sorting of X rays by energy is termed *energy dispersive spectroscopy* (EDS) and can be accomplished by adding a detector and its attendant electronics and computer to a TEM (or STEM) column. This device, with its X-ray detector immersed in the objective lens, can provide elemental analysis of particulate or ultrathin specimens (Fig. 1) (23–27). Elements from sodium to einsteinium can be detected using a standard EDS system. In addition, qualitative data on elemental concentration can be produced, and compositional mapping of elemental location can be conducted.

Specialized detectors and/or systems permit identification of elements to a Z of 5 (boron). Although three common elements in biological samples, carbon, oxygen, and nitrogen are theoretically within the detection range of EDS, in reality, they are not well localized with this technique. Nitrogen and carbon can be detected using WDS, and oxygen can be detected with an additional technique called *electron energy loss spectroscopy* (not covered in this chapter).

The fundamental theory of EDS is based on the production of X rays as one of the resultant signals from specimen-beam interactions. As an electron from the beam interacts with electrons from atoms of the sample (inelastic scattering), it may knock an electron out of a shell. An electron from an outer shell will move into the inner shell and consequently loses energy in the form of an X ray. Each element has characteristic X-ray energies based on the movement of electrons from an M shell to an L shell and an L shell to a K shell. Approximately 1% of the

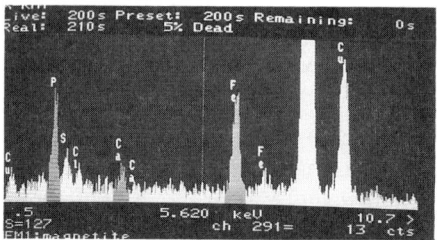

Fig. 1. Energy dispersive X-ray microanalysis spectrum of magnetite granules in the abdomen of a honeybee. The iron (Fe) peaks are from the magnetite granules, phosphorus (P), and calcium (Ca) are from the specimen. The chlorine (Cl) is from the resin, and the copper (Cu) is from the specimen support grid and is also a system peak. Specimens were fixed in glutaraldehyde and embedded in Spurr's firm epoxy resin. Sections were carbon coated before X-ray analysis was conducted in the STEM mode on a JEOL 100CX II equipped with a Link Systems AN 10000 EDS. The authors acknowledge Uko Zylstra, Calvin College, Grand Rapids MI for providing the sections of honeybee abdomen.

X rays produced strike the detector and are then counted and processed by the EDS computer. Thus, EDS on a TEM (or STEM) can determine the specific elements present in a specimen and their relative concentrations. Most EDS units are also capable of constructing compositional maps in order to localize individual elements, as well as the spatial relationship of one element to others (24).

Several caveats must be stated in regard to EDS. One is that for biological samples, the best minimum detectable concentration is approximately 0.004%, with a more realistic limit of ≤0.01% (25) and, in the TEM, spatial resolution is on the order of ≤50nm-100nm, depending on the specific technique and specimen preparation used. Higher resolution is attainable with a dedicated STEM equipped with a field emission electron source and freeze-dried samples. In addition, specimen preparation must be particularly rigorous, since accuracy of detection localization and quantitation depend absolutely on retaining the elements of interest in precisely the same location as in the living state.

Quantitation of X-ray data can be problematic for biological samples, since it is very difficult to prepare suitable standards (26, 27). Accurate quantitation is also difficult, unless it is conducted with consideration for mass loss. It is estimated that proteins may lose as much as 30% of their mass at electron doses common in electron microscopes (8). Although structural detail may be lost, macromolecules may retain enough stability to be analysed using the low-dose electron energy loss. Johnson et al. (25) present an excellent technical discussion of errors in EDS quantitation.

A combination of STEM and either energy dispersive X-ray microanalysis (see below) or electron energy loss can detect low elemental concentrations at the subcellular level. Leapman and Andrews (16) reported the ability to quantitate calcium in Purkinje cell dendrites to an accuracy of $\langle 12$ atoms. A low temperature stage was necessary to reach-this resolution, although, at this writing, a method of transferring frozen samples into the dedicated STEM is soon to be commercially available.

2.4 Specimen Preparation

While correct instrument operation is important, proper specimen preparation accounts for 90% of the quality of the data gathered from the TEM. Hence, the major part of this chapter will be devoted to specimen preparation procedures. It is the diversity of procedures that have been developed to prepare biological specimens for viewing in the microscopes that actually defines their potential applications to biochemical research.

Specimen preparation requires patience, practice, and meticulous attention to cleanliness and detail. While there are many specimen preparation techniques that exist, six will be discussed here, as they are the most useful for biochemical research. These techniques are sufficiently complex and multifaceted that a separate chapter could be written on each. In some cases, specific procedures must be empirically derived for each type of specimen and experimental condition. For these, we refer the reader to excellent reviews of a particular subset of specimen preparation procedures or to specific papers that contain a well-developed methods section. For more generic procedures, we provide a general description of the protocol.

2.4.1. ULTRATHIN SECTIONING

Ultrathin sectioning is used to prepare specimens that are too large to view directly with the electron beam, e.g., animal and plant tissues, fungi, bacteria and cultured cells, and organelle preparations. It is the basis for most of the immunolabelling and *in situ* hybridization techniques. Both the theory and practice of ultrathin sectioning have been well described in a number of texts and laboratory manuals (28–31), so it is sufficient here to describe only the general procedure as a basis for the special applications for biochemical research.

Ultrathin sectioning is the penultimate step in a sequential protocol involving fixation, dehydration, and embedding of a specimen. The protocol itself is devised to preserve the specimen in as lifelike a state as possible, to prepare it for the process of ultrathin sectioning, exposure to the vacuum and electron beam of the microscope, and to impart enough contrast to permit imaging; all this is done with a minimum of introduced artifacts.

The initial step of biological specimen preparation for ultrathin sectioning can be accomplished by one of two methods: *chemical fixation* or *ultrarapid freezing*. Chemical fixation uses a sequential protocol beginning with an aldehyde followed by a second fixation with an organometallic. The method has become quite standardized and requires no special equipment. Ultrarapid freezing generally requires auxiliary equipment, but can result in better-preserved membrane structures and antigen and elemental localization over chemical fixation.

A brief comparison will suffice to define the methodology and uses of these two techniques.

Chemical fixation begins with immersion of small (less than 1mm^3) pieces of tissue or subcellular preparations in a low concentration, 0.5%-4% aldehyde, usually glutaraldehyde or a mixture of glutaraldehyde and formaldehyde (prepared from paraformaldehyde). Glutaraldehyde cross links with both soluble and membrane-bound proteins in the specimen to form complexes of polymers (31). The reaction releases H$^+$ into the solution, therefore, a suitable buffer is necessary to maintain the pH at the physiological level of the specimen. After a number of washes to remove unreacted aldehyde, the specimen is placed into a 0.5%-1% solution of osmium tetroxide, or a similar chemical. Osmium cross-links unsaturated lipids and, because of its high atomic number, imparts a useful degree of contrast to the image.

Ultrarapid freezing is a recently developed alternative to chemical fixation. A number of methods can be employed, including plunge freezing, slam freezing, propane jet freezing, and high-pressure freezing (32-34). With these methods, a small sample is rapidly frozen using a cryogen, such as liquid propane, and stored under liquid nitrogen until further processing. The high-pressure freezer operates by placing the specimen 1n a chamber, raising the pressure to approximately 2100 atmospheres, and spraying the sample with jets of liquid nitrogen. The use of high pressure depresses the freezing point of water to $-22\,°C$ and, thus, greatly reduces ice crystal growth (a common artifact of poorly frozen samples).

Following one of the methods of fixation, the sample is dehydrated. With chemical fixation, conventional liquid phase dehydration is carried out in either anhydrous acetone or alcohol. Ultrarapidly frozen samples are dehydrated by immersing the frozen sample in a bath of solvent (methanol, ethanol, or acetone) held at $-80\,°C$ for several days, a process called *freeze substitution*. For freeze-substituted samples, osmium tetroxide is usually dissolved in the dehydrating solvent in order to aid in fixing the tissue and imparting contrast in the image.

Once dehydrated, the sample is infiltrated and embedded in an epoxy or modified acrylic resin. Most epoxy resins commonly in use can interfere with postembedding immunolabeling, resulting in reduced, although accurate, labeling and have a high content of chlorine that can be detected during X-ray microanalysis. Modified acrylics, such as LR White, can be used for enhanced immunolabeling, or cryo-ultrathin sections can be used (see pg. 89). For X-ray analysis, a standard should be prepared using the same epoxy resin as was used for the sample. There are numerous artifacts associated with the process of ultrathin sectioning, and care must be taken to minimize these, as they can affect image quality and some experimental procedures such as immunolabeling (30, 35).

The final step in the preparation of bulk preparations is to positively stain the ultrathin sections. Positive staining is done with heavy metals, such as uranyl ions and lead, to impart additional contrast to the image for viewing. Straightforward ultrathin sectioning is used to study morphology and anatomy, to make measurements of size and purity of cell fractions (Figure 2), and as a basis for immunolabelling and other techniques as described in the next sections.

2.4.2. TARGET-SPECIFIC LABELING

There are currently many methods used to conduct target-specific labeling and many refinements of each of these methods designed for individual specimens and many experimental variables (36, 37). Target-specific labeling can be conducted using either pre- or postembedding (in resin), employing either an antibody conjugated directly to a particulate probe, e.g., colloidal gold or indirectly via a secondary antibody or Protein A conjugated to gold. The affinity for an enzyme and its substrate, and for a lectin and its specific sugar residue can also be used to perform target-specific labeling at the ultrastructural level (38). These techniques have been used successfully to label a variety of targets at the subcellular level. The highly specific avidin–biotin interaction provides an additional method to label areas of interest (39, 40) and has been used recently with polymerase chain reaction for *in situ* labeling at the LM level (41). Thus, judicious decisions and careful planning at the onset of the experiment are imperative in order to choose the most appropriate method for the specific sample and experimental conditions in order to produce useful data.

As with most specimen preparations for TEM, methods for target-specific labeling are not trivial. Quality results are, however, quite attainable, provided there is meticulous attention to detail. In most cases, decisions concern striking a balance between acceptable morphology and optimum localization of labeling along with the practical considerations of expediency and cost. Antibodies with high specificity for the antigen are an absolute requirement for successful im-

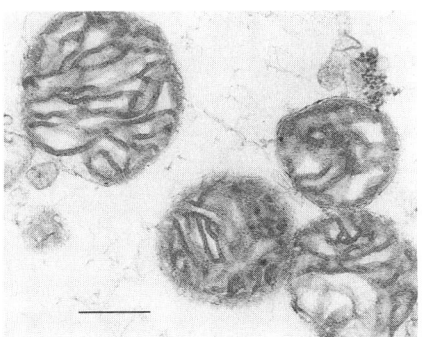

Fig. 2. Transmission electron micrographs of mitochondria isolated from chick heart. Mitochondria were prepared according to standard procedures. Micrographs were used to assess the intactness of the organelles resulting from the isolation technique. Mitochondria were provided by Dr. Peter P. Toth, University of Iowa Hospitals and Clinics. Bar = 1 μm.

munolabeling. Antibody specificity should be assessed by ELISA or Western blot analysis before attempting EM-level labelling.

There is no one correct method for any given sample or experimental circumstance! Only a general outline of procedures can be provided, and this must be varied for each specimen and each set of experimental conditions by empirically derived modifications. Kellenberger and Hayat offer an excellent introduction to the theory of immunolabeling (42). Childs (43) presents an extensive discussion of immunolabeling, and describes the advantages and disadvantages of pre- and postembedding labeling. Bendayan (44–50) has provides numerous reviews of immunolabeling methods, as well.

Protein A-gold immunolabeling is, perhaps, the most commonly used method for localizing antigens with a particulate probe. Protein A can be complexed to colloidal gold which serves as a electron dense marker in the TEM. The colloidal gold can be prepared by reducing chloroauric acid (44) or can be purchased from a number of suppliers either alone or conjugated to Protein A. Protein A-gold can be used in the pre-embedding mode for cell surface antigens and isolated subcellular fractions, and in the post-embedding mode for tissue sections. Protein A can bind to the IgG of most mammalian species and IgM of a few species. Protein A does not bind to mouse IgG1 or IgM; an important factor when choosing a method for use with mouse monoclonal antibodies (51, 52). Protein A will also bind to any endogenous sources of IgG in a specimen and has been reported by Behnke et al. (53) to bind nonspecifically to specimens with a high affinity for the charged surface of the gold particles. For some applications, then, gold conjugated directly to the primary antibody, or to a secondary antibody, may prove more useful. Controls are of fundamental importance for accuracy of labeling interpretation and especially for quantitation (see pg. 88).

2.4.2.1. Pre-embedding General Experimental Procedure. The pre-embedding technique is reasonably straightforward as compared to the postembedding method. It can, however, label only surfaces of the samples, unless a soluble probe, e.g., biotin, is used in a system involving active uptake of the target molecule. While there is no one protocol that will work for all samples and/or experimental conditions, a general procedure would follow these steps: Samples are exposed first to a buffer containing 1% ovalbumin or powdered milk to block nonspecific binding sites. Incubation with the antibody, at an empirically derived concentration, time, and temperature is the next step. A 1:200 dilution of 1 mg/ml IgG for 2 hr at room temperature is a reasonable scheme to start with. The sample is then rinsed in decreasing dilutions of buffer, ending with a solution containing 1% ovalbumin as milk protein. Incubation with the probe (Protein A-gold or secondary antibody-gold) follows. Concentration of the probe, incubation time, and temperature must also be arrived at empirically, although a 1:10 dilution of Protein A for 1–2 hours at room temperature should give results. Samples are given a final rinse and then examined using negative staining (for small particulates) or prepared by fixation, embedding, and ultrathin sections (for cell cultures and other samples requiring sectioning). Controls are identical to those discussed in the section on page 87–88.

2.4.2.2. Postembedding General Experimental Procedure. Labeling using the postembedding method requires that samples be adequately fixed to preserve morphology, but must retain antigenicity (54) (Figure 3). In general, low concentrations (1-4%) of freshly prepared formaldehyde (from paraformaldehyde) or a combination of formaldehyde and 0.1%-1% glutaraldehyde are used. Osmium tetroxide is usually avoided, although there are exceptions to all of these generalizations and final protocols are usually derived empirically. Embedment in LR White or LR Gold (modified acrylic resins) is recommended for TEM and in glycol methacrylate for corroborative LM. Labeling is performed on ultrathin sections that have been collected on nickel grids.

The labeling protocol is similar to that described for preembedding labeling. Before labeling, however, an etching step using 0.4% sodium metaperiodate for 15 min-1 hr may be required to expose antigens, especially if osmium was used as a fixative. Care must be taken not to have the sections dry out, and floating is preferred to immersion in order to minimize background.

Postembedding labeling limits the Ab-Ag interactions to only those at the *surface* of the section. The specimen image in a TEM consists of the entire thickness of the specimen. The *image* may contain antigenic sites that are expected to, but are not labeled, but these may be *within* the section and simply not exposed to the labeling solutions. Hence, image interpretation must be carefully conducted with this in mind. Specific protocols for labeling in plant systems are described by Herman and Melroy (55), and those for labeling enzymes are provided by Sternberger (56) and Bendayan (45).

2.4.2.3. Controls. Controls are of fundamental importance to the accurate and credible interpretation of target-specific labeling experiments. There are three primary controls: 1) incubation with probe (Protein-A gold or secondary anti-

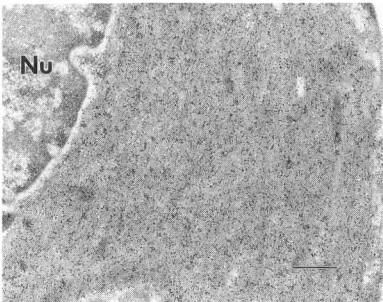

Fig. 3. Transmission electron micrograph of smooth muscle cells of rat small intestine labeled with antiactin antibody followed by gold-labeled goat antirabbit IgG. Specimens were fixed in 2% formaldehyde and 0.1% glutaraldehyde in 0.1M phosphate buffer. Final embedding was in Lowicryl K4M resin. Micrograph courtesy of Dr. Susan J. Hagen, Division of Gasteroenterology, Brigham and Women's Hospital, Harvard Medical School, Boston MA. J. Elec. Micros. Tech. 16:37-44 (1990). Hagen and Trier, H. Histo. and Cyto. 36:717-727 (1988). Nu = nucleus. Bar = 0.5µm.

body-gold) alone to assess nonspecific binding of probe (especially if Protein A is used) to sample or to resin, 2) incubation with primary Ab that has been exposed to an excess of antigen in order to confirm specificity, and 3) incubation with preimmune serum, then probe (Protein A gold or secondary antibody-gold) to assess specificity of the IgG. An additional control that confirms the specificity of IgG and Protein A gold binding can also be performed (44). For plant samples, a control that uses a "nonsense" antibody raised in the same manner as the antibody of interest is recommended. There is some evidence that use of Freund's adjuvant in the preparation of polyclonal antibodies can produce nonspecific binding to plant cell wall components (personal communication, G. de Zoeten, Michigan State University).

2.4.2.4. Quantitation. If comparisons are to be made between areas of a sample or between samples, quantitation should be performed. Subjective evaluations of areas with "more labeling" or "less labeling" are increasingly unacceptable. There are several quantitative methods available (57, 58) that will provide credibility to such comparisons. Griffiths and Hoppeler (57) discuss two approaches to quantitation of immunolabeling. One, which they term relative quantitation, is used to compare amounts of antigen in the same or different organelles or in one organelle under different experimental conditions. These methods can be as simple as counting the gold particles per unit area. Absolute quantitation is used when the number of antigens per unit area or volume of an organelle in a section is desired.

Two major factors influence the accuracy of quantitation: steric hindrance of the antigen–antibody complex and differences in the penetration capacity of the antibody into the surface of the section. Clearly, steric hindrance is a greater problem in samples with a high density of antigens than in those in which the antigens are more dispersed. Penetration into resin sections may be avoided by the use of cryosections. Of course, the compromise here is the difficulty in obtaining a large number of quality cryosections.

Posthuma et al. (58) reports the use of the immunogold technique to quantitate soluble protein in ultrathin sections, using a model system using a matrix of gelatin and cryoultrathin sections. Their choice of embedding in polyacrylamide (PAA) followed by cryoultrathin sectioning demonstrated adequate penetration of the PAA into the gelatin and minimized the problem of matrix density that contributes to error in quantitation of gold cryoultrathin sections.

2.4.2.5. Double-labeling. Double-labeling is also possible with immunocytochemical methods (59). In this case, using Protein A or a secondary antibody bound to gold markers of different sizes can differentiate between antigen sites. An especially refined method of double-labeling the same tissue section with two different antibodies and two sizes of gold probe was described by Bendayan (44). In this method, one side of a grid is floated on the labeling solutions for one antibody of interest, the grid dried and turned over, and floated on solutions for the second antibody of interest. Since the TEM image consists of the entire thickness of the section, both of the different size gold markers, one set on top and the other on the bottom, can be viewed simultaneously. Alternatively, pre-embedding techniques can be used to label one antigen of interest and post-embedding techniques used for the second antigen.

2.4.2.6. Cryosectioning. For some antigens in very low titer, the above techniques may not be successful. Cryoultrathin sectioning, developed by Tokuyasu (60), followed by labeling, may be the method of choice. This method is perhaps the most technically complex and difficult of all TEM specimen preparation techniques. A well-frozen sample is a necessity and this method also requires an ultramicrotome with the cryounit attachment. The specifics of this technique and experimental procedures have been discussed in great detail in two reviews (60, 61).

Often the best approach is to employ several techniques. Ultrathin sectioning and labeling of resin sections provides a straightforward method of preparing many samples, but may have reduced antigenicity. Cryoultrathin sectioning may provide the optimum conditions for maximizing labeling, but is much more difficult in terms of providing many samples for replications. A specific application of this two-pronged approach is presented in the following section.

2.4.2.7. **In Situ *Hybridization Using Gold Probes.*** Another target-specific labeling technique is *in situ* hybridization at the ultrastructural level. Newmark (62) suggested in "An embarrassment of peptides," that one of the next steps in molecular biology should be to determine the physiological function of the numerous peptides that have been defined using molecular biology. In this way, TEM level *in situ* hybridization, sometimes referred to as *hybridization histochemistry*, can be used to localize genes or their transcripts (63–67). Such labeling can be done using radiolabeled nucleotides (see section on autoradiography) or biotinylated probes. The specific protocols for preparing biotinylated probes are detailed by Bayer and Wilchek in a previous volume in this series (39).

Singer and colleagues (68, 69) have successfully visualized cytoskeletal mRNAs and their associate proteins using biotinylated cDNA probes followed by antibodies to biotin and colloidal gold-conjugated antibodies. Their method used *in situ* hybridization followed by whole mount TEM of Triton-extracted chicken embryo fibroblasts. Cytoskeletal mRNAs were found in close proximity to actin protein and further from tubulin filaments. While the whole mount technology does limit the technique to extracted cells, applications to thin sections will allow greater resolution.

Wenderoth and Eisenberg (70, 71) present compelling evidence for using a two-pronged approach for *in situ* hybridization. Their study localized myosin heavy chain mRNA in cardiac tissue and compared frozen and LR White embedded specimens. The LR White embedded samples showed high structural integrity, but lower numbers of gold probe (Fig. 4a), while the cryosections were more highly fragmented, but showed greater numbers of gold probe (Fig. 4b). From their results, it appears that relative concentration of message can be demonstrated, but that current methods do not permit an analysis of absolute numbers.

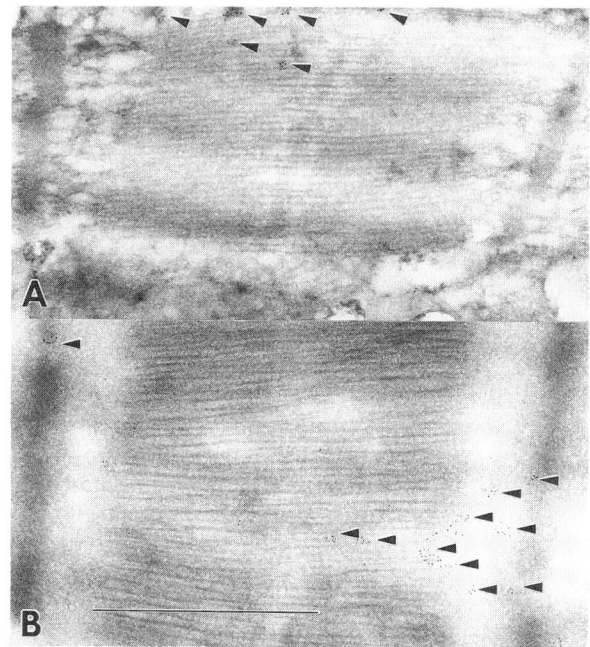

Fig. 4. *In situ* hybridization at the ultrastructural level of localized myosin mRNA at the myofibril borders isn rabbit heart muscle. Tissue was hybridized with biotinated antisense mRNA riboprobe. Biotin was detected with an antibiotin primary antibody and a gold-labeled secondary antibody. A. Ultrathin section from LR White embedded muscle. B. Cryoultrathin section from rapidly frozen muscle. A complete morphometric analysis showed the same distribution for both embedding methods. Cryoultramicrotomy yielded higher absolute numbers of gold clusters, but poor morphological quality. The LR White method was preferred because of good ultrastructural integrity even though there was a lower signal. Micrographs courtesy of Dr. Brenda Russel-Eisenburg, Physiology and Biophysics, University of Illinois, Chicago, IL. Wenderoth and Russell-Eisenberg, J. Histo. and Cyto. 39:1025–1033 (1991). Russell et al. Keystone Conference on Molecular Biology of the Muscle, R339–345 (1992). Bar = 1μm.

McFadden et al. (72) reported a straightforward method of preparing plant specimens in order to use both ^{32}P-labeled DNA probes and biotinylated DNA to localize complementary RNA and rRNA in ultrathin sections of plant tissues. Their method using phase-partition fixation in 4% paraformaldehyde and 0.25% glutaraldehyde, if successfully applied to other cell systems, may overcome the technical problems usually encountered due to mRNAs being difficult to retain in using standard fixation procedures. Such procedures for TEM generally denature and/or mask mRNA rendering the probe ineffective at labeling. Labeling of Vibratome or cryostat sections prior to embedding provide an alternative to minimize this problem. Another avenue to avoid loss of RNA is the use of a low-temperature resin, such as Lowicryl K4M.

Use of biotinylated probes instead of tritium-labeled probes can provide faster signal detection with a similar signal-to-noise ratio. For example, Binder et al (73, 74) used biotinylated dUTP incorporated into mitochondrial rDNA followed by antibiotin antibody conjugated to Protein A-gold and compared that procedure to tritium-labeled rDNA. Their specific protocols were designed for *Drosophila* ovary tissue and are well detailed in their report. As expected, use of the TEM provided superior spatial resolution as compared to LM. In their system, a low concentration of glutaraldehyde (0.1%) with 4% paraformaldehyde was essential for retaining a strong signal.

For some research, Fab fragments can also be used for labeling. These are especially useful for mapping of complex structures such as protein arrays. Such labeling can be used to identify structural domains or can be used to localize functional sites. Aebi et al. (15) illustrated Fab labeling of the giant phage capsid and provided a filtered STEM image to clearly distinguish labeled from nonlabeled capsomeres. Obviously, this technique is limited if the array of interest is labile (damaged) during the labeling protocol.

Thiry (75–77) presents a unique method of localizing DNA on ultrathin sections. He incubated ultrathin sections in terminal deoxynucleotidyl transferase and various nonisotopic biotinylated nucleotide analogues. The biotinylated nucleotide analogues were immunolabeled by either monoclonal antibody coupled to colloidal gold or by antibiotin antibodies to the biotinylated nucleotides. This method appears to be useful with a variety of fixatives and embedding media, thereby maintaining the quality of standard TEM structural and morphological preservation. The exact details of the protocol for this method are well documented by Thiry (75).

Viral RNA can be localized in tissues using biotinylated probes (78, 79), and both light and TEM *in situ* hybridization can localize other mRNAs (80). As should be obvious from the preceeding discussion, each tissue type, probe, and experimental design is unique. There is no "cookbook" labeling protocol.

2.4.3. HIGH-RESOLUTION AUTORADIOGRAPHY

High-resolution autoradiography is an additional method used for target-specific labeling (81–89). This method employs a radioactive label that is incorporated into the specimen prior to fixation and embedding. In general, a low-energy emitter, such as tritium, is the label of choice because of the very thin sections used in TEM. Detecting the label requires the production of metallic silver as a result of the interaction of the product of radioactive decay and a photographic emulsion that has been overlaid on an ultrathin section of the treated sample. Silver scatters electrons from the beam creating the contrast that is necessary for visualization. With careful attention to the specific specimen-source geometry (87), the silver tracks can be correlated to the underlying specimen image.

High-resolution autoradiography requires rather long periods of time for completion as compared to the other target-specific labeling methods described previously. Because the sample bulk in an ultrathin section is so small (less than 100nm thick), only very small amounts of radiolabel are actually present within each experimental section. Even for a low-energy emitter, a monolayer of emulsion is required in order to be able to quantitate the location of the label. Because of these limitations, parallel use of AR at the light level, which uses larger samples ("thick" sections of 1-2 μm), is highly recommended.

Resolution of the TEM-AR method is dependent on the section thickness, emulsion thickness, energy of the radionuclide used, and the developer used to reduce the silver halide to metallic silver. Estimates of resolution range from 70nm to 100nm. This is less resolving power than gold labeling, but is a useful technique when dealing with a target of interest to which antibodies cannot be raised. The flat-substrate procedure is the most common and is described in the next section.

2.4.3.1. Flat-substrate Method. The first step is to coat clean, frosted-end microscope slides with 0.4% collodion. These coated slides are allowed to dry and then are used to support thick sections for light microscope-level autoradiography or ultrathin sections for TEM-level autoradiography. Thick sections may be spaced at intervals along the length of the slide. Ultrathin sections are placed in a row at a distance approximately 5mm from the nonfrosted end of the slide. It is helpful to prepare a cardboard template of the slide with the area to receive the sections well marked, since precision is required when applying the emulsion and when retrieving the sections.

After the sections have been collected on the slides, a thin (3nm-4nm) coating of carbon is applied. Liquid emulsion is applied to the slide according to the manufacturers instructions. For light microscope-level autoradiography, a thicker layer of emulsion is acceptable than is required for TEM-AR. For optimum TEM resolution, a monolayer of emulsion is requisite. One or two blank slides may be coated to use later as a test of the storage conditions. Once the emulsion is dry, the slides are stored in a light-tight box containing a small packet of silica gel or other drying agent. The boxes are wrapped in 1-2 layers of aluminum foil and stored at 4°C. Depending on the initial level of radioactive label, amount of incorporation, and section thickness, the autoradiograms will be ready to develop in 30-60 days. The thick sections for light microscopy can be tested 2-3 weeks after coating with emulsion.

Autoradiograms are developed according to the manufacturer's instructions. One-half strength D-19 or Microdol (both from Kodak) are usually recommended. Agitation while developing, washing, and fixing is not necessary. The light microscope slides can be stained and/or coverslipped before examination. The TEM sections must be released from the slides by floating the plastic with the sections off onto the surface of a dish of water. Hydrofluoric acid can be used to etch the microscope slide and thereby aid in release of the plastic from the slide. A grid can be placed on top of each group of ultrathin sections and the entire sandwich retrieved from the water using paper or a screen.

Once dry, the grids can be picked up, positively stained with uranyl acetate and/or lead citrate and examined in the TEM.

2.4.3.2. In Situ Hybridization. In general, EM-level autoradiography is a compromise between the desire for better resolution than is provided by the light microscope and the need for sufficient efficiency to produce a meaningful Ag signal. While traditional TEM-AR may be useful for some studies, it has lower resolution than most of the immunolabeling techniques. Despite this disadvantage, autoradiography at the ultrastructural level for in situ hybridization using radiolabeled nucleotides is possible (90-92). The method is successful with whole mount chromosomes (65), and for localizing ^{32}P-labeled DNA probes hybridized to complementary RNAs in resin-embedded ultrathin sections of plant samples (72). Tritiated cDNA probes require long exposure times and generally give less-than-desired resolution due to grain size and geometry effects of the AR process, making them less useful than a biotinylated probe (see Section 2g). Raikhel et al. (92) provide a detailed set of procedures for *in situ* hybridization of RNA in plant tissues.

A number of hybridization protocols for light microscopy using radiolabels, including those using synthetic oligonucleotides and intracellular amplification of DNA using polymerase chain reaction, are available (41, 93, 94). Coghlan et al. (95) provide an excellent review of hybridization histochemistry at the LM level that serves as an important foundation upon which to build similar techniques using oligonucleotides and PCR at the TEM level. Included in their review is a comprehensive list of references for hybridization procedures for a variety of tissues at the light level.

At this writing, the technique using PCR has been applied to only a few systems at the TEM level, so it is not possible to prescribe a general protocol. As methods develop in this area, especially those that incorporate biotinylated probes, rather than radioactive probes, the resolution of the technique will improve. Chiu et al. (93) provide the most detailed procedures for such PCR amplification of DNA in tissue sections at the LM level.

2.4.4. NEGATIVE STAINING

Negative staining is, perhaps, the most commonly used TEM preparation technique for biochemical research. The technique derives its name from the specimen's characteristic in which it is surrounded by a contrasting medium. If this medium is more than twice the density of the specimen, the image of the specimen remains electron lucent while the surrounding area is dark (Fig. 5a). Negative stains are salts of heavy metals that exist in a disassociated form in an aqueous medium and dry down to form a fine-grained film.

There are several reasons for the wide use of negative staining. First, the majority of biochemical studies involve, at least in part, macromolecular structures or cell fractions. Subjects in this size range are generally small enough to be viewed directly, if the surrounding area is sufficiently electron dense. Second, as the heavy metal salts dry, they form a cast around the biological material forming a relatively stable structure. The sample seldom requires chemical fixation, and

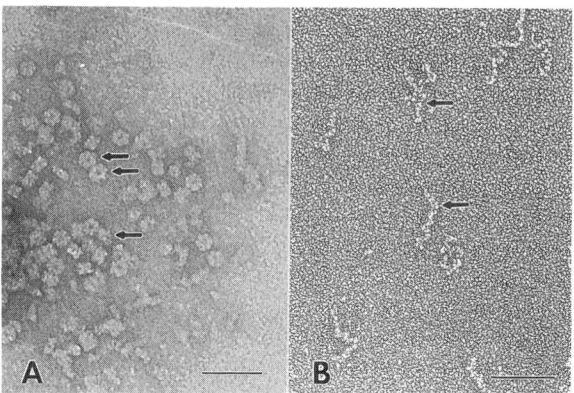

Fig. 5. Glutamine synthetase (*E. coli* (arrows) negatively stained with sodium silicotungstate prepared by the method of Valentine *et al.* (98). This rapid and simple technique, occasionally with some individual modifications, still provides one of the highest resolution molecular images. Bar = 0.1μm. B. Low angle rotary shadowcast of molecules of gum arabic glycoprotein (arrows) printed in reversed contrast. This technique allows easy visualization of extended macromolecules that are difficult to see with negative staining; these preparations are highly resistant to beam damage. Bar = 50nm.

the resulting image is, thus, closer to the living state. Third, the technique itself is generally preformed quickly, with less effort as compared to other preparation techniques. Last, it potentially forms the highest resolution method of imaging an electron lucent specimen.

2.4.4.1. Stain Characteristics. While any element with greater electron scattering power (2×2 $Z_{specimen}$ or higher) than the specimen could theoretically be incorporated into a negative stain, there are a number of other important characteristics. First, in order to be a negative stain, the stain material should not react (*e.g.*, bind to or alter molecular structure) with the sample itself under the conditions of application (*e.g.*, pH, buffers, *etc.*). Second, it should be highly soluble in water so that it dries fine grained and nearly amorphous. Third, it should be able to withstand the electron beam at least long enough to capture an image at the desired level of resolution. Three groups of compounds having these characteristics have been employed routinely since negative staining was first reported (96) for TEM. These are compounds based on molybdenum, tungsten, and uranium.

Molybdenum is used for negative staining in the form of ammonium molybdate $(NH_4)_6MO_7O_{24}$). This stain has been used for a number of membrane studies and yields the greatest ultrastructural details with the least damage (97). Ammonium molybdate (AM) is soluble in 2.3 parts of water but is quite insoluble in alcohol. The pH of a 1% solution is about 5.0 and can be adjusted to about pH 8.0 with NH_4OH.

Tungsten, in the form of a sodium or potassium salt of phosphotungstic acid ($24WO_4 \cdot 2H_3PO_4 \cdot 48H_2O$) or silicotungstic acid (98) ($Na_4(Si[W_3O_{10}]_4) \cdot 20H_2O$), has been used as a negative stain since 1955 (99). These salts are prepared by titration of the acid to the desired pH. They are very beam stable, fine grained, and resist recrystallization. In addition, they will tolerate high buffer concentrations.

Uranium is generally used in the form of uranyl acetate. It can be purchased as such, and stain preparation is merely a matter of dissolving the appropriate quantity in water prior to use.

Uranyl acetate is generally found in the biological EM lab, since it is also used in routine positive staining of ultrathin sections. For negative staining, a saturated solution (about 4%) works well. Uranyl acetate works best at an acidic pH; it is unstable at pH above 6.0. For staining at higher pH, uranyl oxalate may be a better choice where graininess isn't a major factor (97). Uranyl acetate can be used at higher pH values if Na_2EDTA is added to suppress the precipitation of uranyl hydroxide.

There are a number of other high atomic number compounds that have been tried as negative stains; some of the these have been tailored to specific applications (*e.g.*, 100). The best advice for the practitioner embarking on a new project, with no specifics in the literature, is to try one or more of the three types mentioned above at sensible pH and concentrations and to compare results.

2.4.4.2. Support Films. Since the specimens to be negative stained are generally in solution, some support medium is needed to mount the preparation on a grid for observation. While there are a number of choices in this regard, plastic and carbon are the two main compositional categories.

Plastic films can be formed from a number of polymers; two of the most common are collodion (nitrocellulose) and polyvinyl formal (Formvar). A number of other plastic materials can be used for thin films, some with specific advantages (101). Plastic films are usually prepared either by spreading a drop of solution of the plastic (in an appropriate solvent) onto a clean water surface, or by stripping off a film that has been coated onto a clean, flat surface (such as a glass slide) onto the surface of clean water.

In order to cast a film by spreading a drop on to a water surface, the plastic of choice is dissolved in a nonwater soluble solvent, such as isoamyl acetate for collodion or parlodion, at a concentration of 0.5%–4%. Higher concentrations produce thicker films. A single drop [*ca.* 10μl–20μl (microliters)] of this solution is dropped onto the dust-free surface of a dish of distilled water, and the solvent is allowed to evaporate. The water surface can be made scrupulously clean by casting two such films in series, discarding the first with (hopefully) all of the dust particles entrapped.

To cast the film from a flat surface, a similar solution of plastic is made and a clean surface (*e.g.*, glass microscope slide) is dipped into the solution, drained and allowed to dry. One side of the film is detached by scratching the edge with a razor blade, and the slide is gently dipped in to a dish of dust-free water at an angle of about 45°. Fogging the coated slide slightly with one's breath seems to help detach the film, and with a little practice intact films can be removed from both sides of the slide by alternating the angle of inclination during immersion.

After the film is floating on the water, grids are laid carefully out on the surface in rows with about 1mm between grids to facilitate subsequent handling. Generally the grids are positioned with the rough "dull" side down with the thinking that the rough side will make a stronger bond with the film. Areas of the film that have a gray to silver interference color usually have the best combination of strength and thickness (ca. 30nm–50nm) for most applications.

There are a number of ways to retrieve the grid/plastic layer from the water surface. One of the easiest is to pick up the grid/plastic layer with paper laid carefully on top (the paper needs to cover a larger surface then the plastic film for this to work well) and then lifted up off the water surface. Alternatively, a paraffin- (Parafilm) coated slide is brought down across the grid/plastic layer at an angle and then picked up.

In addition to continuous films, plastic films can be made with holes in them. These are useful in viewing certain specimens, especially for crystallography, without the background of a support film. General methods for making them are reviewed by Hayat (29). All plastic films provide strong flexible supports for many negative staining specimens, but they do have some drawbacks. Uncoated plastic films are subject to considerable mass loss under the electron beam. In general, nitrocellulose films seem to be less stable than the Formvar films in this regard. Both types of plastic films can be rendered considerably more stable by evaporating a thin (ca. 5nm) coating of carbon on one surface.

Pure carbon can be used to produce support films and for many applications, this seems to be the best choice. Several negative staining techniques are predicated on the mechanical properties of these films, as well. Although thicknesses of about 10nm are usually used, extremely thin carbon films can be produced at as little as 1nm–2nm, that still retain sufficient strength for subsequent processing steps. In addition to a high mechanical strength in thin dimensions, carbon films are exceptionally beam stable. Carbon films often tend to be hydrophobic, which may interfere with specimen spreading in some negative staining applications. This tendency can be removed by briefly glow discharging them prior to use. Glow discharge equipment is available from most vacuum evaporator manufacturers or can be fashioned in the lab (e.g., 102).

Carbon films can be made by evaporating a layer of carbon over a plastic-coated grid prepared as outlined above and removing the plastic afterwards. This is easily done by placing the grids carbon and plastic side up on a pad of filter paper saturated in the appropriate plastic solvent. Several changes can be approximated by transferring the grids to different areas of the filter paper pad. Several 5-min steps seem to work well for this. Usually, a thicker layer of carbon is applied than would be used for just stabilizing plastic films. A coating that forms a distinct gray color change on a piece of white paper works well.

A second approach to making carbon films is to form them on a smooth surface, then remove them, in an analogous manner to the production of plastic films mentioned above. The support surface of choice is freshly cleaved mica, since this is a crystallinely flat surface. This method also allows the manipulation of the carbon film without the grid attached, which is necessary for some techniques.

To prepare these carbon films, a piece of mica is pried apart by starting an intersheet break by inserting a sharp probe. Once started, the halves are pulled apart easily with fine forceps and placed in the vacuum evaporator, cleaved side up. A carbon film about 5nm–10nm thick is then deposited at high vacuum. If film-coated grids are being made, the film-covered mica is scored into grid-sized squares with a needle and immersed slowly in distilled water at about a 45° angle. The film squares float free and can be picked-up from below with bare copper grids.

One of the advantages of negative staining is the relative ease of specimen preparation. Specimen isolation, to the extent that it has been accomplished for other biochemical techniques, is usually sufficient. Negative staining can be used routinely to image crude clinical samples (*e.g.*, blood, feces, urine, plant tissues) with little or no postcollection processing, as well as to image material purified to the crystalline state.

2.4.4.3. Staining Procedures. There are many methods of applying the specimen and negative stain to the supporting grid. One or more of the following approaches should be suitable for any initial investigation.

The simplest approach is to apply a drop of the specimen in solution to the surface of an appropriately coated grid, wait 20 sec–60 sec, blot off excess solution from the edge of the grid with a piece of absorbent paper, and apply the negative stain solution before the grid dries. The excess negative stain is wicked off to near dryness by touching the edge of the grid to absorbent paper. This is conveniently done by supporting the grid over the edge of a glass slide on one edge of which a piece of low-tack double-stick tape has been applied. The specimen and stain drops can then be accurately applied with an adjustable microliter pipet. This method requires only about 2 µl of sample solution per preparation.

A second approach is to float the coated grid on the surface of a specimen droplet, which has been formed on a hydrophobic surface, such as a paraffin film. The specimen adsorbs to the coated surface of the grid. After about a minute, the grid is transferred directly to a stain droplet on the same hydrophobic surface, left for a few seconds, then picked up and blotted at the edge until near dryness.

The spray approach requires the smallest amount of both specimen and stain. While this can be done with a pharmaceutical nebulizer, it is accomplished most easily by using a modified airbrush, such as that described by Tyler and Brandton (103). This device uses a plastic microliter pipet tip to hold the stain specimen mixture and allows the areosol dispersion of as little as 5 µl–10 µl of the mixture. The target in this case is a coated grid. Since the droplets of stain/specimen are so small (5 µl–20 µm in diameter), they dry-down rapidy, so there is less time for any drying defects to occur. When using either the nebulizer or airbrush approach, one should exercise care in the control of the aerosol overspray, since there is a potential health hazard from all of the negative stains and possibly from the specimen, too.

One of the most useful methods of negative staining of macromolecules is the carbon sheet adsorption aproached used by Valentine (98). In this method, a thin carbon film is generated on a freshly cleaved mica surface, scored with a needle, and then partially submerged in a volume of the specimen solution (a porcelain spot plate works well). This allows the film to detach partially from the mica and to float free. After allowing the sample molecules to attach to the lower surface of the film, the mica and film are withdrawn and the film is floated off on a volume of negative stain. Clean fine-mesh grids are laid on top of the film and then picked up on a piece of paper or Parafilm, as are the coated grids (see above). An interesting variation on this technique uses a pleated carbon film. This is done by pushing the film gently into the edge of the staining dish, which then sandwiches the specimen in a thin "tube" of negative stain (104). This often yields a more uniform preparation (*e.g.*, 105).

2.4.4.4. Beam Damage. As higher- and higher-resolution images are sought, the effect of the electron beam damage on the biological specimen becomes more important. With UA-stained TMV, Williams and Fischer (106) note that beam exposure long enough to expose a micrograph (within 30 sec) causes significant damage: periodicity of the capsomeric subunits is obliterated. Such beam damage could be alleviated to a degree by offsetting the projector lens while scanning and focusing, and returning it to centration for micrography at low beam intensity. Another approach to this is to use low-temperature observation methods (107).

2.4.4.5. Immunosorbent Electron Microscopy. Another variation of negative staining is *immunosorbent electron microscopy* (ISEM). This is based on the specificity of the antigen–antibody interaction and used in conjunction with negative staining of particulate samples (108–110). Using virus as an example, antibody against virus coat protein is applied to the surface of a plastic-coated grid in order to specifically trap the virus in a subsequent step. A second application of antibody can be applied to "decorate" the virus particles and improve the ease of imaging (Fig. 6). Decoration is particularly useful when working with viruses in low titer or in a mixture of virus and virus-like particles. The final step before viewing is to negatively stain the preparation, using any of the stains we just described.

While ISEM is used routinely to monitor virus purification, it may, likewise, have similar application for subcellular fragments. The ISEM preparations are generally free of any extraneous material and successfully trap specimens of interest, even if they are in low concentration.

Negative staining, coupled with image analysis, has provided an increasingly accurate description of the nuclear pore complex (111–113) and ribosomal subunits (114). Well-ordered specimens, such as the nuclear pore complex and virus capsids, have a high enough degree of symmetry to allow processing of a digital image in order to filter out interfering signals.

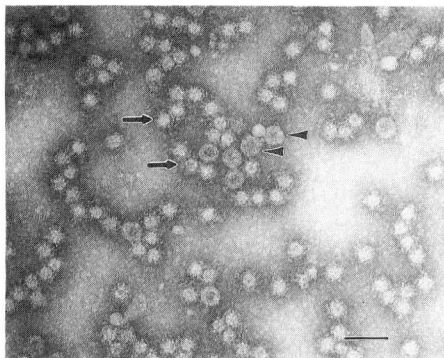

Fig. 6. Transmission electron micrograph of negatively stained mixture of blueberry shoestring virus (BBSSV) (arrows) and cauliflower mosaic virus (CMV) (arrowheads) prepared using the immunosorbent electron microscopy technique with the decoration step added. The antibody used was prepared against coat protein of BBSSV. have a "halo" of antibody molecules surrounding them. The CMV shows no such halo. Viruses were provided by Ms. Jerri Gillett, Botany and Plant Pathology, Michigan State University, East Lansing MI. Bar = 0.1µm.

2.4.5. SHADOWING AND REPLICA TECHNIQUES

Metal shadowcasting, often combined with replica techniques, is another approach to providing contrast to electron lucent samples. It is the oldest method for providing contrast to biological samples for TEM (115). Shadowcasting works by applying a fine vapor of metal atoms at an oblique angle to the surface of a sample. In order for a structure to be differentially shadowed, it must have some physical relief relative to the horizontal surface. The resulting TEM image is, thus, a two-dimensional projection of the three-dimensional structure of the specimen surface (Fig. 5b). However, since the three-dimensional structure is retained by the replica, it can be recovered easily by for viewing by photographing it at various tilts and either digitally or stereo-optically recombining the tilted images.

During the shadowcasting process, the thin, discontinuous metal film accumulates faster on the side of the sample toward the metal vapor source. This causes relatively more electron scattering in areas with a thicker film coating and lighter, shadow regions where little metal is deposited. These light shadows may be disconcerting to the nonmicroscopist. The shadows can be reversed with little additional effort making interpretation easier (e.g., 116). Since the mass thickness of macromolecules or viral-sized biological materials is generally inconsequentially small relative to that of the metal used in shadowing, these small samples can be viewed directly after shadowcasting. Alternatively, the discontinuous metal film can be bound together with an overlying second film, one of amorphous carbon, and this composite replica mounted on the grid.

While small (*ca.* ⩽100nm), particulate samples can be viewed directly after shadowcasting (pseudoreplicas), larger samples require that the biological material be stripped away to form a high resolution replica. Replica techniques are advantageous over other methods, here, in that thicker specimens can be used. With this approach, only the evaporator device dimensions limit primary sample size.

In addition to being able to fit in the evaporator, the sample must tolerate the conditions required for the vacuum evaporation of the metal film and carbon backing. The specimen must be *virtually* dry, *i.e.*, it must not give off appreciable or uncontrolled amounts of water vapor in the evaporator. This can be accomplished by either drying the specimen or keeping the entrained water frozen at a temperature that precludes significant sublimation [discussed under *freeze-etch* techniques (see Section 2.4.5.4)].

2.4.5.1. General Preparation Techniques. For dispersed samples, the general specimen isolation techniques for shadowcasting are generally more rigorous than those required for negative staining. Since the shadowing technique highlights surface structures, care needs to be exercised in the choice of buffers and vehicles, for some techniques, to avoid the obscuration of details by crystallized buffer components (117). Usually, if the specimen will tolerate it, dialysis against distilled water yields a cleaner image (118). If this is not possible, volatile buffers (*e.g.*, ammonium acetate or carbonate systems) can be used (117).

Freeze-drying is one of the first methods used to try to eliminate the damage caused by air drying of the specimen. This is still the method of choice for some surface detail studies (119, 120). Freeze-drying for TEM replica preparation can be accomplished in a number of ways, but by far the easiest is to use the temperature-controlled stage of a freeze-etch machine to keep the specimen temperature below the ice recrystallization temperature during the sublimation process. The exact temperature varies with the solute concentration of the specimen, but generally a temperature of about $-90°C$ should not be exceeded. A vacuum in the 10^{-4} Pa range is usually sufficient, with a liquid nitrogen cold trap (such as the cold microtome arm in the Balzers-type freeze-etch machine). Under these conditions, a thin frozen sample (such as the specimen sprayed or absorbed onto a coated grid or mica surface) should be completely dry in less than half an hour).

Critical point drying (CPD) is a second means of removing fluid phase from a sample prior to replica making. The theory behind CPD is that if the sample can be placed in a solvent that can be brought to its critical point, then the effects of surface tension encountered during dry-down can be avoided. In practice, the fluid of choice is carbon dioxide.

CO_2 exists as a liquid at 5×10^6 Pa at 5 to $10°C$ and can be exchanged with a miscible dehydration solvent in the sample (*e.g.*, ethanol or amyl acetate) in the CPD chamber. After several serial changes, all of the sample volume that was filled with dehydration solvent should be filled with liquid CO_2. The temperature is then raised to above $31°C$ (usually about $40°$), and the pressure is brought up to over 7.4×10^6 Pa. The CO_2 at this point will exist as a gas and can be bled off slowly, thereby avoiding the surface tension problems associated with an evaporative drying.

One of the most widely used methods of preparing biomolecules and other small specimens for shadowcasting is to apply an aerosol of the specimen dissolved in a glycerol solution to the surface of freshly cleaved mica (117, 118). An especially facile and reliable method is to use an artists airbrush, suitably modified to hold a microliter pipet tip filled with the specimen solution. While this method was originally proposed for extended macromolecules, it works well for a range of subjects from globular macromolecules to viruses.

With this technique, 10 µl-20 µl of the specimen in a 10%-50% glycerol solution is sprayed onto the surface of a freshly cleaved mica chip. The mica chip is then placed into a vacuum evaporator, and the vacuum is brought to the evaporation level.

Once a suitable mounting of the specimen has been achieved, the evaporator is pumped down (generally to the low 10^{-4} Pa range) prior to shadowing. The temperature of the sample also can affect the final grain size of the evaporated film and lower temperatures form finer-grained films, due to the suppression of grain migration and coalescence on the specimen surface.

2.4.5.2. Molecular Spreading. Long macromolecules present a unique preparation problem in that an unobstructed view of the entire molecule length is usually desired. In the case of DNA or RNA, this is compounded by a width of about 2 nm, which is at or very near the resolution limit for shadowcasting techniques. The general solution to these problems has been to use molecular spreading techniques that allow the full length of the molecule to be presented in one plane. This technique was first developed by Kleinschmidt and Zahn (121) and has fostered the development of many variant techniques that improve the results for specific classes or condition of molecule (108, 122).

2.4.5.3. Evaporation Techniques. There are a number of evaporation techniques available for the vaporization of the shadowing metal under vacuum (123). The least involved and least expensive is to use *resistance heating.* This is accomplished by heating the metal of choice to the boiling point with a filament heated by an electrical current. Usually, the shadowing metal is placed in a conical coil (basket) formed in the center of the tungsten wire filament. This is heated to incandescence, under vacuum, and the shadowing metal first melts and then is boiled off the filament. While this method can be used for any metal (indeed, the basket itself can be heated to yield a tungsten film), it works better for those metals boiling at temperatures much lower than tungsten. Platinum, under vacuum, boils very close to the tungsten melting point and often alloys with the heater wire. This can be alleviated at a slight cost in resolution, by using a Pt:Pd alloy. A similar resistance approach can be applied to the formation of carbon films. In this case, carbon in the form of a woven strand, is placed between the electrodes and evaporated directly.

A second method of evaporation is to use the heat produced by a carbon arc. In this method, a pair of carbon electrodes are brought into contact in the evaporator by spring or by gravitational force. As a current is run through them, the area where they contact each other becomes exceedingly hot, and an arc develops. If the end of one electrode is wrapped with metal, it will be melted and preferentially evaporated by the arc. This method is used with platinum, with a predetermined length of fine wire being wrapped around the carbon electrode. The metal film thickness can be determined by the length of the wire used once the system has been calibrated. If no metal is provided to the arc, it evaporates the carbon rods. This evaporation allows the production of fine-grained carbon films for replica backing and the other uses we just mentioned.

A more reliable and rapid technique for evaporating both shadowing metals and carbon is via electron beam bombardment. This method makes use of a focused flux of high-energy electrons to provide the necessary power for the evaporation of a small metal or carbon source. Generally, this is used to form thin films of Pt/C, W/Ta, and pure C. The platinum films contain some carbon (usually $\leqslant 5\%$) due to the presence of the carbon support rod. The electron beam gun evaporation approach greatly improves throughput and consistency compared to the other approaches.

High-resolution sputtering has been used to produce fine-grained films for SEM (124). As yet, the utility in TEM is largely unknown and it may prove incompatible with the vacuum and thermal requirements of TEM sample preparation.

The amount of metal, its type, the angle, and the method of shadowing affect the final image information. In general, a low-shadowing angle (*e.g.*, $3°-10°$) is used for small specimens and higher angles for larger specimens (*e.g.*, $45°$). In addition to these "conventional" shadowing conditions, interesting results have been obtained by using near vertical metal shadowing, which yields higher-resolution replicas than any lower angle preparation (125). High resolution replicas of periodic samples have also been made by plain carbon shadowing (126). Carbon-backing films on replicas are generally deposited nearly vertically.

The average metal film thickness can be determined by a number of methods (123). Quartz crystal thickness monitors are generally trouble free and a real advantage in obtaining consistent results. The shadowing-film thickness depends on the contrast and resolution required and on the type of specimen. For specimens with regular features, about 0.5 nm of W/Ta and 1.5 nm of Pt/C give the best result (127).

2.4.5.4. Freeze-etch Techniques. *Freeze-fracture* and *freeze-etch* are two similar techniques that have become the standard methods of revealing high-resolution subcellular details in three dimensions. These techniques have the advantage that the material to be prepared is, generally, subjected to fewer fixation steps than other methods. Indeed, if the sample is ultrarapidly frozen, it holds the potential of being completely "lifelike." In essence, freeze-fracture is a method for forming a replica of a frozen-hydrated sample, which is fractured to expose internal details. The two techniques differ mainly in the postfracture processing. Freeze-fractured material is shadowed directly after fracturing, while freeze-etch material is allowed to sublime for varying periods to expose details that were buried under the fracture-plane surface.

Although there are countless variations of equipment and techniques, there are six basic steps common to any freeze-etch protocol. First, the specimen must be isolated in a manner that does not alter the aspects under investigation. Also, any prefreezing treatments, such as cryoprotecting, mild chemical fixation, or surface labeling need to be completed before freezing. Second, the specimen needs to be frozen rapidly enough to prevent resolvable ice crystals from forming (32). Third, the specimen must be introduced onto the precooled stage of the freeze-etch machine, brought to as low a vacuum as practicable and fractured either by tensile stress or a cryogenic knife. If the specimen is to be etched, the etching process is started as soon as the sample is fractured. This is accomplished by the method described for freeze-drying (above), with the exceptions that the etching temperature is generally lower (e.g., $-100°C$ to $-110°C$) and the time shorter, usually less than 5 min, even for deep etching. When the etching is completed or if it is omitted (in the case of freeze-fracture), the surface is shadowed by one of the methods we just mentioned. Immediately following the application of the shadowing metal, the replica is backed with a thin carbon film. Although it has less effect on the final image contrast or resolution than the thickness of the metal film, it is important to monitor the thickness of the carbon film as well. Too much backing carbon tends to make replicas curl during cleaning and difficult to adhere to the support grid. The last step in the freeze-etch process is cleaning the replica. Unless postfracturing labeling techniques are to be performed (e.g., 128), all remaining sample material should be stripped from the underside of the replica. This is done by floating the replica for several hours on a bleach or chromic acid solution. If the sample is high in silicates (see the following paragraph), an HF bath can be included. After digestion is complete, the replica is transferred by loop to several water washes, then mounted on a grid.

One of the most interesting adaptations of the freeze-etch technique is to use it to prepare samples of macromolecules and macromolecular assemblies carried on finely divided mica flakes (116). In this technique, the sample is briefly incubated with a slurry of mica chips, which are then ultrarapidly frozen, usually by slam freezing. The resulting sample is transferred quickly to the cold stage of a freeze-fracture machine and lightly cut with the microtome knife to expose a fresh surface. The sample is then etched deeply (for 4 min at $-100°C$), rotary shadowed at a low angle, and backed with a thin carbon film. The replica is detached from the mica substrate by digesting it with HF, washed several times, and mounted on a grid. The random orientation of the mica flakes with respect to the fracture face means that some areas of the replica present the molecules at the optimum orientation.

2.4.5.5. Replica Interpretation. While shadowcasting and replica techniques yield specimens that are remarkably stable in the TEM, they do require careful interpretation. Artifacts abound in all shadowing preparations and are especially prevalent in freeze-fracture techniques. A thorough review of these artifacts (e.g., 129) is strongly suggested. Among the artifacts common to all shadowing approaches is *decoration*, which is caused by the preferential nucleation of shadowing metal on other condensible material. This is correlated to physicochemical differences in the specimen surface. Characteristically, decoration provides a "false" structural impression.

One type of decoration that is likely to be inescapable is self-decoration. This occurs during the shadowing phase due to the build-up of shadow cast metal; one shadows the shadow. Self-shadowing is more significant at lower shadow angles. Using high-shadowing angles and tilting the grid with low voltages and using small apertures in the microscope for added contrast also reduce this form of decoration (130).

Glycerol, used in spray mounting techniques may decorate the surface of the specimen on the mica surface. The protein surface structure of the TMV particle, which shows clearly a helical pattern in low-dose negative-stained and freeze-dried shadowcast images (106, 131), is not visible if the particles are applied to mica via glycerol spray.

In freeze-fracture preparations made without consideration of the residual gas composition in relation to specimen temperature, water vapor can condense on the frozen specimen. This can occur nonrandomly on the fracture surface, giving the impression of membrane particles that aren't there or masking those that are.

Interestingly, decoration, while it degrades the apparent topographical image may not degrade the total information in periodic structures. Gold migration on freeze-etched catalase crystals (132) causes lower shadowing resolution but yields higher periodic information, based on the optical transforms.

2.4.5.6. Other Applications. A modification of either negative staining or shadowing is the *Miller chromatin spreading* technique, described in a discussion, "Portrait of a Gene," by Miller and Beatty (133). In this method, nuclear contents can be isolated and separated, and centrifugation can be used to deposit the chromatin on carbon-coated grids. The chromatin can be stained with PTA for imaging proteins or with UA for imaging nucleic acids, or it can be rotary metal shadowed. Osheim and Beyer (134) provide a comprehensive description of these methods.

Snapshot blotting is a technique that allows the transfer of nucleic acids and nucleoprotein complexes from electrophoresis gels to TEM grids for examination (135). In this method, carbon films are attached firmly to TEM grids, subjected to glow-discharge and dipped in spermidine to render the surface hydrophilic. The grids are inserted into ethidium bromide-stained gels, electrophoresed for 30 sec, removed, dehydrated, and shadowed before viewing. Using this method, Jett (135) examines the assembly of transcription complexes. The location of protein bound to DNA matched sites that were predicted from sequence analysis. The method is quick and appears to be a reliable way to examine nucleic acids and protein complexes.

2.4.6. Energy Dispersive X-ray Microanalysis. Specimen preparation for X-ray microanalysis requires exacting protocols, not dissimilar to those for immunolabeling. Preparative techniques must preserve the spatial integrity of the elements of interest and, if quantitation is required, the concentration of those elements. Clearly, bound elements present less of a challenge than a highly mobile, and, of course, very biologically interesting, element, such as calcium.

Accurate preservation of the elements must be accompanied by high-quality preservation of the sample structure. To be useful, the sample must be beam stable and have enough contrast and integrity to produce an image with clearly recognizable structural features. This becomes critical if compositional mapping is to be conducted (Fig. 7).

In general, quick-freezing, by one of the methods mentioned previously, provides the most reliable retention of elements (136–139). Quick-freezing may be followed by freeze-substitution and ultrathin sectioning or by cryosectioning. As with immunolabeling, a combination of these methods may well result in the optimum balance of fidelity of morphology, accuracy of localization, and feasibility of preparation. As with many of the other specimen preparation techniques, there is certainly no one best method for a given sample, or a given set of experimental variables.

Frozen-hydrated samples can be examined using EDS. Image contrast is generally quite low, and the water in the sample contributes to higher background counts and decreased resolution. Radiation damage to frozen-hydrated specimens also reduces the ability to conduct high-resolution X-ray analysis (138). One advantage of frozen-hydrated samples, however, is the ability to examine elements within vacuolar spaces, since elements migrate to the nearest membrane upon drying. In addition, if wet weight data and/or water content is of particular interest, frozen-hydrated samples may be the method of choice. Most EDS, especially for quantitation, is conducted on samples that have been cryosectioned, transferred to the TEM with a cryotransfer device, and then freeze-dried in the microscope column. While EDS may have some application in biochemical research, it is most commonly used to assess elemental distribution at the cell and tissue level and, as yet, has only minor application to analysis of macromolecules, nucleic acids, and enzymes.

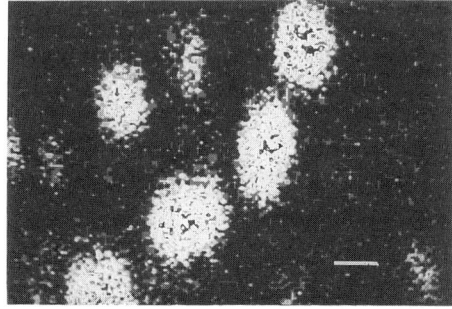

Fig. 7. Compositional map of iron from ultrathin sections of honeybee abdomen (see Fig. 1). The bright pixels represent iron-rich areas and are correlated directly to the magnetite granules seen in the ultrathin sections. Bar = 0.5µm.

3. SCANNING PROBE MICROSCOPY

Scanning probe microscopes (SPMs) are the most recent family of imaging instruments. These microscopes image specimen surface morphology at atomic and near-atomic level resolution (140). The *scanning tunneling microscope* (STM) and the *atomic force microscope* (AFM) are two of the most commonly used. The potential applications of the STM and AFM are just beginning to be applied to biological samples. As Fisher et al. (141) so aptly point out, SPMs have tremendous potential for biology, but the limits of the art and technology, as well as its practical uses, are just beginning to be understood. Hansma et al. (142) present an excellent overview of the potential applications of both STM and AFM. Despite the fact that AFMs have been commercially available for a short time, considerable literature exists, which demonstrates its potential use in biochemical research, especially for examining DNA (143, 144), structural proteins (145), gap junctions (146), and RNA length measurements (147). The SPMs have also been suggested as a new technique for sequencing DNA, but that application is still unproven (148).

Because SPM is a new technology, specimen preparation (Section 3.3.) and reproducibility, and image interpretation, especially for biological specimens, emerge as the predominant challenges to useful application. Image interpretation is, at once, compounded and enhanced by the fact that, unlike other imaging devices with which most researchers are familiar (LM, LSM, TEM), the SPMs use computer-generated images derived from electronic signals representative of quantum mechanical interactions between the sample and the probe. Image processing to enhance detail and remove electronic noise must be conducted judiciously in order to maintain accuracy.

Image interpretation will continue to be a basis for controversy and for progress in applying SPM to biological problems. In just one example, Clemmer and Beebe (149) present evidence that highly ordered pyrolytic graphite, which has been used as a substrate for biomolecules, may actually mimic the "structure" of DNA. In two technical treatments, Stemmer and Engel (150) and Stemmer et al. (151) provide specific information on instrument parameters, tip geometry, and correlation with TEM images that provide a foundation for sound interpretation of topographs. Because scientists are more familiar with interpreting TEM (and SEM) images, these instruments may well find extended use in assessing the developing technology of SPM.

3.1. Image Formation

The STM requires a conductive specimen to operate and is the SPM of choice for examination of conductive material science samples. It produces an image by using the variations in tunneling current, which occur when a fine metal tip is passed very close to a specimen surface. Either the sample or the tip is mounted on a piezoelectric tube that allows for ultrafine movement in X, Y, and Z directions.

The electrons in the atoms on the sample and on the tip form clouds, whose densities decrease exponentially with the distance from the surface. Tunneling current, a quantum mechanical process, occurs when electrons from an atom on the tip of a probe carrying a small electrical current move toward the atoms on the sample surface. Tunneling increases exponentially with decreasing sample-probe tip distance. The tunneling current is, therefore, exceedingly sensitive to the distance between the probe and the atoms on the surface. The variations in this distance are used to generate a computer image of the sample surface at atomic level resolution.

The AFM creates atomic level resolution images of the surfaces of nonconductive specimens, using the quantum mechanical phenomenon of repulsive atomic force. The AFM records the repulsive forces that occur when electron clouds of two atoms, one in the microscope probe and the other at the sample surface, are in close proximity to each other. In most AFMs, the sample is mounted on the piezoelectric tube, so the sample moves in relation to a stationary tip. As the sample is very close to the probe, the fluctuations in electric dipole moment of the interacting atoms create a repulsive action between atoms of the specimen and atoms of the probe. As the probe is deflected by the atoms on the surface of the specimen, its movement is intercepted by a laser beam, which transmits the information to the computer for image generation.

3.2. Advantages and Disadvantages of SPM

The STM has been used to image biological samples, but those must either be "doped" with metal or coated with metal in order to produce the necessary conductivity. Metal-coated samples are more stable than the softer "native-state" sample, and height measurements have been quantified in a few specimens (141, 151). The major disadvantages of this method are that the sample must be dried or frozen, the grain of the metal may interfere with very small surface features, and, of course, the sample is no longer in the native state. Except for some very accurate height measurements, most images are not yet appreciably improved over those produced by high-resolution TEM (141, 151, 152). In addition, for some samples, X-ray crystallographic measurements may differ from STM results, most likely due to artifacts of specimen preparation (150, 151) (see the next paragraph).

Because the AFM does not require a sample to be conductive in order to produce an image, it is used increasingly to examine biological samples. Native state samples can be imaged in their hydrated state. Disadvantages include the difficulty in securing the sample on the specimen support and the probability that the soft sample may be deformed by the tip-sample interaction. The recent availability of high aspect ratio probe tips are expected to increase the resolution in the AFM, but tip interaction with soft biological specimens may remain problematic. However, as with TEM, "disadvantage" may be inherent in a point of view. One unique use of the normally disadvantageous specimen-tip interactions was employed by Hoh et al. (146). They increased the force on the AFM tip to use it as a micromanipulator to dissect the top membrane of isolated gap junctions to expose the array of channels within.

3.3. Specimen Preparation

There are, as yet, no definitely established specimen preparation protocols for SPM of biological samples. Preparation has not become "standard," as in the case for many TEM specimen preparation techniques. The growing body of literature in the area does allow for a discussion of successful methods, but many refinements are to be expected. Fisher et al. (141) provide an excellent overview of methods, although new and modified technique development is occurring at a very rapid pace.

Specimen preparation for SPM differs in the inherent challenges depending on whether or not STM or AFM is used. One challenge common to both is the necessity of preparing a stable, flat sample that is not pushed along the substrate by the probe. In addition, deformation of the sample by probe-tip interactions must be considered.

As stated previously, because STM requires a conductive sample, biological specimens usually undergo some regimen of preparation. For example, Amrien et al. (153) prepared recA–DNA complexes by freeze-drying and then coating with a 1nm layer of platinum-iridium-carbon alloy. Similarly, phage was imaged by STM after coating with platinum-carbon by Garcia et al. (154) and freeze-fracture replicas of biomembranes were imaged by Zasadzinski et al (155). Coating techniques clearly do not allow examination of native material but are often necessary with STM. Native DNA (156) and hydrated bacterial surface proteins (157) have also been imaged by STM with little additional preparation.

Edstrom et al. (158) directly imaged phosphorylase kinase and phosphorylase b in the STM by applying the protein to pyrolytic graphite substrates and drying in nitrogen gas. Their remarkable images, X-ray crystallographic measurements, and TEM images are all consistent. In an additional study (159), they examined the protein complex using both STM and AFM using the same specimen prepartion procedure with similar results. Because proteins are less sensitive to surface tension, freeze-drying or critical point drying (used in SEM specimen preparation) did not offer any advantage. As a result, air drying of proteins was proposed as a useful technique for SPM of proteins.

Imaging with the AFM does not require that the specimen be conductive. For this reason, the AFM offers greater potential for biological samples than does STM, although there are still challenges inherent in the design of the tips and the cantilevers on which they are borne (140). Low-temperature methods, under development, may prove to be as reliable a method of specimen preparation for the AFM as it has been for the TEM.

Zasadzinski et al. (160) imaged polar or headgroup regions of hydrated phosphatidylethanolamine bilayers using AFM. Specimen preparation used the Langmuir–Blodgett technique of deposition onto mica surfaces with the resulting sample imaged under water. They report that their AFM results correlated well with X-ray measurements and electron diffraction of various monolayers. Imaging in water provides the technique most likely to allow accurate imaging of native state lipid and lipid/protein monolayers. Other research (161) also indicates the potential usefulness of Langmuir–Blodgett films for anchoring macromolecules for SPM.

While whole cell surfaces have been imaged by AFM (162), the most useful application for this high resolution technique is likely to be the examination of nucleic acids (163, 164), proteins (156, 159), and subcellular entities such as actin filaments in living glial cells (165).

Specimen preparation techniques for nucleic acids and proteins include various chemical treatments of freshly cleaved mica that permits strong attachment of the specimen to the surface and imaging under alcohol films or in the air. Lyubchenko et al. (147) treated freshly cleaved mica in an atmosphere of 3-aminopropyltriethoxy silane. The mica was then immersed for 2 hr in a buffer of 10mM Tris-HCl, 10nM–20mM NaCl, and 5mM EDTA containing 0.1µg/ml RNA. The mica was washed with dH_2O and imaged in the AFM mode. Hansma et al. (164) report treating mica with magnesium acetate, rinsing, drying, and subjecting it to glow discharge. Single-stranded DNA in 0.5% formaldehyde and 15mM ammonium acetate was applied by floating the mica on drops of the fluid for 2–5 min. Samples were then examined with AFM. Various environmental chambers for specimen observation, such as an electrochemistry cell, can also provide enhanced data collection by adsorbing molecules tightly to a gold surface (166).

4. CONCLUSIONS

Both the transmission electron microscope and the scanning probe microscope (particularly the atomic force microscope) are the highest-resolution-imaging devices available for biochemical research. While knowledge of the instruments is important, the selection of appropriate methods of specimen preparation and the correct execution of those methods are critical for accurate ultrastructural data. In fact, use of more than one method can be quite desirable, especially if alternative methods of data corroboration are not available.

We emphasize again that there is rarely a generic method that can be described for all specimen types and experimental conditions. Indeed, specifics must usually be derived empirically for each research project. While we have discussed the basic guidelines and offer some specific protocols for the most common specimen preparation techniques and have provided references that contain the most detailed methodologies for other techniques, the literature is replete with specifics and should be consulted before beginning any ultrastructural project. As both techniques and instruments improve for electron microscopy and as scientists become increasingly proficient with scanning probe technology, microscopic imaging will continue to provide unique data for biochemical research.

5. REFERENCES

1. J. Pawley (ed.), *The Handbook of Biological Confocal Microscopy*, IMR Press, Madison, WI, 1989, 201 pp.
2. M. T. Postek, K. S. Howard, A. H. Johnson, and K. L. McMichael, *Scanning Electron Microscopy: A Student's Handbook*, Ladd Research Industries, Burlington, VT, 1980, 305 pp.
3. J. I. Goldstein, D. E. Newbury, P. Echlin, D. C. Joy, C. Fiori and E. Lifshin, *Scanning Electron Microscopy and X-ray Microanalysis*, Plenum Press, New York, 1981, 673 pp.

4. A. W. Agar, R. H. Alderson, and D. Chescoe, "Principles and Practice of Electron Microscope Operation," in *Practical Methods in Electron Microscopy* (ed. A. M. Glauert), Elsevier Press, New York, 1974, 345 pp.
5. G. A. Meek, *Practical Electron Microscopy for Biologists*, 2nd ed., John Wiley, New York, 1976, 528 pp.
6. M. A. Hayat, *Correlative Electron Microscopy in Biology: Instrumentation and Methods*, Academic Press, Orlando, 1987, 437 pp.
7. J. F. O'Hanolon, *A User's Guide to Vacuum Technology*, John Wiley, New York, 1980.
8. M. K. Lamvik, *J. Microsc.* **161**, 171 (1991).
9. V. E. Cosslett, *J. Microsc.* **113**, 113 (1978).
10. K.-R. Peters, *Scanning Elec. Microsc.* **4**, 1519 (1985).
11. M. Beer, J. W. Wiggins, C. J. Stoeckert, K. Mrenus, E. Kuhn, and M. Erickson, *Ultramicrosc.* **8**, 207 (1982).
12. A. V. Crewe, *J. Ultrastruc. Res.* **88**, 94 (1984).
13. A. Engel and R. Reichelt, *J. Ultrastruc. Res.* **88**, 105 (1984).
14. C. Colliex, C. Jeanguillaume, and C. Mory, *J. Ultrastruc. Res.* **88**, 177 (1984).
15. U. Aebi, W. E. Fowler, E. L. Buhle, Jr., and P. R. Smith, *J. Ultrastruc. Res.* **88**, 143 (1984).
16. R. D. Leapman and S. B. Andrews, *J. Microsc.* **161**, 3 (1991).
17. J. S. Wall and J. F. Hainfeld, *Ann. Rev. Biophys. Biophys. Chem.* **15**, 355 (1986).
18. R. Freeman and K. R. Leonard, *J. Microsc.* **122**, 275 (1981).
19. A. Engel, W. Baumeister, and W. O. Saxton, *Proc. Natl. Aca. Sci. USA* **79**, 4050 (1982).
20. K. A. Johnson and J. S. Wall, *J. Cell Biology* **96**, 669 (1983).
21. M. W. Mosesson, J. Hainfeld, J. Wall, and R. H. Haschemeyer, *J. Mol. Biol.* **153**, 695 (1981).
22. C. A. Anderson, "An Introduction of the Electron Probe Microanalyzer and Its Application to Biochemistry," in *Methods of Biochemical Analysis*, Vol. 15 (ed. D. Glick), Wiley Interscience, New York, 1967, pp. 147–270.
23. A. P. Somlyo and H. Shuman, *Ultramicrosc.* **8**, 219 (1982).
24. A. P. Somlyo, *J. Ultrastruc. Res.* **88**, 135 (1984).
25. D. Johnson, K. Izutsu, M. Cantino, and J. Wong, *Ultramicros.* **24**, 221 (1988).
26. A. Warley, *J. Microsc.* **157**, 135 (1990).
27. A. P. Reid, W. T. W. Potts, K. Oates, R. Mulvaney, and E. W. Wolff, *Microsc. Res. and Tech.* **22**, 207 (1992).
28. C. Dawes, *Introduction to Biological Electron Microscopy: Theory and Techniques*, Ladd Research Industries, Burlington, VT, 1988, 315 pp.
29. M. A. Hayat, *Principles and Techniques of Electron Microscopy: Biological Applications*, 3rd ed., CRC Press, Boca Raton, 1989, 469 pp.
30. N. Reid, "Ultramicrotomy," in *Practical Methods in Electron Microscopy*, Vol. 3, (ed. A. M. Glauert), North-Holland/American Elsevier, New York, 1975, 353 pp.
31. A. M. Glauert, *Fixation, Dehydration and Embedding of Biological Specimens*, North-Holland/American Elsevier, New York, 1975, 207 pp.
32. J. C. Gilkey and L. A. Staehlin, *J. Elec. Microsc. Tech.* **3**, 177 (1986).
33. R. Dahl and L. A. Staehlin, *J. Elec. Microsc. Tech.* **13**, 165 (1989).
34. A. W. Robards and U. B. Sleyter, "Low temperature methods in biological electron microscopy," in *Practical Methods in Electron Microscopy*, Vol. 10 (ed. A. M. Glauert), Elsevier, New York, 1985.
35. R. F. E. Crang and K. L. Klomparens (eds), *Artifacts in Biological Electron Microscopy*, Plenum Press, New York, 1988, 233 pp.
36. H. Sitte, K. Neumann, and L. Edelmann, "Cryosectioning according to Tokuyasu vs. rapid-freezing, freeze-substitution and resin embedding," in *Immuno-gold Labeling in Cell Biology* (eds. A. J. Verkleij and J. L. M. Leunissen), CRC Press, Boca Raton, FL, 1989, pp. 63–93.

37. J. S. Singer, K. T. Tokuyasu, A. H. Dutton, and W-T. Chen, "High-resolution Immunoelectron Microscopy of Cell and Tissue Ultrastructure," in *Electron Microscopy in Biology*, Vol. 2 (ed. J. D. Griffith), John Wiley, New York, 1982, pp. 55-106.
38. M. Horisberger, "Lectin cytochemistry," in *Immunolabeling for Electron Microscopy* (eds. J. M. Polak and I. M. Varndell), Elsevier, Amsterdam, 1984, pp. 249-266.
39. E. A. Bayer and M. Wilchek, "The Use of the Avidin-biotin Complex as a Tool in Molecular Biology," in *Methods of Biochemical Analysis* (ed. D. Glick), John Wiley, New York, 1980, pp. 1-45.
40. C. Bonnard, D. S. Papermaster, and J.-P. Kraehenbuhl, "The Streptavidin-biotin Bridge Technique: Application in Light and Electron Microscope Immunocytochemistry," in *Immunolabeling for Electron Microscopy* (eds. J. M. Polak and I. M. Varndell), Elsevier, Amsterdam, 1984, pp. 95-111.
41. G. J. Nuovo, "*In Situ* Detection of PCR-amplified DNA and cDNA," in *Amplifications: A Forum for PCR Users*, Perkin Elmer technical publication, Issue 8, Norwalk CT, June 1992, pp.1-3.
42. E. Kellenberger and M. A. Hayat, "Some Basic Concepts for the Choice of Methods", in *Colloidal Gold: Principles, Methods and Applications* (ed. M. A. Hayat), Academic Press, San Diego, 1991, pp. 1-30.
43. G. V. Childs, "Use of Immunocytochemical Techniques in Cellular Endocrinology," in *Electron Microscopy in Biology, Vol. 2 (ed. J. D. Griffith), John Wiley, New York, 1982, pp. 107-179.*
44. M. Bendayan, *J. Elec. Microsc. Tech.* **1**, 243 (1984).
45. M. Bendayan, *J. Elec. Microsc. Tech.* **1**, 349 (1984).
46. M. Bendayan, *J. Elec. Microsc. Tech.* **6**, 7 (1987).
47. M. Bendayan and M. Zollinger, *J. Histochem. and Cytochem.* **31**, 101 (1983).
48. M. Bendayan, A. Nanci, and F. W. K. Kan, *J. Histochem. and Cytochem.* **35**, 983 (1987).
49. M. Bendayan and S. Garzon, *J. Histochem. and Cytochem.* **36**, 597 (1988).
50. P.A. Coulombe, F.W.K. Kan and M. Bendayan, *Europ. J. of Cell Biol.*, **46**, 564 (1988).
51. L.-I. Larsson, *Immunocytochemistry: Theory and Practice*, CRC Press, Boca Raton, FL, 1988, 272 pp.
52. J. Roth, *J. Histochem. and Cytochem.* **31**, 987 (1983).
53. O. Behnke, T. Ammitzboll, H. Jessen, M. Klokker, K. Nilausen, J. Tranum-Jensen, and L. Olsson, *Eur. J. Cell Biol.* **41**, 326 (1986).
54. J. Roth, *J. Microsc.* **143**, 125 (1986).
55. E. Herman and D. Melroy, "Electron Microscopic Immunocytochemistry in Plant Molecular Biology," in *Plant Molecular Biology Manual*, (eds. S. B. Gelvin and R. A. Rachilperoort), B13, Kluwer, Belgium, 1990, pp. 1-24.
56. L. A. Sternberger, "Enzyme Immunocytochemistry," in *Electron Microscopy of Enzymes*, Vol. 1 (ed. M. A. Hayat), Van Nostrand/Reinhold, New York, 1973, pp. 150-191.
57. G. Griffiths and H. Hoppeler, *J. Histochem Cytochem.* **34**, 1389 (1986).
58. G. Posthuma, J. W. Slot, and H. J. Geuze, *J. Histochem. and Cytochem.* **35**, 405 (1987).
59. I. M. Varndell and J. M. Polak, "Double Immunostaining Procedures: Techniques and Applications," in *Immunolabeling for Electron Microscopy* (eds. J. M. Polak and I. M. Varndell), Elsevier, Amsterdam, 1984, pp. 155-177.
60. K. T. Tokuyasu, *Histochem. J.* **12**, 381, (1980).
61. Y-D. Stierhof, H. Schwarz, M. Durrenberger, W. Villiger, and E. Kellenberger, "Yield of Immunolabel Compared to Resin Sections and Thawed Cryosections," in *Colloidal Gold: Principles, Methods and Applications*, Vol. 3 (ed. M. A. Hayat), Academic Press, New York, 1991, pp. 87-95.
62. P. Newmark, *Nature* **303**, 655 (1983).
63. E. H. Herman, *Annu. Rev. Plant Physiol. Plant Mol. Biol.* **39**, 139 (1988).
64. C. Feldherr and S. I. Dworetzky, "Transport of Macromolecules Through the Nuclear Pores," in *Immunogold Labeling in Cell Biology* (eds. A. J. Verkleij and J. L. M. Leunissen), CRC Press, Boca Raton, FL, 1989, pp. 305-315.

65. N. J. Hutchison, "Hybridisation Histochemistry: *In Situ* Hybridization at the Electron Microscope Level," in *Immunolabeling for Electron Microscopy* (eds. J. M. Polak and I. M. Varndell), Elsevier, Amsterdam, 1984, pp. 341-351.
66. R. H. Singer, J. B. Lawrence, F. Silva, G. L. Langevin, M. Pomeroy, and S. Billings-Gagliardi, "Strategies for Ultrastructural Visualization of Biotinated Probes Hybridized to Messenger RNA *In Situ*," in *Current Topics in Microbiology and Immunology: In Situ Hybridization*, Vol. 143 (eds. A. T. Hasse and M. B. A. Oldstone), Springer-Verlag, Berlin, 1989, pp. 55-69.
67. F. Escaig-Haye, V. Grigoriev, G. Peranzi, P. Lestienne, and J-G. Fournier, *J. Cell Sci.* **100**, 851 (1991).
68. F. G. Silva, J. B. Lawrence, and R. H. Singer, "Progress Toward Ultrastructural Identification of Individual mRNAs in Thin Section: Myosin Heavy-chain mRNA in Developing Myotubes," in *Techniques in Immunocytochemistry*, Vol. 4, (eds. G. R. Bullock and P. Petrusz), Academic Press, London, 1989, pp. 127-165.
69. R. H. Singer, G. L. Langevin and J. B. Lawrence, *J. Cell Biology* **108**, 2343 (1989).
70. M. P. Wenderoth and B. R. Eisenberg, *J. Histochem. and Cytochem.* **39**, 1025 (1991).
71. B. Russell, M.P. Wenderoth, and P. H. Goldspink, *Proceedings of Keystone Conference on Molecular Biology of Muscle Development* **R339** (1991).
72. G. I. McFadden, I. Bonig, E. C. Cornish, and A. E. Clarke, *Histochem. J.* **20**, 575 (1988).
73. M. Binder, S. Tourmente, J. Roth, M. Renaud, and W. J. Gehring, *J. Cell Biol.* **102**, 1646 (1986).
74. M. Binder, *Scanning Microscopy*, Vol. 1, **331** (1987).
75. M. Thiry, *J. Histochem. and Cytochem.* **40**, 411 (1992).
76. M. Thiry, *DNA and Cell Biology* **10**, 169 (1991).
77. M. Thiry and D. Dombrowicz, *Biology of the Cell* **62**, 99 (1988).
78. R. A. Wolber, T. F. Beals, R. V. Lloyd, and H. F. Maassab, *Lab. Invest.* **59**, 144 (1988).
79. R. A. Wolber, T. F. Beals, and H. F. Maassab, *J. Histochem. and Cytochem.* **37**, 97 (1989).
80. D. deF. Webster, L. Lamperth, J. T. Favilla, G. Lemke, D. Tesin, and L. Manuelidis, *Histochem.* **86**, 441 (1987).
81. G. C. Budd, "High Resolution Autoradiography," in *Autoradiography for Biologists* (ed. P. B. Gahan), Academic Press, New York, 1972, pp. 95-118.
82. M. A. Williams, "Autoradiography and Immunocytochemistry", in *Practical Methods in Electron Microscopy*, Vol 6 (ed. A. M. Glauert), North-Holland, Amsterdam, 1977.
83. V. M. Pickel and A. Beaudet, "Combined Use of Autoradiography and Immunocytochemistry to Show Synaptic Interactions Between Chemically Defined Neurons," in *Immunolabeling for Electron Microscopy* (eds. J. M. Polak and I. M. Varndell), Elseviser, Amsterdam, 1984, pp. 259-266.
84. A. W. Rogers, *Techniques of Autoradiography, Elsevier*, Amsterdam, 1973, 372 pp.
85. T. C. Appleton, "Autoradiography of Diffusible Substances," in *Autoradiography for Biologists* (ed. P. B. Gahan), Academic Press, New York, 1972, pp. 51-64.
86. M. M. Salpeter and L. Bachmann, "Autoradiography," in *Principles and Techniques of Electron Microscopy: Biological Applications*, Vol. 2 (ed. M. A. Hayat), Van Nostrand Reinhold, New York, 1972, pp. 220-278.
87. M. M. Salpeter, "Sensitivity and Resolution in Electron Microscope Autoradiography," in *Autoradiography of Diffusible Substances* (eds. L. J. Roth and W. E. Stumpf), Academic Press, New York, 1969, pp. 335-348.
88. J. Jacob and G. C. Budd, "Application of Electron Autoradiography to Enzyme Localization," in *Electron Microscopy of Enzymes*, Vol. 4 (ed. M. A. Hayat), Van Nostrand/Reinhold, New York, 1975, pp. 217-266.
89. S. Fakan and J. Fakan, "Autoradiography of Spread Molecular Complexes," in *Electron Microscopy in Molecular Biology* (eds. J. Sommerville and U. Scheer), IRL Press, Washington DC, 1987, 201-214.

90. M. Geuskens, "Autoradiographic localization of DNA in nonmetabolic conditions," in *Principles and Techniques of Electron Microscopy* (ed. M. A. Hayat), Van Nostrand/Reinhold, New York, 1977, pp. 163-199.
91. J. I. Morrell, "Application of *in situ* Hybridization with Radioactive Nucleotide Probes to Detection of mRNA in the Central Nervous System," in *Techniques in Immunocytochemistry*, Vol. 4 (eds. G. R. Bullock and P. Petrusz), Academic Press, London, 1989, pp. 127-146.
92. N. V. Raikhel, S. Y. Bednarek, and D. R. Lerner, *Plant Molecular Biology Manual* **B9**, 1 (1989).
93. K.-P. Chiu, S. H. Cohen, D. W. Morris, and G. W. Jordan, *J. Histochem. and Cytochem.* **40**, 333 (1992).
94. A. T. Haase, E. F. Retzel, and K. A. Staskus, *Proc. Natl. Acad. Sci. USA* **87**, 4971 (1990).
95. J. P. Coghlan, P. Aldred, J. Haralambidis, H. D. Niall, J. D. Penschow, and G. W. Tregear, *Analyt. Biochem.* **149**, 1 (1985).
96. J. L. Farrant, *Biochem. Biophys. Acta.* **13**, 569 (1954).
97. C. E. Hall. *J. Biophys. Biochem. Cytol.* **1**, 1 (1955).
98. R. C. Valentine, B. M. Shapiro, and E. R. Stadtman, *Biochemistry* **7**, 2143 (1968).
99. M. A. Hayat and S. E. Miller, *Negative Staining*, McGraw-Hill, New York, 1990.
100. F. L. A. Buckmire and R. G. E. Murray, *Can. J. Microbiol.* **16**, 1011 (1970).
101. W. Baumeister and M. Hahn, "Specimen supports," in *Principles and Techniques of Electron Microscopy: Biological Applications* (ed. M. A. Hayat), Van Nostrand Reinhold, New York. 1978.
102. U. Abei and T. D. Pollard, *J. Electron Micro. Tech.* **7**, 29 (1987).
103. J. Tyler and D. Branton. *J. Ultrastruct. Res.* **71**, 95 (1980).
104. G. Seegan, C. Smith and U. Shumaker, *Proc. Nat. Acad. Sci. USA* **76**, 907 (1979).
105. C. A. Smith, *Proc. 42nd Annu. Mtg. EMSA*, San Francisco Press, San Francisco, 148-151, (1984).
106. R. C. Williams and H. W. Fisher, *J. Mol. Biol.* **52**, 121 (1970).
107. R. M. Glaeser and K. A. Taylor, *J. Microscopy* **112**, 127 (1978).
108. D. Kay, "Electron Microscopy of Small Particles, Macromolecular Structures and Nucleic Acids," in *Methods in Microbiology*, Vol. 9 (ed. J. R. Norris), 1976, Academic Press, New York, pp. 8-215.
109. K. S. Derrick, *Virology* **56**, 652 (1973).
110. R. G. Milne and E. Luisoni, "Rapid Immune Electron Microscopy of Virus Preparations, in *Methods in Virol.*, Vol. 6 (ed. K. Maramorosch), 1977, pp 265-281.
111. J. E. Hinshaw, B. O. Carragher, and R. A. Milligan, *Cell* **69**, 1133 (1992).
112. M. Jarnik and U. Aebi, *J. Struct. Biol.* **107**, 291 (1991).
113. U. Aebi, W. E. Fowler, and P. R. Smith, *Ultramicrosc.* **8**, 191 (1982).
114. R. Brimacombe, J. Atmadja, W. Stiege, and D. Schuler, *J. Molec. Biol.*, 115 (1988).
115. R. Williams and R. Wycoff, *Science* **101**, 594 (1945).
116. J. Heuser, *J. Elec. Microsc. Tech.* **13**, 244 (1989).
117. J. M. Tyler and D. Branton, *J. Ultrastruct. Res.* **71**, 95 (1980).
118. J. W. Heckman, B. T. Terhune, and D. T. A. Lamport, *Plant Physiol.* **86**, 848 (1988).
119. W. E. Fowler and U. Aebi, *J. Ultrastruct. Res.* **83**, 319, (1983).
120. K. E. Loeser and C. Franzini-Armstrong, *J. Struct. Biol.* **103**, 48 (1990).
121. A. K. Kleinschmidt and R. K. Zahn, *Z. Naturforsch.* **14b**, 730 (1959).
122. C. L. Moore, "The Electron Microscopy of Ribonucleic Acids", in *Electron Microscopy in Biology*, Vol 2 (ed. J. D. Griffith), John Wiley, New York 1982, pp. 67-88.
123. J. H. M. Willison and A. J. Rowe, "Replica, Shadowing and Freeze-etching Techniques", in *Practical Methods in Electron Microscopy*, Vol. 8 (ed. A. M. Glauert), North-Holland, Amsterdam, 1980, pp. 31-55.

124. K. R. Peters, "Penning Sputtering of Ultra Thin Metal Films for High Resolution Electron Microscopy," in *Preparation of Biological Specimens for Scanning Electron Microscopy* (eds. J. A. Murphy and G. M. Roomans), SEM, Chicago, 1980, pp. 173–184.
125. G. C. Ruben, *J. Electron Microsc. Tech.* **13**, 335 (1989).
126. T. Muller, H. Gross, H. Winkler, and H. Moor, *Ultramicroscopy* **16**, 340 (1985).
127. H. Gross, T. Muller, I. Wildhaber, and H. Winkler, *Ultramicroscopy* **16**, 287 (1985).
128. J. E. Rash, T. J. A. Johnson, J. E. Dinchuk, D. S. Duch, and S. R. Levinson, "Labeling Intramembrane Particles in Freeze-Fracture Replicas," in *Freeze-Fracture Studies of Membranes* (ed. S. W. Hui), CRC Press, Boca Raton, 1989, pp. 41–60.
129. U. B. Sleyter and A. W. Robards, *J. Microscopy* **126**, 101 (1982).
130. J. E. Rash and T. Yasumura, *Microsc. Res. Tech.* **20**, 187 (1992).
131. M. Y. Nermut, "Freeze-drying for Electron Microscopy," in *Principles and Techniques of Electron Microscopy*, Vol. 7. (ed. M. A. Hayat), Van Nostrand Reinhold, New York, 1977, pp. 79–117.
132. I. Bachmann, R. Becker, G. Leupold, M. Barth, R. Guckenberger, and W. Baumeister, *Ultramicroscopy* **16**, 305 (1985).
133. O. L. Miller, Jr., and B. R. Beatty, *Cell Physiol.* **74** (Suppl.), 225 (1969).
134. Y. N. Osheim and A. L. Beyer, *Methods in Enzymol.* **180**, 481 (1989).
135. S. D. Jett, *Proceedings of Electron Microscopy Society of America*, Vol 1. 762 (1992).
136. A.J. Morgan, "Preparation of Specimens: Changes in Chemical Integrity," in *X-ray Microanalysis in Biology* (ed. M. A. Hayat), University Park Press, Baltimore, 1980, pp. 65–165.
137. G. M. Roomans and J. D. Shelburne, *Basic Methods in Biological X-ray Microanalysis*, Scanning Electron Microscopy Inc., AMF O'Hare, IL, 1983, 284 pp.
138. K. Zierold, *J. Microsc.* **161**, 357 (1991).
139. I. D. Bowen and T. A. Ryder, "The Application of X-ray Microanalysis to Enzyme Cytochemistry," in *Electron Microscopy of Enzymes*, Vol. 5 (ed. M. A. Hayat) Van Nostrand/Reinhold, New York, 1977, 1984, pp. 187–207.
140. _____, *The Scanning-probe Microscope Book*, Burleigh Instruments, Inc. 1990.
141. K. A. Fisher, K. C. Yanagimoto, S. L. Whitfield, R. E. Thomson, M. G. L. Gustafsson, and J. Clarke, *Ultramicrosc.* **33**, 117 (1990).
142. P. K. Hansma, V. B. Elings, O. Marti, and C. E. Bracker, *Science* **242**, 209 (1988).
143. M. Radmacher, R. W. Tillmann, M. Fritz and H. E. Gaub, *Science*, **257**, 1900 (1992).
144. H. G. Hansma, J. Vesenka, C. Siegerist, G. Kelderman, H. Morrett, R. L. Sinsheimer, V. Elings, C. Bustamante, and P. K. Hansma), *Science* **256**, 1180 (1992).
145. N. M. D. Brown and H. X. You, *J. Struc. Biol.* **107**, 250 (1991).
146. J. H. Hoh, R. Lal, S. A. John, J.-P. Revel, and M. F. Arnsdorf, *Science* **253**, 1405 (1991).
147. U. L. Lyubchenko, B. L. Jacobs, and S. M. Lindsay, *Nucleic Acids Res.* **20**, 3983 (1992).
148. S. M. Lindsay and M. Philipp, *GATA* **8**, 8 (1991).
149. C. R. Clemmer and T. P. Beebe, Jr., *Science* **251**, 640 (1991).
150. A. Stemmer and A. Engel, *Ultramicros.* **34**, 129 (1990).
151. A. Stemmer, A. Hefti, U. Aebi, and A. Engel, *Ultramicrosc.* **30**, 263 (1989).
152. H. Arakawa, K. Umemura, and A. Ikai, *Nature* **358**, 171 (1992).
153. M. Amrein, R. Durr, A. Stasiak, H. Gross, and G. Travaglini, *Science* **243**, 1708 (1989).
154. R. Garcia, D. Keller, H. Panitz, D. G. Bear, and C. Bustamante, *Ultramicros.* **27**, 367 (1989).
155. J. A. N. Zasadzinski, J. Schneir, J. Gurley, V. Elings, and P. K. Hansma, *Science* **239**, 1013 (1988).
156. T. P. Beebe, Jr, T. E. Wilson, D. F. Ogletree, J. E. Katz, R. Balhorn, M. B. Salmeron, and W. J. Siekhaus, *Science* **243**, 370 (1989).
157. R. Guckenberger, W. Wiegrabe, An. Hillebrand, T. Hartmann, Z. Wang, and W. Baumeister, *Ultramicros.* **31**, 327 (1989).

158. R. D. Edstrom, M. H. Meinke, X. Yang, R. Yang, and D. F. Evans, *Biochemistry* **28**, 4939 (1989).
159. R. D. Edstrom, M. H. Meinke, X. Yang, R. Yang and V. Elings, *Biophys. J.* **58**, 1437 (1990).
160. J. A. N. Zasadzinski, C. A. Helm, M. L. Longo, A. L. Weisenhorn, S. A. C. Gould, and P. K. Hansma, *Biophys. J.*, **59**, 755 (1991),
161. A. L. Weisenhorn, M. Egger, F. Ohnesorge, S. A. C. Gould, S.-P. Heyn, H. G. Hansma, R. L. Sinsheimer, H. E. Gaub, and P. K. Hansma, *Langmuir* **7**, 8 (1991).
162. H.-J. Butt, E. K. Wolff, S. A. C. Gould, B. Dixon Northern, C. M. Peterson, and P. K. Hansma, *J. Struct. Biol.* **105**, 54 (1990).
163. F. Zenhausern, M. Adrian, B. Ten Heggeler-Bourdier, R. Emch, M. Jobin, M. Taborelli, and P. Descouts, *J. Struct. Biol.* **108**, 69 (1992).
164. H. G. Hansma, R. L. Sinsheimer, M.-Q. Li, and P. K. Hansma, *Nucleic Acids Res.* **20**, 3585 (1992).
165. E. Henderson, P. G. Haydon, and P. S. Sakaguchi, *Science* **257**, 1944 (1992).
166. L. A. Nagahara, T. Thundat, P. I. Oden, S. M. Lindsay, and R. L. Rill, *Ultramicroscopy* **33**, 107 (1990).

Quantitative Fluorescence Imaging Techniques for the Study of Organization and Signaling Mechanisms in Cells

MARGARET H. WADE, ADRIAAN W. DE FEIJTER, MELINDA K. FRAME,
Meridian Instruments, 2310 Science Pkwy, Okemos, Michigan
AND MELVIN SCHINDLER, *Department of Biochemistry, Michigan State University, East Lansing, Michigan*

1. Introduction
2. Instrumental Design and Probes
    2.1. Fluorescence Microscopy for Imaging and Analysis
        2.2.1. Passive Reporter Groups
        2.2.2. Active Probes
    2.2. Probes—Natural Fluorophores and Fluorescent Indicators
3. Applications and Methods
    3.1. Quantitation of Molecular Components and Parameters in Single Cells or Cell Populations
        3.1.1. Quantitation of Glutathione in Living Cells in Tissue Culture
        3.1.2. Quantitation of Oncogene Expression in Tissue Culture Cells
        3.1.3. Measurements of pH in Mammalian Cells
        3.1.4. Measurements of Membrane Potential in Cells Growing in Tissue Culture
    3.2. Kinetic Analysis of Cellular Activities
        3.2.1. Ca^{++} Oscillations in Cultured Cells
        3.2.2. Diffusion of Membrane and Cytoplasmic Components
        3.2.3. Intercellular Communication
    3.3. Fluorescence *In Situ* Hybridization (FISH)
    3.4. Confocal Microscopy and 3-D Reconstruction
4. Conclusion
5. References

Bioanalytical Instrumentation, Volume 37, Edited by Clarence H. Suelter.
ISBN 0-471-58260-3 © 1993 John Wiley & Sons, Inc.

1. INTRODUCTION

The rapid growth in our understanding of biochemical reactions and mechanisms during the past century, in large part, depended on the use of physical methods to systematically disassemble cellular organization to yield purified biochemical components. Plant and animal tissues were subjected to homogenization, sonication, centrifugation, and other disruptive physiochemical techniques in order to prepare sufficient biological material for examinations within the controlled environments of a test tube or cuvette. Investigations of cellular organization using electron microscopy clearly indicated that compartmentalization and sequestration within the cell are important to control the function and activity of specific cellular components. However, it is difficult to use this technical approach to integrate "test-tube" biochemical investigations with *in situ* cellular experiments.

Although measurements in a cuvette make it easier to modify salt concentration and pH, it would be more appropriate, biologically, for such measurements to be performed in a controlled, quantifiable manner within a cell. Could the cell itself be transformed into an analytical chamber to pursue holistic experimental approaches for biochemical examinations within the context of cellular environments? To perform these types of *in situ* investigations, at least three issues require close attention in designing analytical approaches and instrumentation; namely, specificity, sensitivity, and resolution. To pursue such experiments on specific molecular species within the complex biochemical milieu of the cell, one requires appropriate sensors or probes. Either a probe is required that is unique for a protein whose function and location is to be examined, or a probe must serve as a reporter molecule, providing quantitative information about the steady-state or fluctuations in biological parameters (e.g., Ca^{++} concentration, pH, membrane potential). Since most investigations on individual cells are signal-limited, it is necessary to develop methods that have high sensitivity for measurements of subcellular localization and biological activity. Finally, approaches and instrumentation must be designed to provide resolution at micron and submicron levels to demonstrate localization of structure and activity within the dimensions of the cell. The recent introduction of a large variety of fluorescent probes and sophisticated imaging systems now provides the technical ability to both visualize and quantitate biochemical events in living cells. It is the aim of this chapter to demonstrate that fluorescence-based whole-cell analytical approaches can meet the demanding requirements of specificity, sensitivity, and resolution that are necessary for the examination of biochemical components and mechanisms within a single cell or tissue.

2. INSTRUMENTAL DESIGN AND PROBES

2.1. Fluorescence Microscopy for Imaging and Analysis

Fluorescence is the property of a molecule in an excited electronic state to emit light as a result of returning to the ground state. The emission spectrum of fluorescent molecules is red-shifted in relation to the excitation spectrum. Measurable parameters of this process are the excitation and emission spectra (1–6), fluorescence intensity (1–6), fluorescence lifetime (7–9), and degree of polarization (10–12). To take advantage of these measurable fluorescent properties for cellular analysis and imaging, optical instruments have been designed consisting of the following components:

- A microscope to both focus the excitation source onto the sample and then focus the emitted light onto a photo-detection system (e.g., photomultiplier, charge-coupled device (CCD), film, retina) is a central requirement for all imaging systems.
- A source of high intensity, coherent illumination that can be focused to a small beam diameter (approximately 1μm) to excite a wide variety of fluorescent probes. Most analytical systems employ an Argon or Krypton laser to provide illumination between 360nm and 600nm.
- Optical detectors in the form of a photomultiplier tube or CCD to convert fluorescence intensity into electronic pulses that can be manipulated by computers.
- Light modulation devices such as acousto-optic modulators (AOM) or neutral density filters to provide attenuation for the excitation source.
- Galvanometric mirrors or two-dimensional stages to offer addressable, precise beam positioning on the sample.
- A computer for system control and management, experimental organization, data collection, manipulation, and analysis.

Two instrument configurations that demonstrate these features are shown in Figs. 1 and 2. The ACAS 570 Interactive Laser Cytometer (Meridian Instruments, Inc.) (Fig. 1) utilizes a high speed two-dimensional stage to move the sample across a stationary excitation beam that is focused through the center of the optical path of the objective. The ACAS is comprised of an argon laser, two-dimensional (X–Y) scanning stage, inverted microscope, and photomultiplier tubes, all under computer control (13). The laser beam passes through an acousto-optic modulator (AOM) and is then focused by a microscope objective. The AOM provides rapid and precise control of the intensity and duration of the laser illumination. The fluorescent signals detected by the photomultiplier tube(s) are digitized and presented in a variety of formats by the computer. The pixel-resolution of the image can be varied by adjusting the step size between data points (from 0.1μm to 100μm). The same cells can be monitored over a period of time to provide kinetic data following biochemical manipulations. The fluorescence cytometer can be used in confocal mode by adding optics that reduce the laser spot to a diffraction limited size (approximately 0.3μm) in conjunction with a focusing lens and pinhole positioned in front of the photomultiplier tube(s). Confocal microscopy (discussed in more detail below) is an imaging feature that can significantly enhance axial resolution and, to a lesser extent, lateral resolution.

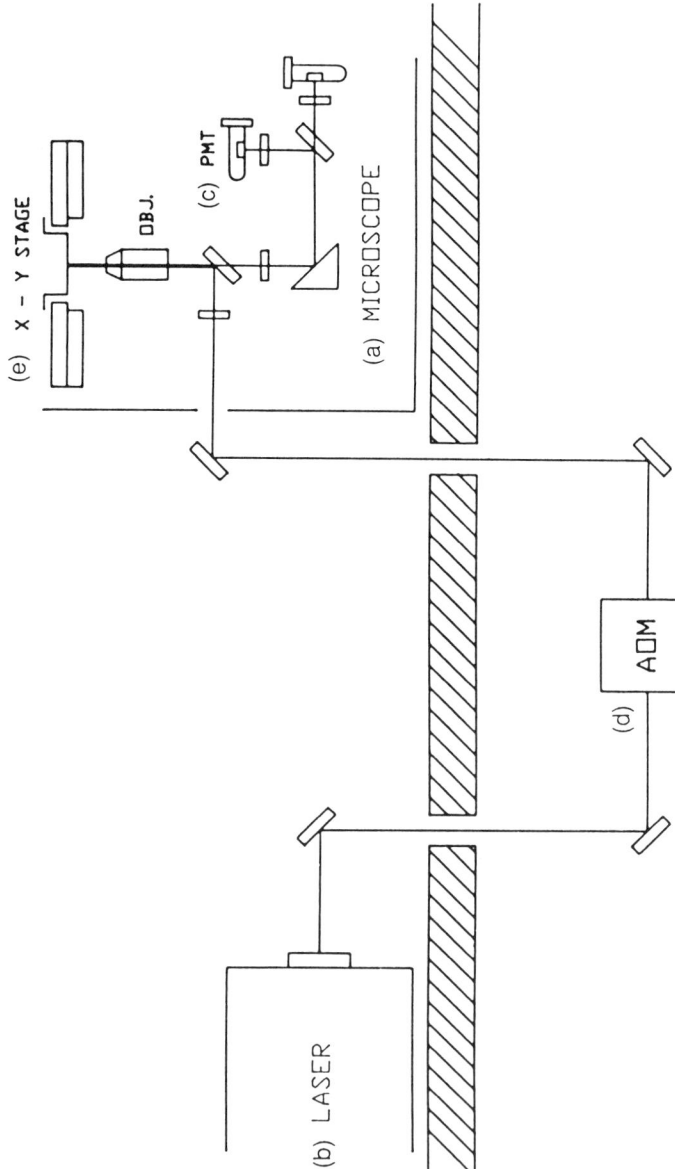

Fig. 1. Diagram of the optical path of a fluorescence-imaging system (ACAS 570 Interactive Laser Cytometer). This system consists of a tunable 5W Argon ion laser, an acousto-optic modulator (AOM) to control the duration and intensity of excitation, an inverted microscope with an X–Y scanning stage, a Z-axis stepper motor, and two photomultiplier (PMT) tubes. A computer performs command, control, and analytical routines.

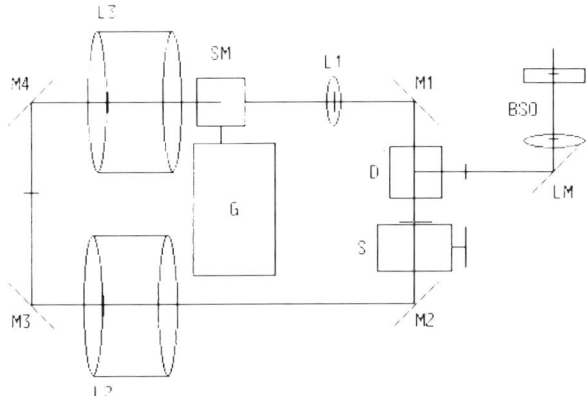

Fig. 2. Diagram of the optical path of the bilateral laser scanning confocal microscope (InSIGHT). Real-time confocal imaging is obtained through the use of a double-sided mirror which simultaneously scans the sample and the oculars or detector at video rates. BSO, beam shaping optics; LM, laser mirror; D, dichroic; M4, mirrors; SM, scanning mirror; G, galvonometer; L1-L3 lenses; S, variable slit.

The bilateral laser scanning confocal microscope (Fig. 2) is another type of fluorescence imaging system that utilizes a CCD as a detector and a scanning light cursor to excite fluorescence in cells. A double-sided, galvanometer-driven mirror simultaneously scans the specimen and the detector, building a confocal image (14, 15). This configuration utilizes a slit at the detector rather than a pinhole. The instrument can provide "real-time" confocal views of fluorescence that can be observed through the oculars.

In conventional fluorescence microscopy, emitted light from above and below the plane of focus can degrade image contrast and resolution. Confocal imaging significantly reduces out-of-focus fluorescence by introducing an aperture (slit, pinhole, or spinning disk) in the emission path, restricting light observed at the detector (Fig. 3). In both systems described here, the degree of confocality can be adjusted to optimize for a particular sample thickness and fluorescence intensity of the sample. The advantages of this technology for a variety of analytical approaches will be discussed in more detail later in this chapter.

2.2. Probes—Natural Fluorophores and Fluorescent Indicators

2.2.1. PASSIVE REPORTER GROUPS

Although a number of biological molecules are inherently fluorescent, e.g., aromatic amino acids, flavins, vitamin A, chlorophyll, and NADH, most of these endogenous fluorophores can only be excited in the ultraviolet (UV) region (between 260nm and 360nm). For both technical (special UV optics) and biological (photodamage) reasons, the use of these chromophores for imaging has been

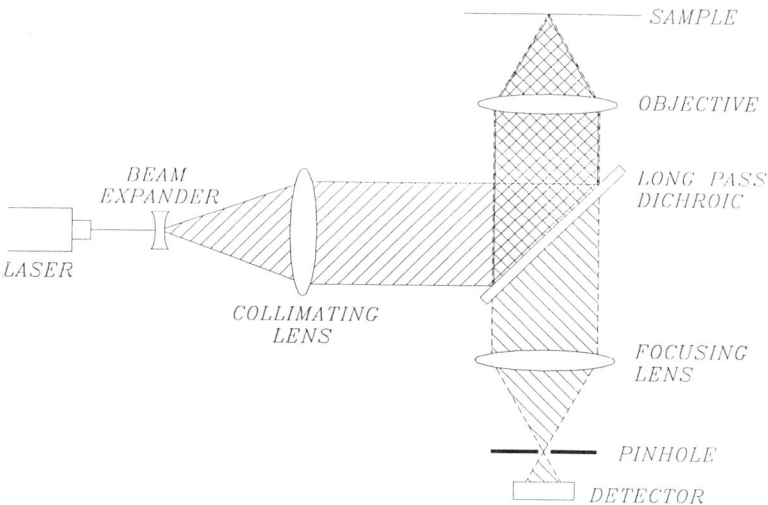

Fig. 3. Simplified diagram of the confocal components of the ACAS 570. The laser beam is expanded to provide a diffraction limited spot at the sample. A focusing lens in front of the detection system directs the fluorescence from the focal plane through an adjustable pinhole. In the bilateral laser scanning confocal microscope, a similar technique is employed with a cursor rather than a point for illumination and detection.

limited to relatively few investigations (16, 17). The bulk of cellular fluorescence measurements has been performed with dyes that are incorporated either biosynthetically or by covalent modification of appropriate site-specific or function-specific molecules (18-23). Perhaps the two best examples of such probes are antibodies derivatized with either fluorescein isothiocyanate (FITC) or rhodamine isothiocyanate (TRITC) to visualize specific antigen cellular localization and phospholipids derivatized with fluorescent dyes to demonstrate lipid environments within the cell (18-23). The most comprehensive description of commercially available fluorescent probes and references discussing their use may be found in *Molecular Probes—Handbook of Fluorescent Probes and Research Chemicals* (24).

In general, fluorescent probes offer several physical characteristics that may be exploited by the investigator to pursue cellular investigations.

1. Fluorescence is very sensitive to the chemical environment and may be utilized to provide information about the microenvironment surrounding the probe (5, 6). The fluorescence intensity (quantum yield) (1-3), the maximum emission wavelength (1-3), the fluorescence lifetime (6-9), or the polarization (10-12) may all be monitored for specific changes that are induced as a result of changes in polarity, pH, ion concentration, membrane potential, or ligand binding.
2. Measurements of specific fluorescence quenching processes may also be utilized to provide environmental information, specifically with regard to

the distance between macromolecules within or on the surface of the cell. The two types of quenching phenomena that have demonstrated their value for cellular analysis are the long-range (~80nm) nonradiative process of resonance energy transfer (25–27) and the quenching that results from intermolecular collisions (6, 28). In the case of resonance energy transfer, energy is transferred from the excited singlet state of a donor to the excited singlet state of an acceptor. The amount of transfer or resonance is given by the overlap of the fluorescence emission spectrum of the donor and the absorption spectrum of the acceptor. Utilizing the Förster relationship (29),

$$R = R_0 \left(\frac{1 - E_T}{E_T}\right)^{1/6},$$

where R_1 is the intermolecular distance, R_0 is a constant for each donor–acceptor pair and E_T is the efficiency of transfer, intermolecular distances may be determined between donor–acceptor pairs, e.g., FITC and TRITC. Collisional quenching may be utilized to examine accessibility of fluorophores and has been used more selectively to assess exposure of fluorophores on the cell surface (6, 28).

3. Measurements of fluorescence polarization have also been most useful in determining macromolecular organization and dynamics in cells. When a fluorophore is excited with plane-polarized light and the fluorescence is observed through analyzing polarizers, the degree of polarization of fluorescence usually decreases. This is called *fluorescence depolarization* and is a result of the random movement of the fluorophore. If the fluorophores, however, become constrained by associating with macromolecular assemblies or with the membrane, the degree of fluorescence polarization increases. In this manner, the polarization value can serve as an indicator for degree of immobilization (30–32). In addition to these steady-state fluorescence analytical methods, there are a number of time-resolved approaches to explore fluorophore environments (33, 34). Although very powerful, the use of these methods has been limited by the absence of commercial instrument packages to image time-related fluorescence in cells under growth conditions.

2.2.2. ACTIVE PROBES

In the previous section, we discussed the general properties of fluorophores that serve as passive reporter molecules for cellular organization and biological activity. Recently, a variety of photoactivatible or "caged" probes have become available that are either nonfluorescent or biologically inert prior to activation by the appropriate exciting wavelength. These fluorescent dyes have been modified chemically to quench their fluorescence or biological activity until the probe is "uncaged" by incident illumination (24, 35–39). Utilizing these probes, light may be used to initiate intracellular biological activity and then to follow the biochemical consequences of this optical triggering by the use of appropriate passive probes.

3. APPLICATIONS AND METHODS

Analytical approaches utilizing fluorescence imaging may be categorized as follows:

- Quantitation of molecular components in single cells or cell populations
- Topologically defined kinetic analyses of cellular activities
- Fluorescence *In Situ* Hybridization (FISH)
- Confocal microscopy and three-dimensional reconstruction

An additional advantage of these approaches is that repetitive measurements on the same field(s) of cells can be performed during the course of an experiment. Cells can be maintained under growth conditions and optically probed in a continuous manner. Computer control of experiments permits the sequential and/or multiparametric analysis of hundreds of cells. Results may then be presented in a variety of formats giving detailed information about cell structure and function of specific metabolic and signaling pathways. This is difficult with other single-cell analytical methods such as flow cytometry (40–42). In the following section, we will present a series of model investigations to highlight a number of analytical approaches. Unless otherwise noted, the data were obtained with the ACAS 570 Interactive Laser Cytometer.

3.1. Quantitation of Molecular Components in Single Cells or Cell Populations

3.1.1. QUANTITATION OF GLUTATHIONE IN LIVING CELLS IN TISSUE CULTURE

Glutathione (GSH) and glutathione-S-transferase (GST) protect cells against free radical damage and chemical injury (43, 44). GSH is found at high levels in most cell types and is the most prevalent cellular thiol. GSH and GST levels in cells are important in predicting drug resistance and cellular integrity (45, 46). Monochlorobimane (MCB) is incorporated readily into live cells and becomes fluorescent when conjugated to GSH by GST (44, 45). Therefore, utilizing fluorescence imaging approaches, it is possible to directly measure GSH concentration on an individual cell basis.

In the experiments shown in Figs. 4 and 5, NIH 3T3 fibroblasts grown in Dulbecco's Modified Eagle's Medium (DMEM) containing 10% calf serum at 5% CO_2, were loaded with 1μM MCB for 30 min at 37°C. The fluorescence was quantitated by exciting with the argon ion laser (351–363nm) and collecting emitted light above 390nm. Manipulations to block GSH synthesis, such as addition of buthionine disulfoxide (BSO), dramatically reduce GSH levels (Fig. 5).

Similar reductions in GSH levels can be seen with the sulfhydryl-reactive mycotoxin patulin in granulosa cells (47). Figure 6 depicts a series of images showing the dose response of GSH activity after 2 hr incubation with patulin in granulosa cells. Doses of 0.1μM and 1μM patulin reveal dose-dependent de-

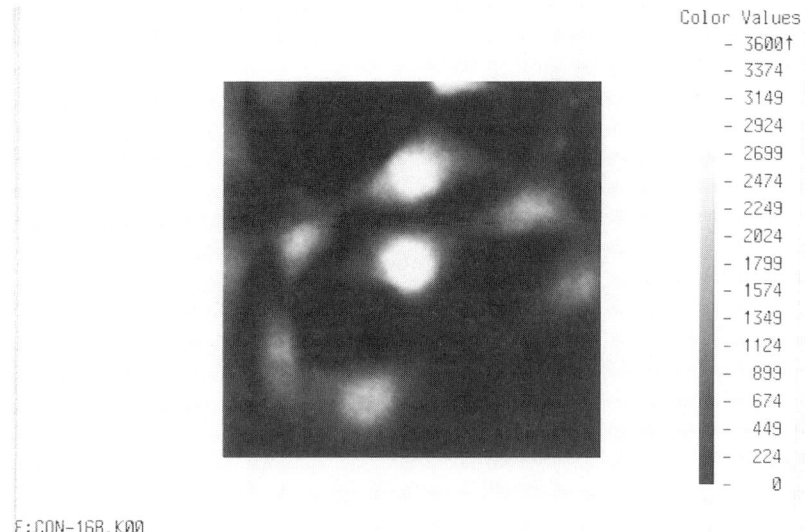

Fig. 4. GSH fluorescence in NIH 3T3 cells labeled with 1µM monochlorobimane for 30 min.

pletion of GSH, while 10µM and 100µM patulin overlap with the background values, indicating total depletion of GSH.

These results demonstrate a number of advantages provided by imaging approaches for quantitation. The concentration of an important bioactive mole-

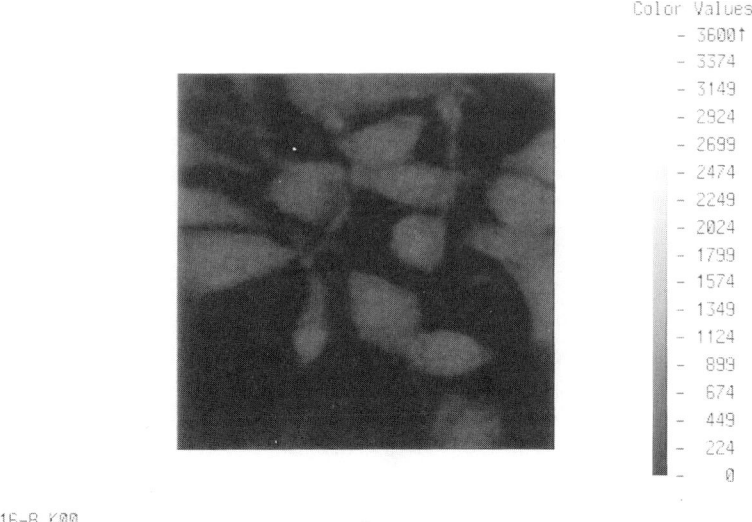

Fig. 5. GSH fluorescence in NIH 3T3 cells incubated overnight with 10µM buthionine disulfoxide and then labeled with 1µM monochlorobimane.

cule can be measured both spatially and temporally within living cells. It is also possible to examine large populations of cells to determine cellular heterogeneity in relation to responses to modifying drugs. Experiments may be performed on small numbers of cells, limiting the need for large amounts of tissue.

3.1.2. QUANTITATION OF ONCOGENE EXPRESSION IN TISSUE CULTURE CELLS

Abnormal expression of the ras oncogene has been correlated to tumor formation and carcinogenesis (48, 49). Using a fusion gene prepared with a metallothionein (MT) promoter and the ras T24 structural gene (49–51), ras T24 protein (p21) expression in rat liver epithelial MTR6 cells could be manipulated by altering the zinc concentration in the growth media, and quantitated using fluorescence cytometry (51). It has been shown that increased expression of the ras gene causes tumors in rats (48, 49), and a correlation between expression of the oncogene and down regulation of gap junction function has been established (51).

Untreated cell cultures (i.e., no additional zinc) and cultures exposed to 100µM ZnSO$_4$ for 2 days were fixed and stained with a streptavidin-fluorescein antibody against p21 (for detailed method, see 51). The average fluorescence in several hundred single cells was determined and displayed in a histogram showing control (no additional zinc) and 100µM zinc-treated cells (Fig. 7). Zinc

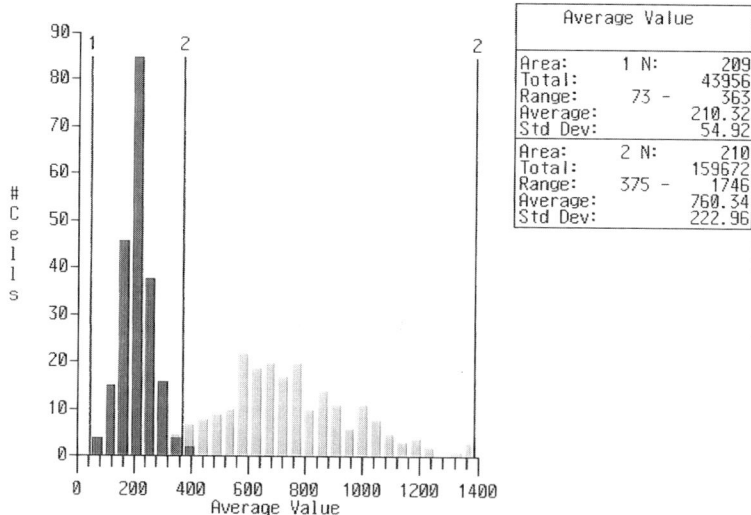

Fig. 7. Histogram of the average fluorescence intensity of rat liver epithelial cells carrying a zinc-inducible MTrasT24 fusion gene. Control cells (dark gray) and cells treated with ZnSO$_4$ (100µM, 48 hr; light gray) were fixed, probed with FITC-antibody against the ras T24 protein (p21), and analyzed on a single cell basis. Notice the marked increase in fluorescence in the treated cells, reflecting the enhanced accumulation of ras T24 protein.

induces the expression of ras T24, and the result is the observed fluorescence increase due to increased FITC-antibody binding. The MTR6 cells show an increase of about fourfold in fluorescence intensity following exposure to zinc. In a similar manner, fluorescence immunoassays may be pursued at the single-cell level. Although these measurements were performed on fixed cells, microinjection techniques may be employed in conjunction with fluorescently labeled antibodies to pursue similar investigations in living cells.

3.1.3. MEASUREMENTS OF pH IN MAMMALIAN CELLS

The strict regulation of intracellular pH is vital for cell growth and function. Variations in pH affect both biosynthetic and metabolic pathways (52–56). A number of growth-related biochemical factors have been demonstrated to require changes in intracellular pH for their activity (57–59). To pursue such measurements in a noninvasive manner in living cells, it is possible to exploit the environmental sensitivity of fluorescent probes to monitor pH. The ester of SNAFL calcein AM (24) is one of a variety of probes that may be utilized for this purpose (60–62). This molecule is initially nonfluorescent and can passively diffuse across the plasma membrane of cells. Within the cell cytoplasm, endogenous esterases cleave the ester-linked groups producing SNAFL calcein, a strongly fluorescent molecule now trapped within the cell cytoplasm. The calcein derivative of SNAFL serves to bind the SNAFL to intracellular glutathione and inhibit dye leakage (24), a significant problem with previous pH probes. SNAFL calcein is optimally excited by the 514nm line of the argon laser, and emissions at 570nm and 630nm are collected simultaneously and ratioed. Figure 8 shows the ratiometric pH profile within 3T3 cells. The pH values have been extrapolated from a standard curve generated following exposure of cells to the proton ionophore nigericin (63) (10μg/ml) and subsequent incubation with a series of buffers of known pH.

Although there is considerable heterogeneity in pH values observed within a cell, whole cell pH values are between 7.1 and 7.2, in agreement with previously reported values for 3T3 cells (64). This same type of intracellular heterogeneity exists in cells labeled with BCECF (another ratiometric pH probe) and may be in part due to a chemical interaction of the fluorescent probe with the cells. This may represent a biologically relevant feature not observed previously for measurements performed on cells in suspension. With BCECF, some of the heterogeneity can be diminished by including glucose in the buffer and treating the cells with the mitochondrial poison CCCP (M. Wade, unpublished results).

3.1.4. MEASUREMENTS OF MEMBRANE POTENTIAL IN CELLS GROWING IN TISSUE CULTURE

The use of potential-sensitive fluorescent probes to monitor the electrical potential across a cell membrane permits an accurate, noninvasive measurement of membrane potential changes in a wide variety of cells, vesicles, and organelles without the external electrical or mechanical manipulation required by microelectrode techniques (65, M. K. Frame *et al.*, manuscript in preparation). Fluorescence-imaging techniques can quantitate membrane potential changes in single cells or groups of cells equilibrated with the potential-sensitive bisoxonol dye $DiBAC_4(3)$ (66, 67), which exhibits a potential-dependent partitioning across the plasma membrane. Since $DiBAC_4(3)$ fluorescence increases significantly in the cell compared to extracellular $DiBAC_4(3)$ as a result of dye binding to intracellular components (predominantly proteins), cell depolarization results in an increase in cell fluorescence (67). Figure 9 shows several images of a field of human teratocarcinoma (HT) cells grown in DMEM plus 10% FCS at 37°C in 5% CO_2 and subsequently labeled with 5μM $DiBAC_4(3)$ in DMEM for 20 min at 37°C. The $DiBAC_4(3)$ remained in the media throughout the measurements at 37°C. After establishing a baseline, 50mM KCl was added to depolarize the cells. The normalized plot, reflecting about 35% enhancement in fluorescence following the addition of KCl is also shown in Fig. 9.

Figure 10 shows the results of a stepwise addition of increasing concentrations of KCl to a group of cells. A clear dose-dependent increase in fluorescence is apparent as the cells become more depolarized. The actual changes in membrane potential can be estimated by using the Nernst equation to calculate membrane potential at various external potassium concentrations (68).

3.2. Kinetic Analysis of Cellular Activities

Cellular function depends on the movement of molecules within membranes (18, 20, 23, 69), through cytoplasmic compartments (13, 21, 70), and between

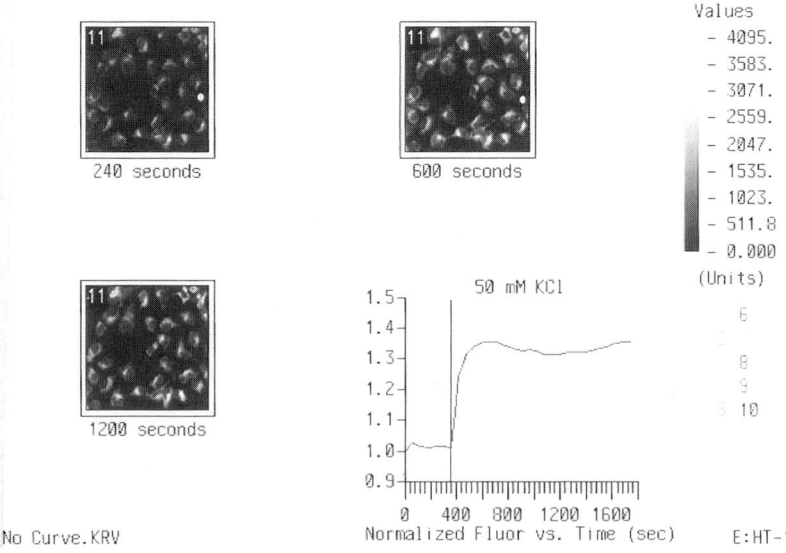

Fig. 9. Fluorescence of HT cells labeled with 5μM $DiBAC_4(3)$ and depolarized by 50 mM KCl. The time plot is normalized against the first measurement.

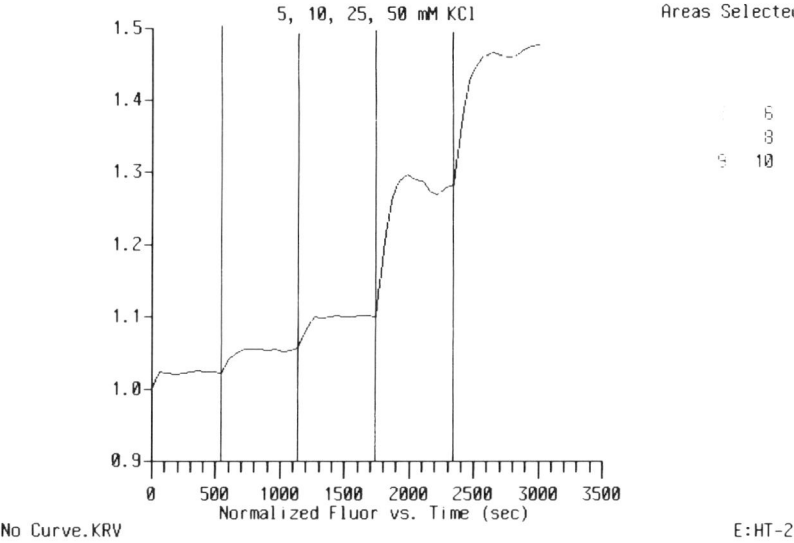

Fig. 10. Sequential addition of KCl to DiBAC$_4$(3)-labeled HT cells.

cells (intercellular communication) (71–73). Rapid changes in ionic fluxes, Ca^{++} concentration, and membrane potential are important controllers of these dynamic cellular processes. Quantitative fluorescence microscopy has made significant contributions to our understanding of rapid dynamic processes that are necessary for signaling mechanisms within and between cells. The following experimental approaches have been utilized to measure such transient fluorescence changes.

3.2.1. Ca^{++} Oscillations in Cultured Cells

Small changes in free intracellular calcium can affect gap junction function, cell division, muscle contraction, neutrophil stimulation, and other essential cell functions (74–77). Recently, several fluorescent probes to measure intracellular free calcium have been developed (24, 78–80). Indo-1 and Fura-2 were developed as dual emission or dual excitation calcium-sensitive ratiometric probes. The ability to ratio emissions or excitations can eliminate or mitigate artifactual fluorescence changes resulting from variations in dye concentration, photobleaching, and dye leakage that may exist within or between cell populations. However, both Indo-1 and Fura-2 require UV excitation. More recently, the single excitation/emission probes, Fluo-3 and Rhod-1, have become available (24, 78). Fluo-3 can be excited by the 488nm line of the argon laser and emits in the visible spectrum at a wavelength similar to fluorescein (530nm). The advantage of this probe is that UV excitation is not required. A major disadvantage is that it is difficult to quantitate the absolute calcium concentration in cells because ratioing is not possible. The relative size and temporal resolution of

Ca^{++} transients, however, are relatively easy to determine in cells. Further quantitation can be pursued with the calcium ratio probe, Indo-1.

A potentially useful approach to provide better quantitation with Fluo-3 would be to label cells with a combination of Fluo-3 and Fura Red, a long wavelength calcium indicator that displays an intensity decrease as calcium values increase (81). Theoretically, the ratio of the emissions of these two probes could be correlated to calcium concentration by using a standard curve. Preliminary findings indicate that such a ratio of emissions at 530nm (Fluo-3) and 630nm (Fura Red) can be measured. HT cells were labeled with both Fluo-3 and Fura Red and excited with 488 nm light from the argon ion laser. The emissions at 530 nm and 630 nm were simultaneously detected and ratioed. Figure 11 depicts images from both detectors before addition of ionophore. Figure 12 shows the same field several seconds following addition of 3 µM ionomycin. Ionomycin is a calcium ionophore that not only opens membrane channels to extracellular calcium, but also causes significant internal release of calcium stores. Note the increase in signal in Detector 1 (Fluo-3) and the drop in signal in Detector 2 (Fura Red), which indicates that an increase in intracellular calcium has occurred. The plot of free Ca^{++} change, displayed in Fig. 13, shows the Fluo-3/Fura Red emission ratio as the ordinate and time as the abscissa. An initial rise in calcium is clearly seen, followed by a return to near baseline levels. The photobleaching characteristics of cells labeled with Fluo-3 or Fura Red were determined by repetitively scanning the same area of cells and plotting the integrated value as a function of scan number. Following 50 scans, the fluorescence intensity had only decreased by 5% (data not shown). The minimal photobleaching is a result of the use of a highly focused laser beam (approximately 1µm in diameter) and an acousto-optic modulator (AOM). The AOM pulses the laser such that the sample is only illuminated for a brief period of time over a discrete area.

The bilateral laser scanning confocal microscope permits real-time measurements of changes in intracellular calcium. Freshly isolated rat heart cells were loaded with 5µM Fluo-3. Fluorescence images were collected every 0.2 seconds by a frame grabber, or in real-time (30 frames/second) on video tape using the INSIGHT. Figure 14 shows three images and a time plot illustrating the calcium oscillations seen in these beating cells as a function of time. This type of temporal resolution requires the rapid detection capability obtained by a real-time imaging system.

3.2.2. DIFFUSION OF MEMBRANE AND CYTOPLASMIC COMPONENTS

The measurement of fluorescence redistribution after photobleaching (FRAP) is a unique, noninvasive method for directly analyzing dynamic processes in living cells (13, 18–23, 69–73). FRAP experiments have provided important insights into membrane structure (18, 20), mechanisms of hormone action (82, 83), nucleocytoplasmic communication (13, 21), cytoplasmic organization and structure (84), actin and tubulin assembly (85, 22), cell-cell communication (13, 65, 67–69), cell differentiation and proliferation (86), parasite membrane structure (87), and bacterial membrane biosynthesis (88–89).

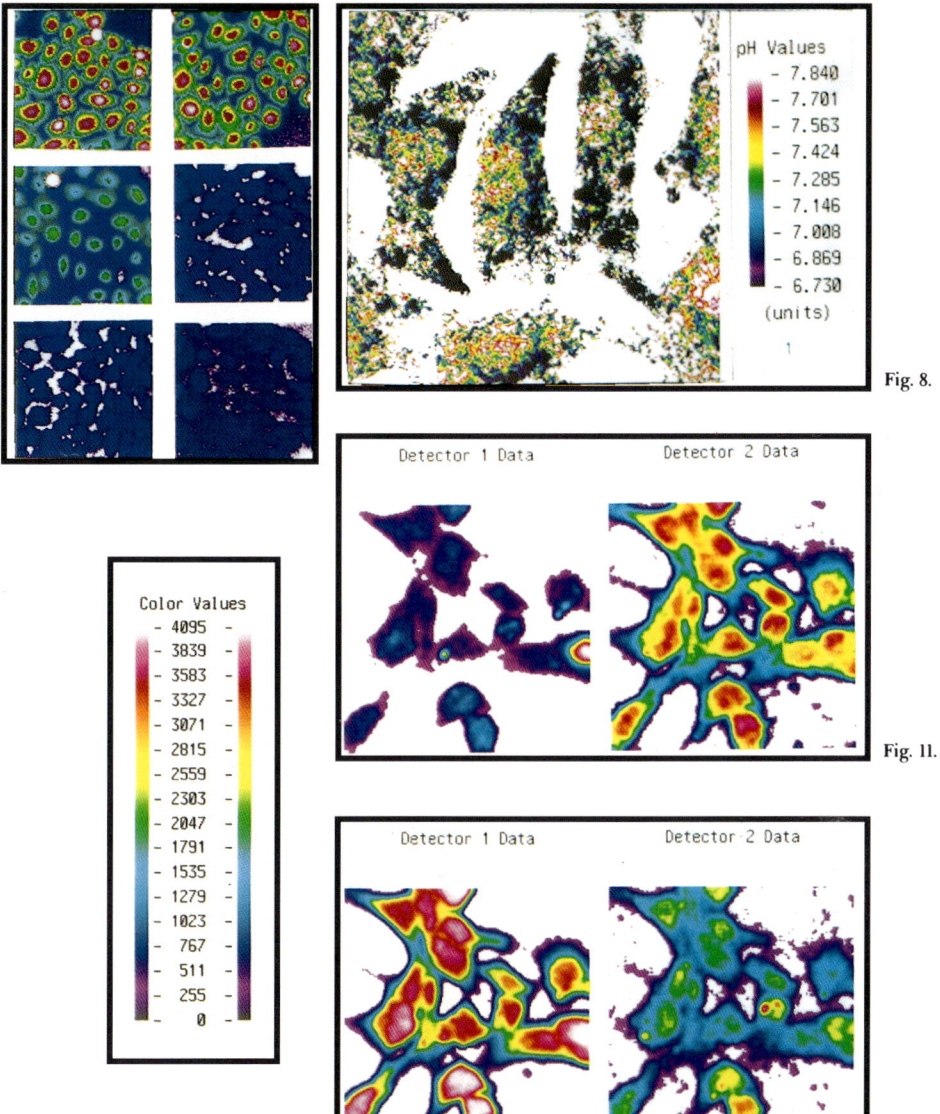

Fig. 6. GSH fluorescence in granulosa cells stained with 40μM monochlorobimane after 2-hr incubation with various concentrations of patulin. Reading from left to right: (top) control (0μM), 0.1μM; (middle) 1.0μM, 10μM; (bottom) 100μM and background control without MCB labeling. (Data courtesy of Dr. Robert C. Burghardt, Texas A & M University, College Station, Texas.)

Fig. 8. SNAFL calcein fluorescence ratio map in fibroblasts. Ratio of 570nm and 630nm emissions was calculated and plotted as a function of pH, which was independently determined in nigericin-treated cell cultures.

Fig. 11. HT cells labeled simultaneously with 3μM Fluo-3 (detector 1) and 5μM Fura Red (detector 2).

Fig. 12. The same field as Fig. 11 following the addition of 1.5μM ionomycin. Note the increase in signal in detector 1 and the decrease in detector 2.

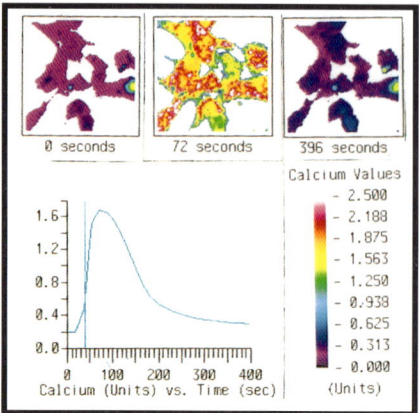

Fig. 13.

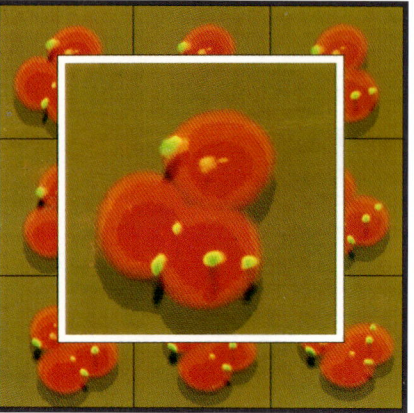

Fig. 22.

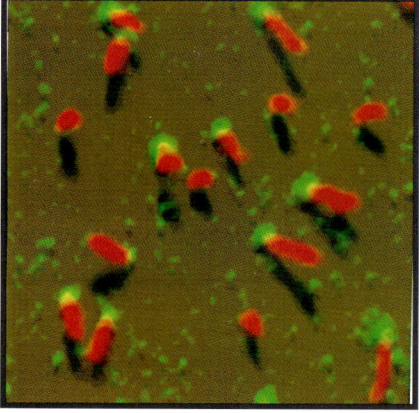

Fig. 25.

Fig. 13. Time plot summarizing the ratio change as a function of time. The vertical line denotes the addition of the ionomycin.

Fig. 22. Male peripheral blood mononuclear cells labeled with biotin-#1 chromosome probe/avidin (FITC; green) and counterstained with propidium iodide (red). Seventeen optical sections (27 x 27μm each) were imaged at 0.3μm intervals in the Z-axis, and reconstructed at various viewing angles with the SFP algorithm. (Sample courtesy of Dr. James F. Leary, University of Rochester Medical Center, Rochester, N.Y.)

Fig. 25. Bacteria (*Bradyrhizobium japonicum*) labeled with propidium iodide to define the bacterial volume (red) and a fluorescent antibody against BJ38 (FITC; green). Twenty-two optical sections (18 x 18μm each) were imaged at 0.3μm intervals in the Z-axis and reconstructed with the SFP algorithm. Notice the polar organization of BJ38, a protein that is thought to play a role in attachment to soybean root hairs.

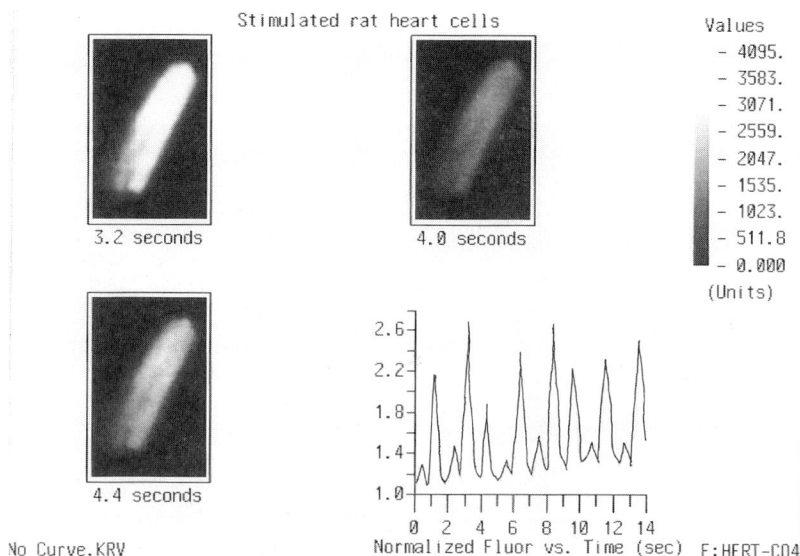

Fig. 14. Measurements of Ca^{++} oscillations in adult rat heart cells labeled with 3µM Fluo-3. These cells beat in culture with electrical stimulation. (Sample courtesy of Dr. Scott Henry, Parke-Davis, Ann Arbor, MI). Data were obtained with an INSIGHT confocal microscope.

FRAP measurements on membranes are made by photobleaching a microscopic area (1µm-2µm in diameter) with a short, intense pulse of light from an argon ion laser. Recovery of fluorescence within the bleached area due to the lateral diffusion of neighboring intact fluorophores is subsequently measured by repetitive scanning across the cell surface with an attenuated laser beam. The rate and extent of recovery permits calculation of the diffusion coefficient (mobility) of the fluorescent species and the fraction, which is mobile (termed the recoverable percent). In a representative investigation, NIH 3T3 fibroblasts were labeled with 1-acyl-2-(N-nitrobenzo-2-oxa-1,3-diazole)-aminocaproyl-phosphatidylcholine (NBD-PC) (20µg/ml at 4°C for 20 minutes), a fluorescent derivative of a naturally occurring membrane phospholipid. A gray-scale image of fluorescence is presented to visualize the distribution of the fluorophore across the entire cell (Fig. 15). Prebleach scans are made across the cell surface to establish the initial fluorescence intensity profile. An area on the membrane is then targeted for photobleaching (denoted by an X on the image). Postbleach scans monitor the recovery of fluorescence (Fig. 16). Final data from an experiment include mobility (in cm^2/sec), standard deviation from the fit curve, percent fluorescence recovery, and the initial bleach percentage (Fig. 17). The use of FRAP techniques is based upon the introduction of fluorescent probes into either the cytoplasm, specific cytoplasmic compartments, or the plasma membrane. The total time needed to photobleach, monitor, and analyze fast diffusing species in the membrane is approximately 2 to 3 min. Typically, NBD-PC and

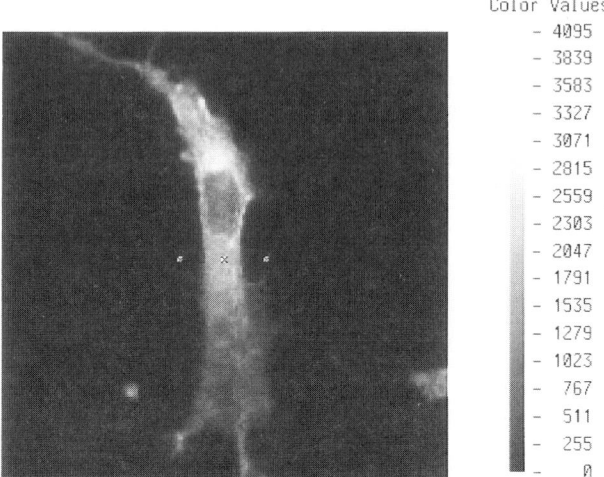

Fig. 15. Image of a fibroblast, fluorescently labeled with NBD-PC (20µg/ml for 20 min). The image displays the start (.), end (.), and bleach point (X) for the FRAP experiment.

other lipid probes display a recovery percentage of 60%–90% and a mobility of 1.5×10^{-9} to 6×10^{-9} cm^2/sec. Proteins usually recover less than 50% and move at rates that are a factor 10 to 100 slower. FRAP experiments with proteins therefore require more time than measurements with lipids.

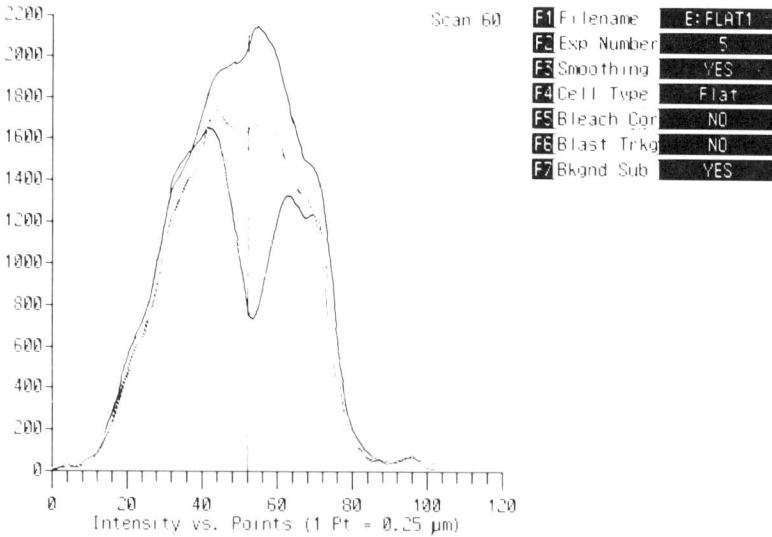

Fig. 16. Composite plot displaying the gradual recovery of NBD-PC fluorescence intensity (Y-axis) along a cross section of a 3T3 cell (X-axis) within 50 sec after photobleaching a spot on the cell membrane at the position indicated by the vertical line.

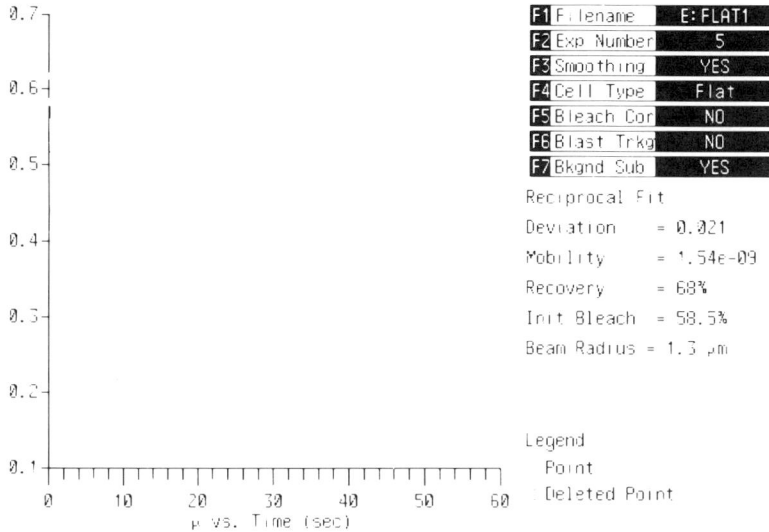

Fig. 17. Final result of the FRAP analysis of a 3T3 cell. The fluorescence recovery, indicated by reciprocally decreasing μ-value (Y-axis), is plotted versus time after photobleaching (X-axis). Several FRAP parameters, are displayed next to the graph.

3.2.3. INTERCELLULAR COMMUNICATION

Intercellular communication measurements are another way in which photobleaching and fluorescence recovery can be employed to monitor movement of fluorescent molecules. In this approach, FRAP is used to measure dye transfer between contacting cells (71–73). Molecules less than 1000 dalton in molecular weight (MW), such as carboxyfluorescein (MW 370), can freely transfer through gap junctions. The measurement of gap junctional communication has been examined by loading cells with carboxyfluorescein diacetate (1μg/ml for 15 min @ 37°C) and then selectively bleaching whole cells. Cells that are in contact and have functional gap junctions are able to reestablish the fluorescence equilibrium by transfer of carboxyfluorescein from unbleached neighboring cells to the bleached cell. Numerous studies have documented the utility of this technology for communication measurements (69, 71–73, 93). Rates have been calculated, and it has been shown that the rate of communication is proportional to the number of contacting cells (94). In rat liver epithelial cells that contain a zinc-inducible ras T24 gene, a reduction in intercellular communication was correlated with increased expression of the ras p21 protein [see above and (51)].

3.3. Fluorescence *In Situ* Hybridization (FISH)

Fluorescence *In Situ* Hybridization is a technique that utilizes fluorescently modified DNA probes to hybridize to specific sequences in metaphase chromosome preparations or interphase nuclei. These probes can be used to localize single-gene copies, teleomeres, centromeres, and specific whole chromosomes

(95–97). A more specialized use of this technique with great promise for clinical evaluations of chromosomal abnormalities has been termed chromosome painting. The DNA probes for this type of analysis are prepared from a chromosome-specific library. Reciprocal translocations within chromosomes are detected rapidly by fluorescence imaging, reducing the need for the more involved labor-intensive procedures of chromosome banding. Movements of chromosomal pieces are observed as shifts in the fluorescence signal from the hybridized chromosome. Figure 18 shows a metaphase spread in shades of gray from a mouse, human hybrid in which all chromosomes are stained with propidium iodide (dark gray) and only the human chromosomes are stained with FITC (light gray). The overlap of the two signals is evident.

3.4. Confocal Microscopy and 3-D Reconstruction

Laser scanning confocal microscopy can significantly enhance the details of subcellular structures labeled with fluorescent probes. In addition, spatial resolution is improved significantly, permitting topically defined investigations of cellular activity using fluorogenic reagents or reporter molecules. The advantage of confocal microscopy is that out-of-focus fluorescence from above or below the plane of focus is greatly reduced (98). Three-dimensional reconstructions of subcellular structures, cells, or tissue sections can be prepared from a series of optical slices (theoretical Z axis resolution is approximately 0.6μm–0.7μm) acquired

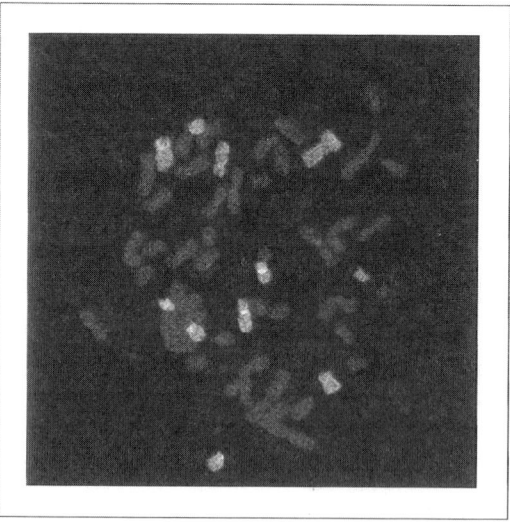

Fig. 18. Metaphase image of a mouse/human hybrid bearing multiple human chromosomes, probed with FITC-biotinylated human DNA (light gray) and counterstained with propidium iodide (dark gray). The human chromosomes show up in light gray as a result of a combined signal of FITC and PI. (Sample courtesy of Dr. Roger A. Schultz, University of Maryland School of Medicine, Baltimore, MD.)

as the plane-of-focus is automatically moved in small increments through the sample. A significant advantage of this technique is that it results in nondestructive, high-resolution optical sections of living tissues so that dynamic processes may be followed and cellular variations in organization can be analyzed temporally. In addition, it provides the axial resolution that is not available in conventional transmission electron microscopy without recourse to the difficult technical process of preparing serial sections. Recently, several confocal studies have been published that implicate differences between nuclear and cytoplasm calcium values in smooth muscle and neuronal cell types (99, 100). These data were generated using the UV-excitable dye, Indo-1, and confocal optics to localize the calcium response accurately.

The advantages of confocal microscopy for observing fluorescent structures can be seen in Fig. 19. A composite photograph is presented of a confocal image and a standard nonconfocal fluorescence image of the same astrocyte labeled with FITC conjugated anti-tubulin; note the blurring in the nonconfocal image.

In Fig. 20 various optical sections are depicted from a neuron microinjected with Lucifer Yellow, which was located within a thick (>100μm) brain section. The various sections do not show much information, but when they are combined in a reconstruction as shown in Fig. 21, the three-dimensional nature of the structure becomes clear. Similar detailed images are observed for interphase nuclei which have been hybridized with FITC derivatized oligonucleotide probes. Figure 22 displays a reconstruction in which it is quite evident that two

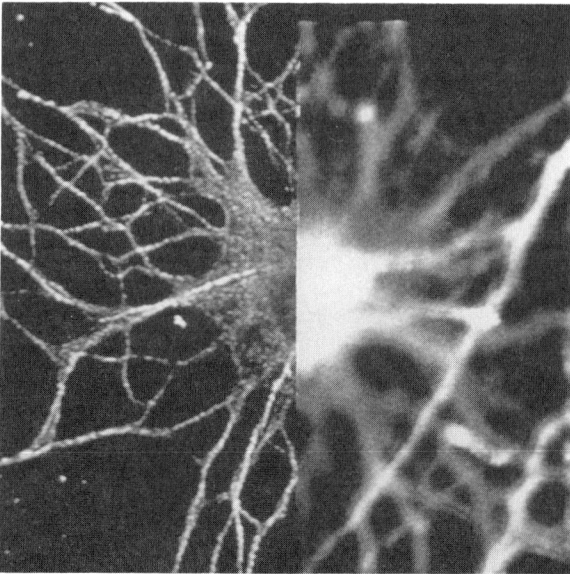

Fig. 19. Composite photograph showing confocal (left) and non-confocal (right) images of an astrocyte labeled with FITC anti-tubulin. (Sample courtesy of Dr. Charissa Dyer, Wayne State University, Detroit, MI).

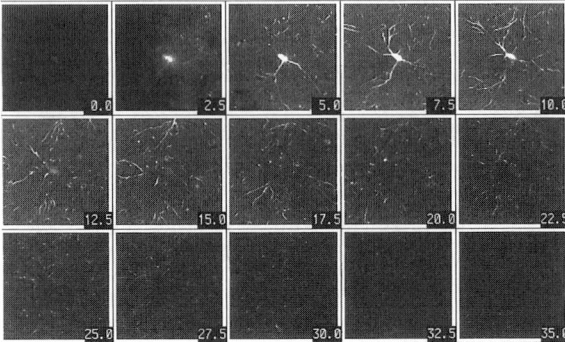

Fig. 20. Tissue section from the ventral posteromedial nucleus of rat thalamus. A neuron was microinjected with Lucifer Yellow and 15 optical sections (180 × 180µm each) were imaged at 2.5µm intervals in the Z-axis. (Sample courtesy of Dr. Robert Rhoades, Medical College of Ohio, Department of Anatomy, Toledo, OH.)

copies of the number 1 chromosome exist (FITC, green) within each nucleus (propidium iodide, red).

Lily pollen was stained with acridine orange and imaged using the bilateral laser scanning confocal microscope (INSIGHT). Figure 23 shows a three-dimensional reconstruction of 49 separate slices, using a modified version of the SFP

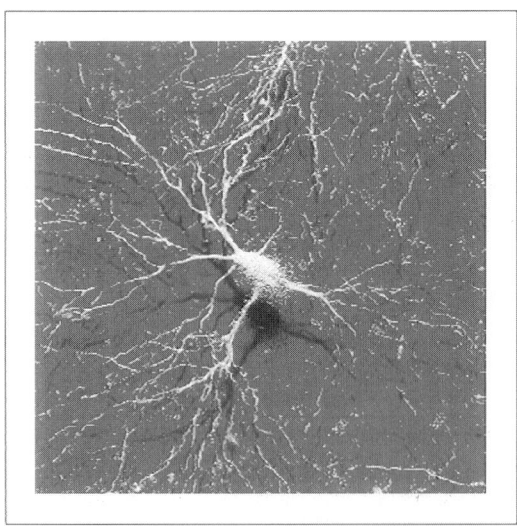

Fig. 21. Three-dimensional reconstruction of the optical sections displayed in Fig. 20. Notice how the many small and separate areas of fluorescence that were observed in most of the sections combine into long extensions from a single cell. (Sample courtesy of Dr. Robert Rhoades, Medical College of Ohio, Department of Anatomy, Toledo, OH.)

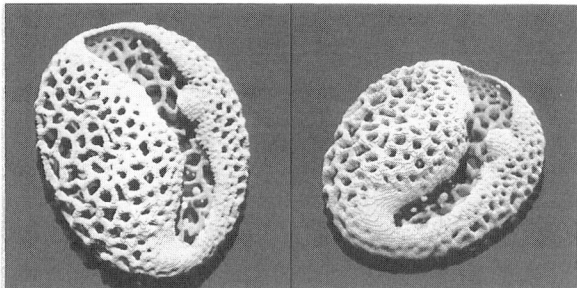

Fig. 23. Lily pollen stained with acridine orange. Forty-nine optical sections (86 × 86μm each) were imaged with the INSIGHT at 0.5μm intervals in the Z-axis and reconstructed with the SFP algorithm at two different viewing angles.

algorithm (101–103). The SFP algorithm combines multiple sections into one 3-D reconstruction using a modified volume rendering technique known as *ray-casting* combined with a depth-shaded function. Shown here are two different views reconstructed from two angles. Figure 24 shows a 3-D reconstruction of an insect isolated from Italian bacon and labeled with acid fuchsin. The entire insect was scanned in a confocal mode with a high numerical aperture objective.

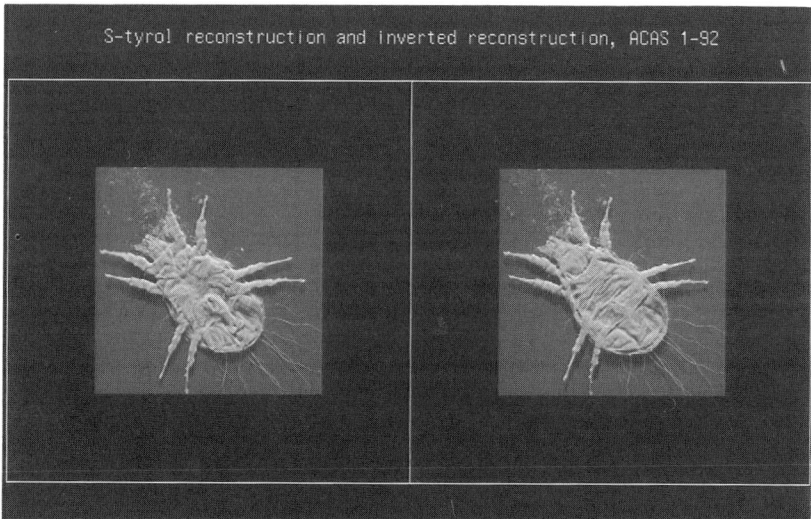

Fig. 24. Macerated male mite (Tyrophagus Longer), stained with acud fycgsin. Forty optical sections (720 × 720μm each) were imaged at 0.5μm intervals in the Z-axis, and reconstructed with the SFP algorithm. Views from the bottom (left panel) and top (right panel) of the sample are displayed. (Sample courtesy of Dr. Manfred G. Walzl, University of Vienna, Austria.)

A particularly important future role for confocal microscopy and D reconstruction may be as a means to attain the imaging capability usually reserved for scanning electron microscopy. To minimize the harsh preparation procedures necessary for electron microscopy, confocal imaging was attempted on bacteria stained with two distinct fluorescent probes: the DNA-binding dye propidium iodide and a fluorescein tagged antibody against a bacterial antigen, BJ38 (Fig. 25). Reconstruction of the optical sections demonstrates a unique polar organization for this protein localized to a loose tuft of material projecting away from the bacterial outer membrane. The reconstructed image of the bacteria demonstrates the power of this technology for examining the surface organization of bacterial components without the need for scanning electron microscopy. The significant advantages of this technology are that: a) multiple fluorescent probes can be viewed simultaneously; b) a variety of fluorescence-based analytical approaches may be utilized; c) no dehydration or vacuums are required for sample preparation resulting in less damage to the antigen or membrane organization; and d) microbial organisms such as bacteria or yeast may be used to obtain such images.

4. CONCLUSION

The instrumentation and probes for fluorescence analysis and imaging are now available to perform multiple biochemical analyses in living cells. The interplay between cellular structures and differentiated cell function can now be dynamically visualized and monitored to provide topologically specific information about the biology of the cell.

5. REFERENCES

1. M. Schindler, Y. Assaf, N. Sharon, and D. M. Chipman, *Biochemistry* **16**, 423–431 (1977).
2. M. Schindler and M. J. Osborn, *Biochemistry* **18**, 4425–4430 (1979).
3. A. S. Waggoner and L. Stryer, *Proc. Natl. Acad. Sci. USA* **67** 579–589 (1970).
4. D. Gabel, I. Z. Steinberg, and E. Katchalski, *Biochemistry* **10**, 4661–4669 (1971).
5. J. R. Lakowicz, *Principles of Fluorescence Spectroscopy*, Plenum, New York, (1983).
6. M. R. Eftink, *Methods of Biochemical Analysis* **35**, 127–205 (1991).
7. A. Griswald and I. E. Steinberg, *Biochem. Biophys. Acta* **427**, 663–678 (1976).
8. D. M. Jameson and T. L. Hazlett, in *Biophysical and Biochemical Aspects of Fluorescence Spectroscopy* (ed. T. Gregory Dewey), Plenum, New York, 1991, pp. 105–134.
9. J. Yguerabide, *Methods Enzymol.* **26**, 498–578 (1972).
10. G. Weber, *Biochem. J.* **51**, 155–167 (1952).
11. V. I. Teichberg and M. Shinitzky, *J. Mol. Biol.* **74**, 519–531 (1973).
12. M. Shinitzky, A-C. Dianoux. C. Gitler, and G. Weber, *Biochemistry* **10**, 2106–2113.
13. M. Schindler, L-W. Jiang, M. Swaisgood, and M. H. Wade, *Methods in Cell Biol.* **32** part B, 423–446 (1989).
14. G. J. Brakenhoff and K. Visscher, *MICRO* **90**, 247–252 (1990).

15. C. J. Koester in *Handbook of Biological Confocal Microscopy* (ed. J. B. Pawley), Plenum Press, New York, 1990, pp. 207–214.
16. I. K. Lichtscheidl and W. G. Url, *Eur. J. of Cell Biol.* **43**, 93–97 (1987).
17. A. L. Plant, D. M. Benson, and L. C. Smith, *J. Cell Biol.* **100**, 1295–1308 (1985).
18. D. Koppel, M. Sheetz, and M. Schindler, *Proc. Nat. Acad. Sci. USA* **78**, 3576–3580 (1981).
19. T. N. Metcalf III, J. L. Wang, K. R. Schubert, and M. Schindler, *Biochemistry* **22**, 3969–3975 (1983).
20. T. N. Metcalf III, J. L. Wang, and M. Schindler, *Proc. Natl. Acad. Sci. USA* **83**, 95–99 (1986).
21. L-W. Jiang and M. Schindler, *J. Cell Biol.* **106**, 13–19 (1988).
22. Y-Li Wang, F. Lanni, P. L. McNeil, B. R. Ware, and D. L. Taylor, *Proc. Natl. Acad. Sci. USA* **79**, 4660–4664 (1982).
23. L. S. Barak and W. W. Webb, *J. Cell Biol.* **95**, 846–852 (1982).
24. R. Naugland, *Molecular Probes—Handbook of Fluorescent Probes and Research Chemicals* (ed. K. D. Larison), 5th ed., 1992–1994.
25. L. Stryer, *Biochem. Biophys. Acta.* **35**, 242–244 (1959).
26. T. G. Dewey, *Biophysical and Biochemical Aspects of Fluorescence Spectroscopy* (ed. T. G. Dewey), Plenum, New York, 1991, pp. 197–230.
27. B. K. Fung and L. Stryer, *Biochemistry* **17**, 5241–5248 (1978).
28. M. R. Eftink in *Biophysical and Biochemical Aspects of Fluorescence Spectroscopy* (ed. T. G. Dewey), Plenum, New York 1991, pp. 1–41.
29. T. Förster, *Z. Naturforsch A* **4**, 321–323 (1949).
30. M. Shinitzky and M. Inbar, *Biochem. Biophys. Acta.* **433**, 133–149 (1976).
31. J. M. Collins, R. N. Dominey, and W. M. Grogan, *J. Lipid Res.* **31**, 261–270 (1990).
32. J. M. Collins and W. M. Grogan, *Cytometry* **10**, 44–49 (1989).
33. L. Seveus, M. Vaisala, S. Syrjanen, M. Sandberg, A. Kussisto, R. Harju, J. Salo, I. Hemmila, H. Kojola, and E. Soini, *Cytometry* **13**:329–338 (1992).
34. C. G. Morgan, A. C. Mitchell, and J. G. Murray, *J. of Microscopy* **165**, 49–60 (1992).
35. S. Gilroy, N. D. Read, and A. J. Trewavas, *Nature* **346**, 769–771 (1990).
36. J. A. McCray and D. R. Trentham, *Annu. Rev. Biophys. Chem.* **18**, 239–270 (1989).
37. A. Minta, J. P. Y. Kao, and R. Y. Tsien, *J. Biol. Chem.* **264**, 8171–8178 (1989).
38. J. A. Theroit and T. J. Mitchison, *J. Cell Biol.* **119**, 367–377 (1992).
39. J. W. Walker, A. V. Somlyo, Y. E. Goldman, A. P. Somlyo, and D. R. Trentham, *Nature* **327**, 249–252 (1987).
40. H. S. Kruth, *Anal. Biochem.* **125**, 225–242 (1982).
41. K. Muirhead, P. K. Horan, and G. Poste, *Biotechnology* **3**, 337–356 (1985).
42. M. R. Melamed, P. F. Mullaney, and M. L. Mendelson, *Flow Cytometry and Sorting*, John Wiley, New York, 1986.
43. A. J. Meister, *Biol. Chem.* **263**, 17205–17208 (1988).
44. G. C. Rice, E. A. Bump, D. C. Shrieve, W. Lee, and M. Kovacs, *Cancer Res.* **467**, 6105–6110 (1986).
45. D. C. Shrieve, E. A. Bump, and G. C. Rice, *J. Biol. Chem.* **263**, 14107–14114 (1986).
46. J. A. Cook, S. N. Iype, and J. B. Mitchell, *Cancer Res.* **51**, 1606–1612 (1991).
47. R. C. Burghardt, R. Barhoumi, F. H. Lewis, R. H. Bailey, K. A. Pyle, B. A. Clement, and T. D. Phillips, *Toxicol. and Appl. Pharm.* **112**, 235–244 (1992).
48. A. W. de Feijter, J. S. Ray, C. M. Weghorst, J. E. Klaunig, J. I. Goodman, C. C. Chang, R. J. Ruch, and J. E. Trosko, *Mol. Carcinog.* **3**, 54–67 (1990).
49. T. Seyama, A. K. Godwin, M. DiPietro, T. S. Winokur, R. M. Lebovitz, and M. W. Lieberman, *Mol. Carcinog.* **1**, 89–95 (1988).

50. Y. Li, T. Seyama, A. K. Godwin, T. S. Winokur, R. M. Lebovitz, and M. W. Lieberman, *Proc. Natl. Acad. Sci. USA* **75**, 348 (1988).
51. A. W. de Feijter, J. E. Trosko, D.B. Krizman, R. M. Lebovitz, and M. W. Lieberman, *Mol. Carcinog.* **5**, 205-212 (1992).
52. W. B. Busa and R. Nuccitelli, *Amer. J. Physiol.* **246**, R409-R438 (1984).
53. A. Kurkdjian and J. Guern, *Annu. Rev. Plant Physiol. Plant Mol. Biol.* **40**, 271-303 (1989).
54. J. C. Chambard and J. Pouyssegur, *Exp. Cell Res.* **164**, 282-294 (1986).
55. D. J. Yamashiro and F. R. Maxfield, *Trends in Pharm. Sci.* **9**, 190-193 (1988).
56. I. Mellman, R. Fuchs, and A. Helenius, *Annu. Rev. Biochem.* **55**, 663-700 (1986).
57. R. J. Lee, J. M. Oliver, G. G. Deanin, C. D. Troup, and R. F. Stump, *Cytometry* **13**, 127-136 (1992).
58. P. E. J. van Erp, M. J. J. M. Jansen, G. J. deJongh, J. B. M. Boezeman, and J. Schalwijk, *Cytometry* **12**, 126-132 (1991).
59. G. R. Bright, J. E. Whitaker, R. P. Haugland, and D. L. Taylor, *J. Cell Phys.* **141**, 410-419 (1989).
60. E. Musgrove, C. Rugg, and D. Hedley, *Cytometry* **7**, 347-355 (1986).
61. S. Bassnet, L. Reinisch, and D. C. Beebe, *Am. J. Physiol.* **258**, C171-C178 (1990).
62. Z. Wang, G. L. Chu, W. C. Hyun, H. A. Pershadsingh, M. J. Fulwyler, and W. C. Dewey, *Cytometry* **11**, 617-623 (1990).
63. J. A. Thomas, R. N. Buchsbaum, A. Zimiak, and E. Racker, *Biochem.* **18**, 2218 (1979).
64. G. R. Bright, G. W. Fisher, J. Rogowska, and D. L. Taylor, *J. Cell Biol.* **104**, 1019-1033 (1987).
65. A. S. Waggoner, *Annu. Rev. Biophys. Bioeng.* **8**, 47-68 (1979).
66. H. J. Apell and B. Bersch, *BBA* **903**, 480-494 (1987).
67. T. Brauner, D. F. Hulser, and R. J. Strasser, *BBA* **771**, 208-216 (1984).
68. I. Holopainen, M. O. K. Enkvist, and K. E. O. Akerman, *Neurosci Lett.* **98**, 57-62 (1989).
69. M. Schindler, P. K. Gharyal, and L-W. Jiang, in *Biophysical and Biochemical Aspects of Fluorescence Spectroscopy* (ed. T. G. Dewey), Plenum, New York, 1991, pp. 261-281.
70. K. Luby-Phelps, D. L. Taylor, and F. Lanni, *J. Cell Biol.* **102**, 2015-2022 (1986).
71. O. Baron-Epel, D. Hernandez, L-W. Jiang, S. Meiners, and M. Schindler, *J. Cell Biol.* **106**, 715-721 (1988).
72. M. H. Wade, J. E. Trosko, and M. Schindler, *Science* **232**, 525-528 (1986).
73. M. Schindler, J. E. Trosko, and M. H. Wade, *Methods in Enzymol.* **141**, 439-459 (1987).
74. A. K. Campbell, *Intracellular Calcium*, John Wiley, New York (1983).
75. J. Szollosi, B. G. Feurstein, W. C. Hyun, M. K. Das, and L. J. Marton, *Cytometry* **12**, 707-716 (1991).
76. P. H. Cobbold and T. J. Rink, *Biochem. J.* **248**, 313 (1987).
77. A. P. Somlyo and B. Himpens, *FASEB* **3**, 2266-2276 (1989).
78. A. Minta, J. P. Y. Kao, and R. Y. Tsien, *J. Biol. Chem.* **264**, 8179-8184 (1989).
79. J. P. Y. Kao, A. T. Harootunian, and R. Y. Tsien, *J. Biol. Chem.* **264**, 8179-8184 (1989).
80. G. Grynkiewicz, M. Poenie, and R. Y. Tsien, *J. Biol. Chem.* **260**, 3440-3450 (1985).
81. N. Kurebayashi, *Biophys. J.* **61**, A160 (1992).
82. Y. Schechter, L. Hernaez, J. Schlessinger, and P. Cuatrecasas, *Nature* **278**, 835-838 (1979).
83. G. Ven Katakrishman, C. A. McKinnon, A. H. Ross, and D. E. Wolf, *Cell Regul.* **1**, 605-614 (1990).
84. K. Jacobson and J. Wojcieszyn, *Proc. Natl. Acad. Sci. USA* **81**, 6747-6751 (1984).
85. P. Wadsworth and E. D. Salmon, *J. Cell Biol.* **102**, 1032-1038 (1986).

86. E. Balint, A. Aszalos, and P. M. Grimley, *BBRC* **157**, 808–815 (1988).
87. P. Johnson, T. B. Garland, T. Campbell, and J. R. Kusel, *FEBS LETT.* **141**, 132–135 (1982).
88. M. Schindler, M. J. Osborn, and D. E. Koppel, *Nature* **283**, 346–350 (1980).
89. M. Schindler, M. J. Osborn, and D. E. Koppel, *Nature* **285**, 261–263 (1980).
90. J. J. Anders, *GLIA* **1**, 371–379 (1988).
91. D. W. Bombick, *In Vitro Tox.* **3**, 27–30 (1990).
92. L. S. Stein, G. Stoica, R. Tilley, and R. C. Burghardt, *Cancer Research* **51**, 696–706 (1991).
93. B. V. Madhukar, S. Y. Oh, C. C. Chang, M. H. Wade, and J. E. Trosko, *Carcinogenisis* **10**, 13–20 (1989).
94. A. W. de Feijter, J. E. Trosko, and M. H. Wade, in *Fluorescent and Luminescent Probes for Biological Activity: A Practical Guide to Technology for Quantitative Real-Time Analysis* (ed. W. T. Mason), Academic Press, London, 1993, pp. 378–388.
95. M. J. Dicke-Evinger, C. Kuszinski, G. A. Perry, J. Sun, and Z. Wang, *Blood* **78**,335a (1991).
96. J. Bresser and M. J. Evinger-Hodges, *Gene Anal. Technology* **4**, 89–104 (1987).
97. R. A. Cardullo, S. Agrawal, C. Flores, P. C. Zamecnik, and D. E. Wolf, *Proc. Natl. Acad. Sci. USA* **85**, 8790–8794 (1988).
98. D. M. Shotton, *J. Cell Sci.* **94**, 175–206 (1989).
99. B. Himpens, H. DeSmedt, G. Droogmans, and R. Casteels, *Am. J. Phys.* **263**, C95–C105 (1992).
100. D. A. Przywara, S. V. Bhave, A. S. Bhave, T. D. Wakade, and A. R. Wakade, *The FASEB Journal* **5**, 217–222 (1991).
101. G. J. Brakenhoff and K. Visscher, *J. Microscopy* **165**, 139–146 (1992).
102. H. T. M. VanderVoort, G. J. Brakenhoff, and M. W. Baarslag, *J. Microscopy* **153**, 123–132 (1989).
103. R. C. Hallgren and C. Buchholz, *J. Microscopy* **166**, RP3–RP4 (1992).

BIOANALYTICAL INSTRUMENTATION   VOLUME 37

Automated Enzyme Assays

JOHN A. LOTT, PH.D., *Professor of Pathology, The Ohio State University, Starling Loving M-368, 320 W. 10th Avenue, Columbus, OH 43210-1240*

DANIEL A. NEALON, PH.D., *Director, Reference, Laboratory Group, Eastman Kodak Company, Clinical Diagnostics Division, 1600 Lexington Avenue, Rochester, NY 14652-4149*

1. Introduction
2. What is Automation?
3. Solution-Based, Absorption Spectrophotometric Methods
    3.1. Specimen Identification
    3.2. Specimen Preparation
    3.3. Specimen Integrity
    3.4. Pipeting Specimens
    3.5. Temperature Control
    3.6. Absorbance Accuracy
    3.7. Wavelength Accuracy
    3.8. Time Control
    3.9. Cuvet Control
    3.10. Control of Interferences
        3.10.1. Keto Acids
        3.10.2. Ammonium Ions
        3.10.3. Glycerol
        3.10.4. Oxidized Enzymes
        3.10.5. Amino-Alcohol Buffers
    3.11. Photometric Accuracy
    3.12. Extracting Enzyme Activities from Spectrophotometric Data
        3.12.1. Lag Phase
        3.12.2. Search for the Linear Reaction-Rate Curves
4. Solution-Based Enzyme Assay Methods Other Than Absorbance
    4.1. Fluorescence
    4.2. Bioluminescence and Chemiluminescence
    4.3. Turbidimetry and Nephelometry
    4.4. Fluorescence Polarization
5. Nonliquid Reagent Systems
    5.1. Advantages of Nonliquid Systems
    5.2. Evaluations of Nonliquid Systems
    5.3. The Ektachem Systems
        5.3.1. Fluid Dynamics on the Ektachem

Bioanalytical Instrumentation, Volume 37, Edited by Clarence H. Suelter.
ISBN 0-471-58260-3   © 1993 John Wiley & Sons, Inc.

        5.3.2. Spectrophotometry on the Ektachem
        5.3.3. Processing of Enzyme Data on the Ektachem
6. Automated Enzyme Assays in Other Fields
    6.1. Animal Testing
    6.2. Environmental Testing
7. Quality Control and Standardization of Enzyme Tests
    7.1. Quality Control
    7.2. Standardization: Now and Proposed
    7.3. Reference Materials for Enzymes
    7.4. Standardization of Enzymes as Proteins
8. The Ideal Automated Enzyme Analyzer
    8.1. General Requirements
    8.2. Mechanical and Computer Needs
    8.3. Analytical Specifications
9. Appendix 1
10. Appendix 2
11. References

1. INTRODUCTION

The automated assay of enzyme activities is a major activity of clinical, food, environmental, research, and other analytical laboratories. The assay of enzymes in body fluids is the authors' area of expertise. Earlier reviews of laboratory automation are by Alpert (1) and Schwartz (2); the latter described the early days of continuous flow, i.e., Technicon AutoAnalyzer™ systems with brief mentions of alternate automated devices. Perez-Bendito and Silva (3) describe continuous flow as applied to rate methods, and Roodyn (4) in his book, describes the topic of automated enzyme analyzers up to about 1970 and gives detailed information on enzyme assays on the Technicon AutoAnalyzer including a FORTRAN program for data reduction; he also provides a detailed bibliography to about 1969. He discusses other early instruments for enzyme assays such as the Bausch and Lomb "Zymat 340," the Joyce Loebl "Enzymat," the LKB "Reaction Rate Analyzer," the AGA Medical "Autochemist," the Vickers "Multichannel 300," the Warner-Chilcott "Robot Chemist," the Beckman "Kintrac VII," the Gilford "Multiple Sample Absorbance Recorder," and the Smith Kline "Eskalab." The degree of automation of the AutoAnalyzer was superior to that of the other devices; this was surely the reason for the dominant position of Technicon equipment in clinical laboratories in the late 1960s and early 1970s. The addresses of the manufacturers of the above devices are given by Roodyn (4). This information is largely of historical interest and is included here for completeness. The above automated enzyme analyzers have nearly all disappeared from clinical chemistry laboratories.

The development and application of centrifugal analyzers, a major advance in automated enzyme assays, is described by Tiffany et al. (5). This device was the first to use computerized process control and data reduction. There are many reports in the literature that evaluate or compare specific instruments. Some examples of such studies are those by Valcárcel and Luque de Castro (6) describing the Technicon RA-100, BMD Hitachi 705, and Beckman Astra-8 analyzers (7), the Abbott Spectrum (8, 9, 10), the DuPont aca (11, 12), the DuPont Dimension (13, 14), the Hitachi 704 (15, 16), the Hitachi 717 (17, 18), the Hitachi 736 and 737 (19, 20), the Kodak Ektachem 700 (21, 22, 23), and the Corning ACS 180 (24). Other reports make comparison of various instruments, e.g., the Hitachi 737 versus the Ektachem 700 (25), and the Baxter Paramax versus the Ektachem 700 (26). These studies deal with enzymes and other tests, comparability of the

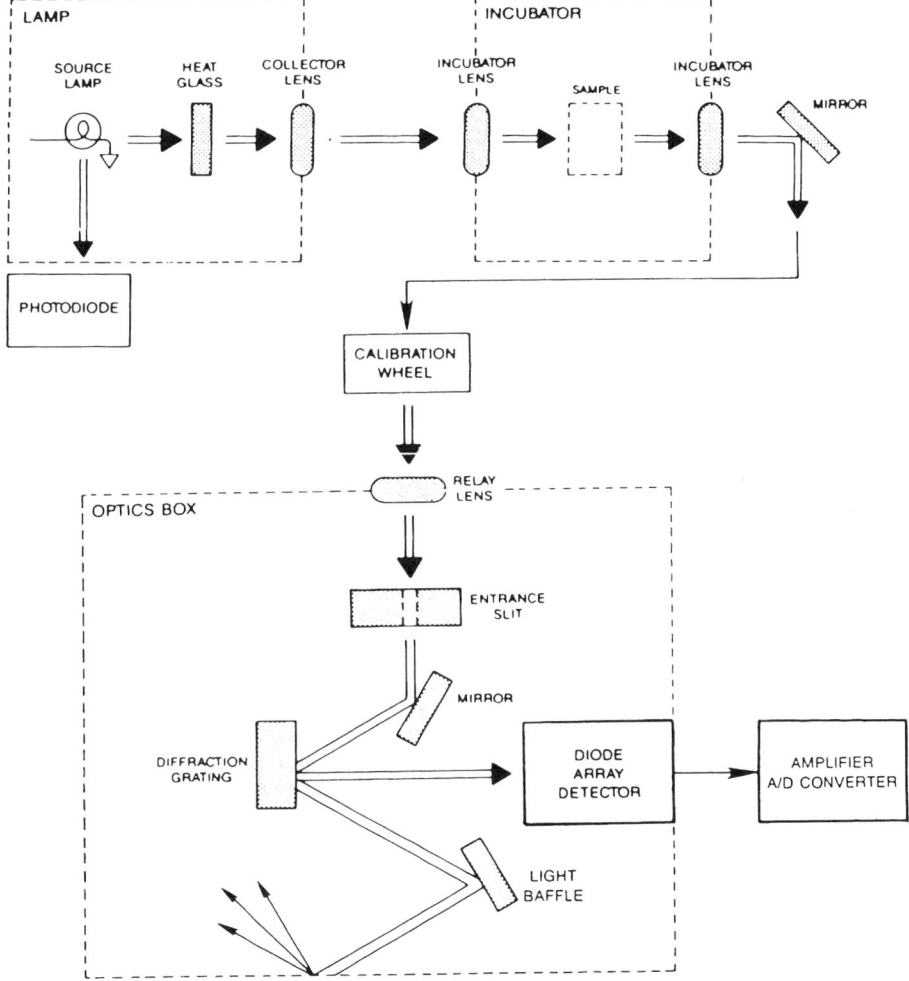

Fig. 1. Part of the optical system of the Abbott Spectrum. The instrument uses a grating to isolate various wavelengths, and simultaneous measurements at multiple wavelengths are made with the diode array detector.

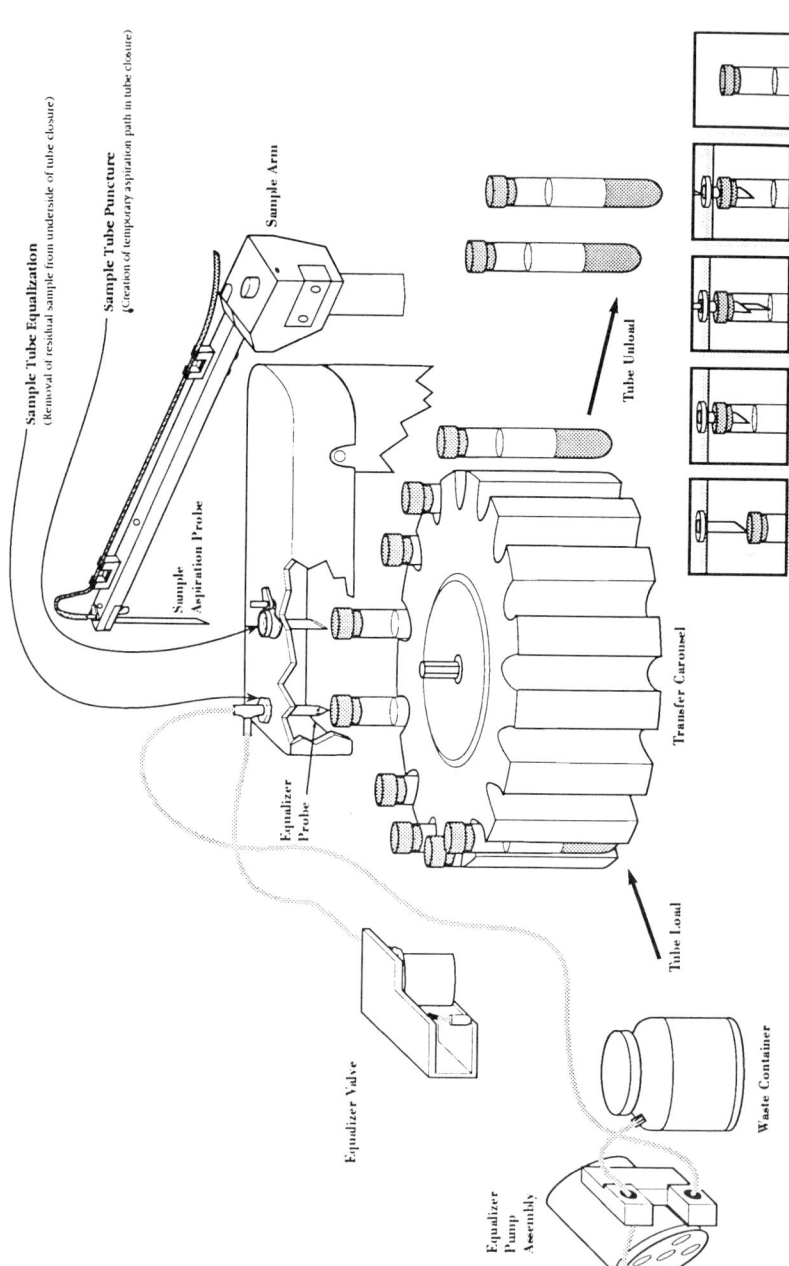

Fig. 2. Detail of the Baxter Paramax analyzer showing the handling of specimens. The instrument pierces the rubber septum to gain access to serum. The cuvets are provided as long spools for the continuous feeding into the instruments.

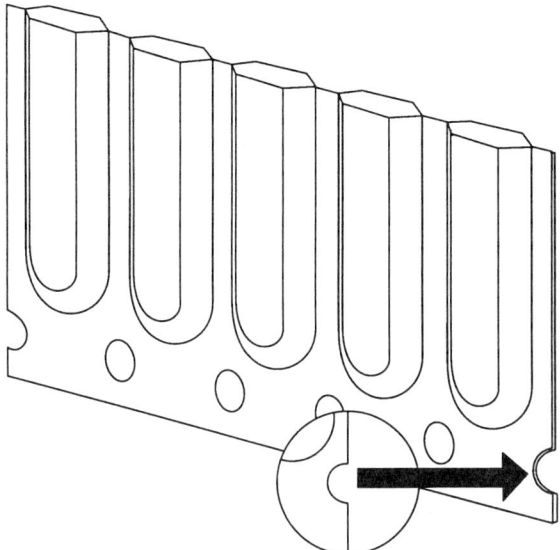

Fig. 3. View of the disposable cuvets used in the Baxter Paramax. The cuvets are provided as a long chain and are cut off by the instrument.

results with those from accepted methods for many analytes, quality control data, and other performance characteristics. Unfortunately, these reports become obsolete quickly owing to model changes, improvements of all types, reagent reformulations, and the like. Figures 1 to 7 show some representative details of contemporary automated enzyme analyzers with some special or unique features.

The distinction between "batch" and "discrete" analyzers is an important one. The AutoAnalyzer is a batch instrument; groups of specimens are analyzed together along with standards and quality control specimens or knowns. It is inefficient to analyze one specimen on the AutoAnalyzer owing to the required calibration with every run. Contemporary analyzers such as the DuPont aca and Kodak Ektachem are extremely stable, and only infrequent calibration is needed; these devices meet the needs of clinical laboratory analyses: any test at any time. Assaying specimens as they arrive in the laboratory provides better and more timely testing services rather than accumulating specimens and analyzing them in groups.

Our review covers the major developments in automated enzyme assays since about 1975. The well-documented advantages of automation are several: reductions in imprecision, increased productivity of the staff, standardization of methods, faster assay of specimens and reporting of results, and fewer human blunders.

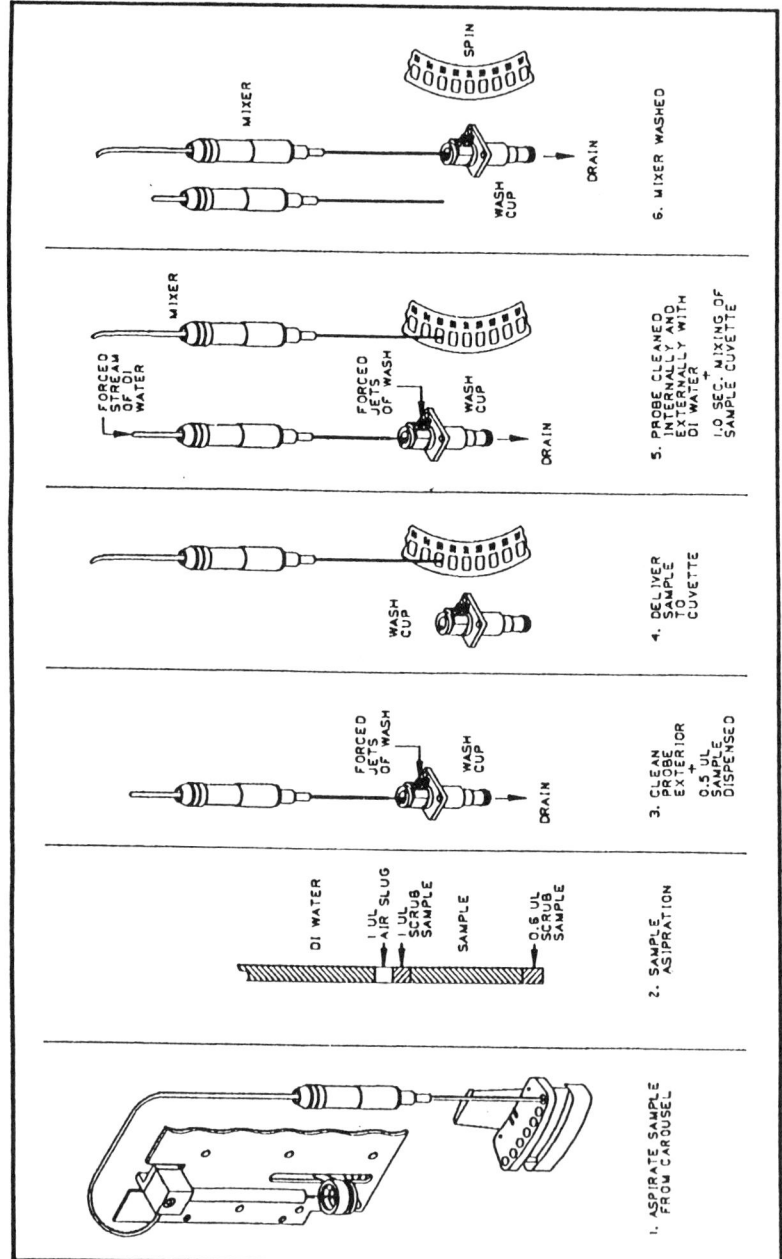

Fig. 4. Detail of the Beckman CX 4/5 sample probe sequence showing the aspiration of specimen, delivery to the cuvet, and washing of the probe. The washing system is designed to minimize carry-over of specimen.

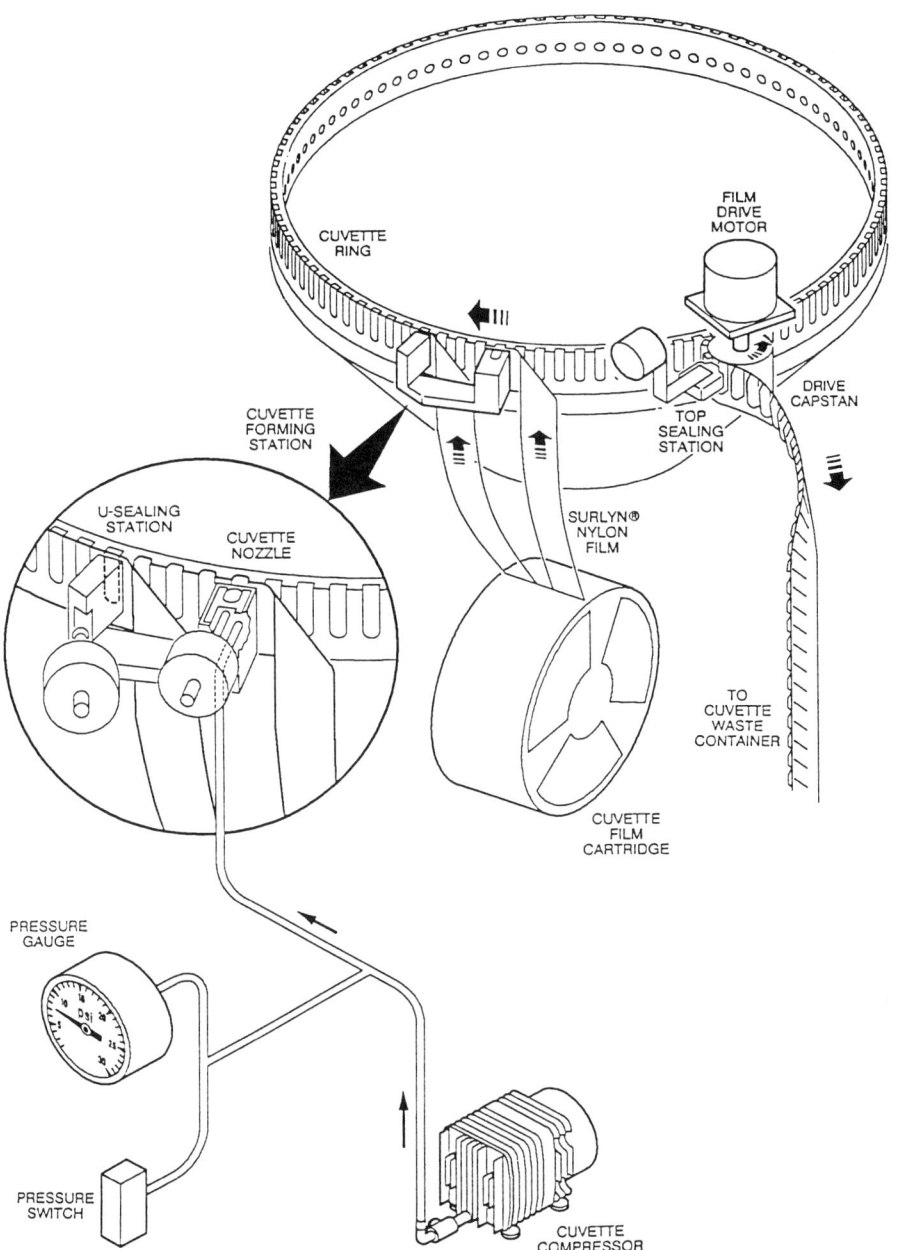

Fig. 5. Detail of DuPont Dimension Analyzer. Cuvets are formed on the instrument from two bands of nylon film, and absorbance measurements are made on these cuvets.

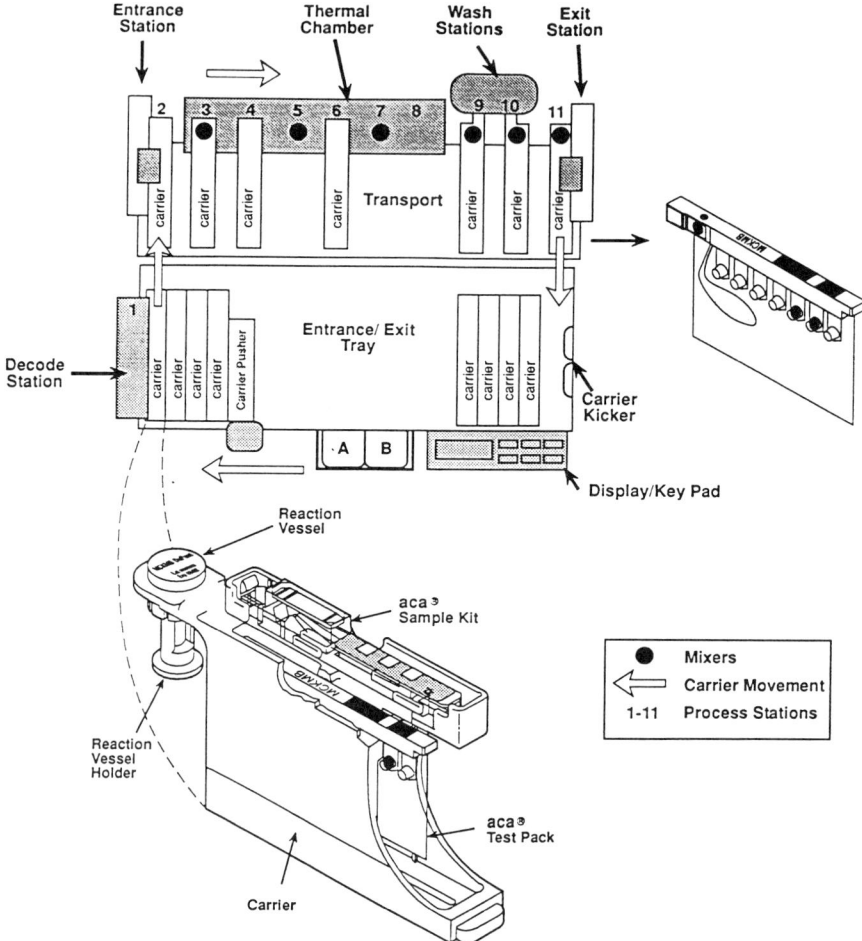

Fig. 6. Detail of the preprocessing device, the "Plus," for the DuPont aca analyzer. Separation and pretreatment steps are carried out here, and then the treated sample is added to an aca cup for analysis by the DuPont aca.

2. WHAT IS AUTOMATION?

As is discussed by others (6), "automation" is often poorly defined, and there are clearly many levels of automation (27). Modern, computer-controlled instruments require less-and-less human intervention. But is it desirable to have the instrument (or robot) replicate every human manipulation? The answer is a qualified yes, and we discuss here the rationale and precautions of various steps in the analytical chain and the safeguards that must be built into the instrument to maintain the integrity of the results.

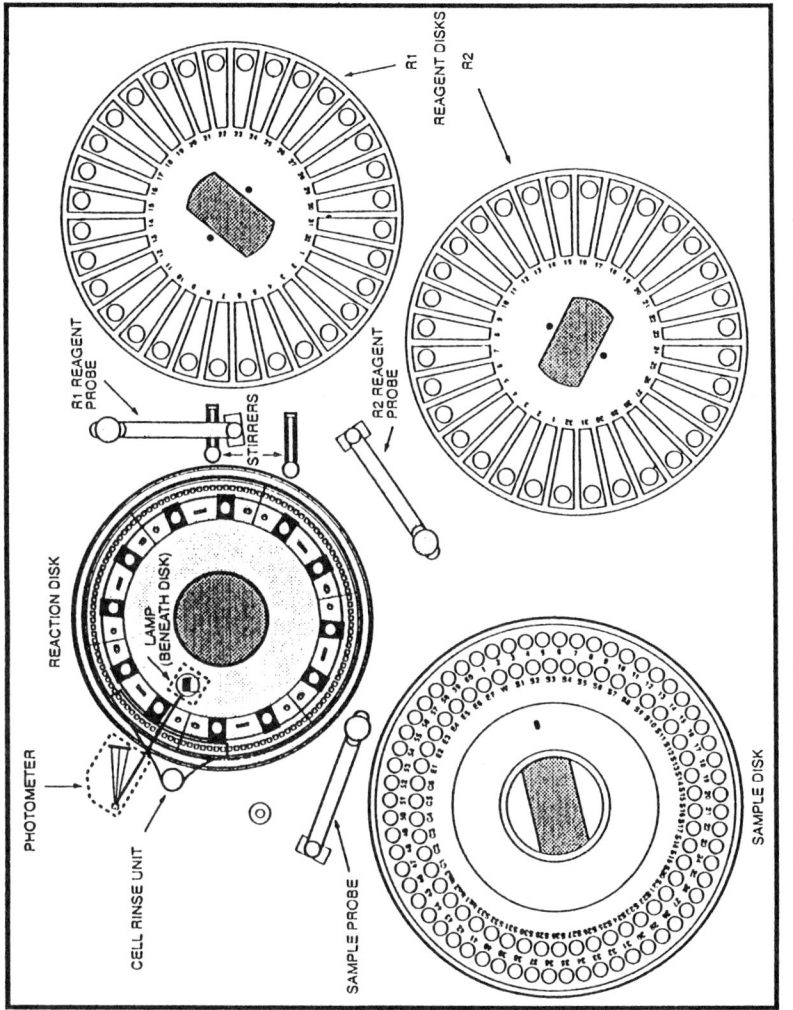

Fig. 7. Detail of the Hitachi 717 analyzer. Reagents are stored in "R1" and "R2." Specimens and reagents are automatically pipetted into the "Reaction Disk" for incubation and photometry.

A comparison of some currently available automated enzyme analyzers is given in Table 1. Two caveats apply: The selection of instruments is based on those that have large peer groups in the College of American Pathologists (28) proficiency testing surveys, and continuing updates of equipment makes it highly likely that some of the instruments will have features not shown in the table.

3. SOLUTION-BASED, ABSORPTION SPECTROPHOTOMETRIC METHODS

Ultraviolet and visible absorption spectrophotometry are still the most widely used techniques for enzyme assays in aqueous systems. Nephelometry or turbidimetry are used for a few special situations, e.g., assay of serum or urine amylase with a starch solution as the substrate where the clearing of the solution is measured, or similarly, the assay of serum lipase where the clearing of an olive oil emulsion is followed. Table 2 shows the current list of clinically important enzymes and their substrates. Automated analyzers rely largely on "self-indicating" substrates such as NAD(P)H, p-nitrophenylphosphate, phenolphthalein monophosphate, thymolphthalein monophosphate, derivatives of 4-nitrophenyl maltopentaoside, and so on. These instruments generally do not permit a separation step to complete the analysis; the reaction vessel is also the cuvet or measuring container. The reaction chemistries may be quite complex; however, the specimen handling and spectrophotometry must be adapted to the mechanical requirements of the instrument. Microwell readers can serve nicely to measure enzyme activities simultaneously in 96 wells (29). Flow-injection analysis of enzyme is a variant of continuous flow spectrophotometry and has been applied to the assay of enzymes (30).

3.1. Specimen Identification

Identification is a major problem in clinical laboratories, and serious untoward events can occur with misidentified specimens. Unambiguous identification is possible today with bar-coding and similar machine-labeling techniques (31). The model discussed here is for testing patient-derived materials in a clinical laboratory; the model can, of course, be extended to other applications. A machine-readable label on every specimen is the contemporary standard of modern equipment. Keying in identifiers to an instrument is less desirable owing to the inevitable human errors. In our experience with bar-code readers, they "read" the label correctly or don't work at all. The topic, automated specimen identifications, is described in more detail in Section 8.2 here (32).

3.2. Specimen Preparation

Preparation for analysis ranges from a mundane step such as centrifugation to sophisticated procedures like pretreatment by chromatography, extraction, reaction with antibodies, and so on. The DuPont aca (see Table 1) uses some reagent packs that incorporate pretreatment columns in the packs, e.g., those for the assay of the CK-MB isoenzyme of creatine kinase (CK, EC 2.7.3.2). A pretreatment module for the aca, the aca **plus**, provides additional capabilities to carry out separation and isolation steps so that the treated specimens can then be analyzed in the usual way on the aca.

The Kodak Ektachem 250 uses reactions on thin films and a "radial wash" technique as a pre-analysis step for certain analytes. The Baxter Stratus instrument uses a similar strategy. The Corning ACS 180 employs a double-antibody technique to measure CK-MB and a fully automated separation step using antibody coated on magnetic particles; nothing is required of the operator except loading the reagents and bar-coded specimens into sampling wheels.

Removing the stopper from a tube of blood can be eliminated—and the occasional aerosol that is produced during opening—by piercing the stopper's rubber septum with the probe as is done on the Baxter Paramax or as has been suggested by Columbus and Palmer (33) in a special blood collection device that is centrifuged axially and then sampled through a port in the plastic tube. Others (34) have devised a plastic cone that is inserted into the rubber stopper that serves as an opening to the specimen and as a guide for the sampling probe. Safety concerns in the handling of potentially infectious specimens is stimulating more development in this area of automation.

3.3. Specimen Integrity

Unattended assays have unique requirements that manual techniques do not. On-instrument stability is an issue, particularly if the environment of the instrument is hotter than the laboratory bench, the typical case. For example, CK is unstable at 37°C, is light sensitive, and is easily oxidized during storage; such concerns impact the validity of CK assays (35). Specimen evaporation can be a serious problem, particularly for small specimens held in containers with large, exposed surfaces for prolonged periods of time (36, 37). Evaporation lids are desirable, but they cannot always be used owing to instrument constraints on the specimen probe.

3.4. Pipeting Specimens

Specimen sampling has special needs in automated instruments. The analyzer must detect an incomplete or "short" specimen, an air bubble, and any solid material such as a fibrin clot in a serum specimen or the silicon separator gel used in many blood collection tubes; failures of any of these invalidates the results, and the operator must be alerted. Use of anticoagulants is common in hospitalized patients, and therefore delayed clotting is a major problem in clinical laboratories. The instrument must sense when excessive force is needed to aspirate the specimen. Because it is impossible for the analyst to monitor the pipeting of specimens, these checks for possible failures must be in place, particularly with the typical 2µl to 20µL specimens in common use. The metering of tiny specimens is an art that is beyond the discussion here.

TABLE 1
Comparison of Automated Enzyme Analyzers

	Abbott Spectrum	Baxter Paramax	Beckman CX7	BMD Hitachi 717	DuPont Dimension AR	Kodak Ektachem 700 XR
Minimum sample vol., µL[1]	1.25–10	5–20	3	1	14–40	11
Minimum total vol., µL[2]	50	50	50	100	44–70	50
Closed-container sampling[3]	N	Y	N	N	N	N
Cuvet type[4]	Disposeable	Disposeable	Glass	Semi-disposeable	Disposeable	Dry slides
Auto reassay[5]	Y	Y	Y	Y	Y	N
Open channels[6]	Y	Y	Y	Y	Y	N
Wash water usage[7]	3 L/H	1 L/4500 tests	7 L/h	20 L/h	2 L/h	None
Wastestream[8]	Cups, cuvets, wash	Wash, cuvets	Cups, wash	Cups, wash, cuvets	Cuvets, cups, wash	Slides, pipette tips
Liquid-sensing probe[9]	N	Y	Y	Y	Y	Y
Air-sensing probe[10]	N	N	N	N	N	Y
Clot-sensing probe[11]	N	N	N	N	N	Y
Optics[12]	Grating	Filters	Grating	Filters	Filters	Filters
Time to perform one test (min)[13]	10	10	8	12.5	3–7	7
Enzyme assays/hr[14]	200	700	225	600	300	250
Calibration stability[15]	Days	Weeks	Days	Days	Weeks	Weeks

Note. All instruments had the capability to make measurements in the visible or ultraviolet region of the spectrum, all had temperature thermostatting, all had the capability to interface to a host laboratory computer with bidirectional communications and to retransmit data, all had direct-tube sampling and bar-code identification of the specimen, all had reagent identification to the instrument, and usually with bar codes.

[1] The minimum amount of specimen that is used for an enzyme test.
[2] The minimum volume for one test **PLUS** the "dead" volume in the specimen container that represents unusable specimen.
[3] The instrument opens the specimen container and samples the contents.
[4] Cuvet types can be plastic and replaced after a single assay or plastic and can be reused. Glass (or quartz) cuvets are multiuse and are washed between specimens.
[5] The instrument senses when a specimen is out of range and automatically, without operator intervention, reassays the material on a dilution or on a smaller sample or by some other means.
[6] An "open channel" is one the user can configure for their own particular needs or tests.
[7] Water volume used for washing the cuvets and/or probe or for diluting specimens or reagents.
[8] All instruments have specimen tubes in the wastestream; the above are in addition to this.
[9] A probe determines if there is liquid in a tube.
[10] The probe can tell if an air bubble has been aspirated.
[11] The probe can tell if excessive force is needed to aspirate a specimen.
[12] The device used to isolate wavelengths in the spectrophotometer.
[13] The total time from inserting a specimen into the instrument to the production of the result.
[14] The maximum number of assays per hour if only enzymes were assayed.
[15] The approximate time between required recalibration.

TABLE 2

Specificity of Enzyme-Catalyzed Reactions

Enzyme (Abbreviation, IUB Group, EC No.)	Substrate	Comments
Alpha-amylase (hydrolase, EC 3.2.1.1)	α-1,4 links in D-glucose polymers, e.g., polyglucose (G6, G7, etc.), many starches, glycogen, dextrins	Does not cleave β-1-4 links as occur in cellulose, α-1-6 links at branches of starches, maltose
Acetylcholinesterase (hydrolase, EC 3.1.1.7)	Acetylcholine acetyl thiocholine	Moderately specific
Pseudocholinesterase (ChE, hydrolase, EC 3.1.1.8)	Many aliphatic esters of choline	Choline is $HOCH_2CH_2N(CH_3)_3^+$, a quaternary amine. The OH-group can be esterified with many groups; esters are substrates.
Acid phosphatase (ACP, hydrolase, EC 3.1.3.2)	Hydrolyses many phosphate esters, e.g., pNPP, G-6-P, phenyl-P, β-glycerophosphate, phenolphthalein-P, thymolphthalein-P, naphthol-P	Cleaves phosphate esters like ALP but at pH of about 5.
Alkaline phosphatase (ALP, hydrolase, EC 3.1.3.1)	Hydrolyses same phosphate esters as ACP	ALP has unusual pH optimum of about 9. Optimum pH varies with substrate.
Angiotensin converting enzyme (ACE, hydrolase, EC 3.4.15.1)	Splits peptides at certain sites, e.g., between GLY and PHE, also hippuryl-L-histidyl-L-leucine is substrate	A nonspecific hydrolase; also acts on angiotensin I bradykinin, met-enkephalin, leu-enkephalin
Aspartate aminotransferase (AST, transferase, EC 2.6.1.1)	L-aspartate + 2-oxoglutarate or L-glutamate + oxaloacetate	Absolutely specific; only reacts with L-aspartate or L-glutamate. Enzyme is stereo-specific.
Alanine aminotransferase (ALT, transferase, EC 2.6.1.2)	L-alanine + 2-oxoglutarate or L-glutamate + pyruvate	Absolutely specific. Only reacts with L-alanine or L-glutamate. Enzyme is stereo-specific.
Aldolase (ALD, lyase, EC 4.1.2.13)	Fructose-1,6-di-P	Highly specific
Ceruloplasmin (oxidoreductase, EC 1.16.3.1)	Many oxidizable substrates e.g., phenylenediamine	Nonspecific. Capable of oxidizing many substrates
Creatine kinase (CK, transferase, EC 2.7.3.2)	Creatine + ATP or creatine − P + ADP	Highly specific
Gamma-glutamyl transferase (GGT, transferase, EC 2.3.2.2)	Transfers a gamma glutamyl group from many "donor" peptides to an "acceptor" peptide	Nonspecific but requires gamma-glutamyl group.

TABLE 2
(Continued)

Enzyme (Abbreviation, IUB Group, EC No.)	Substrate	Comments
Glutamate dehydrogenase (GDH, oxido-reductase, EC 1.4.1.3)	2-oxoglutarate + NH_4^+ + NADH or glutamate + NAD	Moderately specific
Lipase [tiracylglycerol lipase] (hydrolase, EC 3.1.1.3)	Emulsified long-chain tri-, di-, and monoglycerides. Also slowly splits soluble esters and esters in micelles.	Nonspecific but requires carboxylic acid-alcohol ester group.
Lactate dehydrogenase (LD, oxido-reductase, EC 1.1.1.27)	Pyruvate and other ketoacids + NADH. Also lactate and other α-hydroxy acids + NAD	Moderately specific
Hexokinase (HK, transferase, EC 2.7.1.1.)	ATP + D-glucose, ATP + D-fructose, ATP + D-mannose, or ATP + 2-deoxy-D-glucose	Moderately specific
Pyruvate kinase (PK, transferase, EC 2.7.1.40)	Phosphoenol pyruvate + ADP or pyruvate + ATP	Abolutely specific

3.5. Temperature Control

Temperature control to maintain the reaction solution within +/−0.1°C of the desired temperature is required for enzyme assays. The rule-of-thumb is that the rate of enzyme-catalyzed reactions doubles for every 10°C or about 7% increase in the rate for every degree increase. Copeland et al. (38) described the effects of temperature on the activity of purified alkaline phosphatase (ALP, EC 3.1.3.1) from human liver, intestine, placenta, and porcine kidney. All the enzymes exhibited linear Arrhenius (log activity vs. temperature) relationships, and the activity at 30°C was about 1.2-times that at 25°C. The activity at 37°C was about 1.7-times that at 25°C. The choice of temperature depends of course on the assay. For clinical work, 30°C is a compromise owing to the instability of some enzymes at 37°C, but the latter has the advantage of giving faster rates (39).

The National Institutes of Standards and Technology [NIST, formerly the National Bureau of Standards (40)] sells a high-purity gallium standard in a Teflon and nylon container [Standard Reference Material (SRM) no. 1968] for use as a temperature standard. The melting point is 29.77°C and can be used to calibrate thermometers and other temperature-sensing devices.

In the past, 25°C was the commonly used temperature for enzyme assays; this temperature has the disadvantage of requiring cooling of the reaction chamber or cuvets and of course slower enzyme reaction rates as compared to 30°C or 37°C. A number of professional societies (e.g., International Federation of Clinical Chemistry (IFCC), National Committee for Clinical Laboratory Standards

(NCCLS)] have recommended that 37°C be used as the standard temperature for enzyme assays, and this setting is now widely used in automated analyzers.

3.6. Absorbance Accuracy

In cuvet-based analyzers, absorbance accuracy is checked readily with solutions of either $K_2Cr_2O_7$ or $(NH_4)_2Co(SO_4)_2$, both in dilute H_2SO_4. Solutions of KNO_3 are useful in the ultraviolet region (see Table 3). For assays using NAD(P)H, $K_2Cr_2O_7$ is particularly convenient; its peak absorbance wavelength, 350nm, is close to that of NAD(P)H, 340nm. $K_2Cr_2O_7$ in 5 mmol/L H_2SO_4 is stable for years in well-closed, hard-glass containers. Such a solution containing 0.9350 g/L $K_2Cr_2O_7$ has an absorbance of 10.0 at 350nm, and suitable dilutions can be prepared with 5 mmol/L H_2SO_4 to check absorbance accuracy at 350nm, to see if the instrument is giving a linear absorbance response with concentration, and to check for stray light with a solution having a nominal absorbance of about 3.0 A. NAD(P)H can be used to check for linearity and stray light, but it is unsuitable for checking absorbance accuracy owing to its variable water content. Other useful compounds to check linearity are: oxyhemoglobin, p-nitrophenol, cyanomethemoglobin, and biuret-protein complex prepared with BSA or HSA (41). Glass filters that fit 1-cm cuvet holders are available from the NIST as SRM 9301 for checking absorbance accuracy.

Grating and prism spectrophotometers should recover the theoretical absorptivities shown in Table 3; interference-filter photometers typically produce lower values, and glass-filter photometers produce lower values still. As a general rule, the wider the band-pass of the filter, the lower the absorptivity. In any case, even filter photometers should show constant absorptivities with time. Because many automated enzyme assays do not use a standard but rely on the absorptivity of the chromophore to determine the activity, low values of the absorptivity owing to a wide band-pass filter will yield lower enzyme activities.

Instructions on the use of $K_2Cr_2O_7$ solutions for checking absorbance accuracy are given in Appendix 1.

3.7. Wavelength Accuracy

Wavelength accuracy, i.e., does the instrument's setting agree with the actual wavelength of the light, affects the accuracy of spectrophotometric assays. The emission lines of mercury or deuterium permit very precise wavelength settings; less satisfactory but still useable are holmium oxide or didymium filters, e.g., NIST SRMs 2009, 2010, 2013, and 2014. The problem with most automated enzyme analyzers is that the above techniques cannot be used owing to the configuration of the optics and cuvets, inability to insert a filter in the measuring chambers, and so on. Korzun and Miller (42) proposed a methyl red solution measured at two pH values to judge if the wavelength setting is correct. Their method does not establish wavelength accuracy but can be used to monitor shifts in wavelength.

TABLE 3

Reagents for Checking Absorbance Accuracy (179)

Material (Formula weight)	Wavelength, nm	Concn., g/L	a, L/g-cm	Note
$K_2Cr_2O_7$ (294.12)	257	0.0500	14.38	In 5 mmol/L H_2SO_4
	350	0.0500	10.72	In 5 mmol/L H_2SO_4
	257	0.1000	14.45	In 5 mmol/L H_2SO_4
	350	0.1000	10.74	In 5 mmol/L H_2SO_4
$(NH_4)_2Co(SO_4)_2$ (287.14)	400	10.538	0.3406	In 1% H_2SO_4 (10 mL conc. H_2SO_4/L)
	450	10.538	2.1062	In 1% H_2SO_4 (10 mL conc. H_2SO_4/L)
	500	10.538	4.4550	In 1% H_2SO_4 (10 mL conc. H_2SO_4/L)
	510	10.538	4.7466	In 1% H_2SO_4 (10 mL conc. H_2SO_4/L)
	550	10.538	2.1117	In 1% H_2SO_4 (10 mL conc. H_2SO_4/L)
	600	10.538	0.3733	In 1% H_2SO_4 (10 mL conc. H_2SO_4/L)
	400	21.105	0.3537	In 1% H_2SO_4 (10 mL conc. H_2SO_4/L)
	450	21.105	2.0680	In 1% H_2SO_4 (10 mL conc. H_2SO_4/L)
	500	21.105	4.4354	In 1% H_2SO_4 (10 mL conc. H_2SO_4/L)
	510	21.105	4.7075	In 1% H_2SO_4 (10 mL conc. H_2SO_4/L)
	550	21.105	1.9864	In 1% H_2SO_4 (10 mL conc. H_2SO_4/L)
	600	21.105	0.2993	In 1% H_2SO_4 (10 mL conc. H_2SO_4/L)
KNO_3 (101.10)	302	11.70	0.0703	In H_2O

3.8. Time Control

Control of the reaction monitoring time is important for accurate enzyme assays. Electronic time control is used almost universally, but a discussion of this is beyond the scope of our chapter.

3.9. Cuvet Control

Disposable plastic or permanent glass cuvets are being used in automated enzyme analyzers. Plastic cuvets can be manufactured to close tolerances for the solution path length and be sufficiently transparent in the near-ultraviolet spectrum. A disadvantage is cost and disposal. Permanent glass cuvets have better dimensional stability, and quartz cuvets have excellent transparency in the ultraviolet region to about 200nm. Their disadvantage is the necessary flushing, washing, and drying after each specimen and the build-up of protein on the walls when serum specimens are analyzed. Periodic replacement is necessary even for "permanent" cuvets, a significant cost item. On balance, plastic cuvets are probably superior for enzyme assays. Some automated analyzers monitor the cleanliness of cuvets prior to each use, an advantage. Standard 1-cm cuvets are available from the NIST as SRM 932.

3.10. Control of Interferences

With permanent cuvet systems, carryover of reagents and (or) specimens can be a problem. For example, if the same cuvet is used for a lipase assay and a triglyceride substrate and is then used to measure triglycerides in serum, a potential problem of triglyceride carryover exists. Turbidity of any cause, e.g., lipemia, can cause serious errors owing to the high background absorbance and the small change in absorbance therein for a specimen with low enzyme activity. Many automated analyzers will flag starting absorbances that are too high. A detailed examination of instrument- and method-associated interferences owing to icterus, hemolysis, and turbidity is available in the monograph by Glick and Ryder (43).

Interferences that occur in enzyme assays are numerous. The list of enzyme inhibitors prepared by Zollner (44) is invaluable. The work by Young (45) details many effects on enzyme results (or absence of same) caused by drugs and in-vivo changes or chemical, in-vitro interferences.

Another issue with sampling is possible cross contamination of adjacent specimens or specimen carry-over. At least two options exist: The instrument must wash or clean the probe between specimens, a sometimes fallible procedure because large wash volumes must be used if a low enzyme activity specimens follows one with tremendous activity. Alternately, a new pipette tip or probe is used with each specimen as is done on the Kodak Ektachem 700 avoiding the problem of contamination but adding the cost of pipette tips.

If the same probe selects different reagents for successive assays, then probe washing is an issue as is possible contamination of the reagent by the probe. In cuvet chemistries, mixing of the reagent with the specimen and/or the reaction-triggering agent must produce a uniform solution.

Some examples of problems found in the assay of serum enzyme are probably quite general: Many enzymes are not substrate specific but give interfering reactions with analytes present in serum. A few instances of nonspecificity are given as follows.

3.10.1. KETO ACIDS

Keta acids, such as acetoacetic or β-hydroxybutyric, present in the blood of patients with diabetes mellitus or after a prolonged fast, interfere in alanine aminotransferase (ALT, EC 2.6.1.2) assays and will give falsely increased values. The rate-limiting reaction is:

$$\text{Alanine} + \text{2-oxoglutarate} \xleftarrow{\text{ALT}} \text{L-glutamate} + \text{pyruvate}.$$

The indicator reaction is

$$\text{Pyruvate} + \text{NADH} \xleftarrow{\text{LD}} \text{Lactate} + \text{NAD}.$$

A reaction that interferes is

$$\text{Keto acids} + \text{NADH} \xleftarrow{\text{LD}} \text{NAD} + \text{products}.$$

In the design of the ALT assay, the LD (EC 1.1.1.27) is usually allowed to "burn out" the keto acids before the rate measurements are made. LD reagent must be free of ALT activity, and a water-blank specimen will reveal reagent contamination. In general, the reagent enzymes must always be checked for contamination with the enzyme being measured or any other enzymes that might interfere.

3.10.2. AMMONIUM IONS

Ammonium ions interfere in the assay of aspartate aminotransferase (AST, EC 2.6.1.1). The rate-limiting reaction is

$$\text{Aspartate} + \text{2-oxoglutarate} \xleftarrow{\text{AST}} \text{oxaloacetate} + \text{L-glutamate}.$$

The indicator reaction uses MDH (EC 1.1.1.37) and is

$$\text{Oxaloacetate} + \text{NADH} \xleftarrow{\text{MDH}} \text{malate} + \text{NAD}.$$

However, endogenous glutamate dehydrogenase (GDH, EC 1.4.1.3) will allow the following reaction to occur yielding falsely increased AST values:

$$\text{2-oxoglutarate} + \text{NH}_4^+ + \text{NADH} \xleftarrow{\text{GDH}} \text{glutamate} + \text{NAD}.$$

Obviously, reagents containing NH_4^+ ions must be avoided.

3.10.3. GLYCEROL

Glycerol is present at low concentration in serum; occasionally, patients have hyperglyceridemia. Endogenous glycerol must be consumed before triacylglycerol lipase (EC 3.1.1.3) is measured to avoid falsely increased lipase values.

3.10.4. OXIDIZED ENZYMES

CK has thiol groups that are oxidized easily; this is most noticeable in specimens that have been stored. Oxidized CK can be restored with thiol-reducing agents, but this generally takes more time than what is allowed in the lag phase, and the enzyme activity will be underestimated. This problem is even more acute in lyophilized products that are being used as calibrators or control materials. Grossly erroneous calibration can occur if the CK is not fully reduced, i.e., fully activated, prior to use.

3.10.5. AMINO-ALCOHOL BUFFERS

Amino-alcohol buffers are used commonly for alkaline phosphatase (ALP, EC 3.1.3.1) assays. Some sources of these buffers are contaminated with a potent ALP inhibitor, an obvious interferant (46).

3.11. Photometric Accuracy

The accuracy needed in the measurement of absorbances is related inversely to the measuring interval. The DuPont aca uses a 17-sec measuring time for kinetic enzyme assays. Thus for a specimen with an AST activity of 20 U/L and measured using NADH in the indicator reaction, then the absorbance change in 17 sec is only 0.035 A units. A trivial error of 0.004 A introduces an 11% error. Clearly, highly accurate spectrophotometry is needed here and in other automated enzyme analyzers that use very short measuring intervals.

3.12. Extracting Enzyme Activities from Spectrophotometric Data

Data reduction of spectrophotometric readings must consider the extreme situations that are encountered with patients' specimens. The kinetics of enzyme-catalyzed reactions are described in the excellent volume by Bergmeyer and Gawehn (47). The discussion here assumes solution-based reactions occurring in cuvets; kinetic, dry-film reactions are described elsewhere in this review. The factors discussed here include the lag phase, finding the linear portion of the rate curve, substrate depletion, highly turbid specimens, and highly active specimens. An example for calculating enzyme activities from spectrophotometric data is in Appendix 2.

3.13.1. LAG PHASE

The *lag phase* is the time for the enzymatic reaction to reach maximum velocity; the latter is the rate that is zero-order in substrate, and substrate-saturation kinetics are assumed. At the beginning of the reaction, changes in concentration of the substrate are assumed to be trivial, and the substrate is present in large excess. The lag phase depends on the specimens' activity and the complexity of the reaction scheme being used. For example, the determination of LD with the pyruvate-to-lactate reaction has little or no lag phase, even with serum specimens having low activity:

$$\text{Pyruvate} + \text{NADH} \longleftrightarrow \text{LD} \longrightarrow \text{lactate} + \text{NAD}.$$

We can infer that the mechanism of the above reaction is simple and that the intermediates form and decay at a rate that is greater than the overall rate.

A coupled reaction is used for the assay of AST:

Aspartate + 2-oxoglutarate ⟵ AST ⟶ oxaloacetate + L-glutamate

Oxaloacetate + NADH ⟵ MDH ⟶ malate + NAD.

To estimate the AST activity, the first reaction must be rate limiting, and for this to be so, the oxaloacetate must reach a fairly high steady-state concentration; the time for this to occur is the lag phase. Generally, a large excess of MDH is used so that the rate of the second reaction (above) is always much greater than that of the first. If the AST activity is extremely high, the MDH may be insufficient to obtain accurate results. More MDH can be used, but the reaction may then be too fast for the measuring device to provide meaningful enzyme data.

Bergmeyer (48) gives some estimates of the required enzyme activity in the indicator reactions. The AST reaction scheme can be restated as:

$$A \xrightarrow{k1} B \xrightarrow{k2} C,$$

where k1 and k2 are the rates of the first and second reactions. Table 4 illustrates the effect of the relative values of k1 and k2 on the overall rate. Obviously, k2 must be more than 100-times k1 for the assay to be useful. Increasing the indicator enzyme activity to give a k2 of 1000 is not necessary unless many highly active specimens must be analyzed with good accuracy. For many enzyme assays, the latter yields too few measurement points to give a satisfactory rate curve necessitating dilution and reassay.

A three-step reaction scheme is used for the assay of CK (49). The reactions are:

Creatine-P + ADP ⟵ CK ⟶ creatine + ATP

ATP + glucose ⟵ hexokinase ⟶ ADP + glucose-6-phosphate

G-6-P + NAD(P) ⟵ G-6-PDH ⟶ NAD(P)H + 6-P-gluconolactone.

TABLE 4

Effect of the Rates of the First and Second Reaction on the Overall Rate in Coupled Systems of Two Reactions

k1	k2	Overall Rate
1	0.3	<0.2
1	1	0.46
1	10	0.74
1	100	0.96
1	1000	0.993
1	Infinite	1.00

The CK assay has a lag phase of several minutes for specimens with a normal CK activity even in the presence of a large excess of the activities of hexokinase (EC 2.7.1.1) and glucose-6-P-dehydrogenase (G-6-PDH, EC 1.1.1.49). A long lag phase for the CK assay is predictable based on the above discussion for a two-step assay. Specimens with increased CK activities generally have shorter lag phases.

3.13.2. SEARCH FOR THE LINEAR REACTION-RATE CURVES

Algorithms exist for curve searching; many use only two points at a fixed time after the start of the reaction, an unsatisfactory approach for accurate work. At a minimum, a curve-searching algorithm must be implemented to trap certain errors, to identify when the lag phase is over, to be certain that there is an adequate linear portion of the curve, and that the region of substrate exhaustion has not been reached. With highly active specimens, substrate exhaustion may occur during the lag phase, and the absolute absorbance can be used to trap such results. Highly turbid specimens with low or high enzyme activities and with high starting absorbances present special problems. Usually centrifugation to remove lipids or dilutions solves the problem. The change in absorbance (ΔA) of adjacent points can be compared until the ΔAs are constant within defined limits. At least six ΔAs over a period of 1–6 min should be obtained; the linearity of the rate curve can be tested with a standard linear regression algorithm and tested for outliers with the standard deviation of the residuals, i.e., $S(x,y)$, that is the SD of the points' distance from the least-squares, straight-line region in the Y direction. The value of $[S(x,y)/\text{mean } \Delta A] \times 100$ should be less than 10%; a greater value indicates excessive scatter or noise about the line and (or) loss of linearity owing to substrate exhaustion and hence unsatisfactory results.

4. SOLUTION-BASED ENZYME ASSAY METHODS OTHER THAN ABSORBANCE

4.1. Fluorescence

The classic text by Udenfriend (50) initiated the surge in interest in the use of fluorescence by analytical and clinical chemists. This approach was of interest because of its potential for significant increases in analytical sensitivity. In fact, the fluorescence signal by the chromophore is 100 to 1000 times greater than the sensitivity of absorbance measurements. Because of this level of analytical sensitivity, fluorescence has been applied almost exclusively to the measurement of compounds at low concentrations.

Despite its advantage of increased analytical sensitivity, fluorescence measurements have drawbacks: concentration effects (the so-called inner filter effect), background effects owing to Rayleigh and Raman scattering, solvent effects (for example, interfering nonspecific fluorescence and quenching), and, most commonly, sample matrix effects such as light scattering, interfering fluorescence absorption, and photodecomposition. Nevertheless, successful applications of fluorescence measurements to automated enzyme assays have been described (51).

Instruments that measure fluorescence are termed either *fluorometers* or *spectrofluorometers*. The distinction between the two types is based on the approaches that are used to separate the excitation and emission light into monochromatic light. Fluorometers use interference or glass filters, and spectrofluorometers use gratings or prisms.

4.2. Bioluminescence and Chemiluminescence

These approaches differ from fluorescence and other luminescent techniques such as phosphofluorescence in that the excitation event is caused by a chemical reaction rather than photolumination. There are a number of reviews that detail the development and applications of luminescent technology (52, 53, 54, 55, 56, 57, 58) and instrumentation (59, 60). Particularly outstanding is the recent detailed review by Stanley (61) where more than 90 luminometers (manual, automatic, microtiter plate, HPLC, LC, GLC, imaging, and others) from more than 60 sources are described.

Bioluminescence is a unique type of chemiluminescence found in biological systems; these reactions can be classified as either pyridine- or adenine-nucleotide linked systems or enzyme-substrate systems. In clinical enzymology, the most commonly used system is the firefly luciferin–luciferase system for the measurement of ATP:

$$LH2 + E + ATP \xrightarrow{Mg^{++}} E.LH2\text{-}AMP + Products$$

$$E.LH2\text{-}AMP + O_2 \longrightarrow E + oxyluciferin + CO_2 + AMP + light,$$

where LH2 is firefly luciferin and E is firefly luciferase. As is evident from the above reaction, all enzymes and metabolites that participate in "ATP reactions" or that are linked to such reactions could be assayed by the luciferase bioluminescent assay, and indeed it has been applied to the direct measurement of adenylate kinase, pyruvate kinase, and hexokinase. Of greatest interest has been the application of the above reaction to the measurement of CK and the B-subunit activity of the CK-B isoenzymes (62, 63). The use of chemiluminescence has been limited severely by the general lack of a fully automated luminometer with a rate of specimen throughput comparable to that of existing major automated enzyme analyzers. One of the few luminometers that may provide the advantages of a fully automated system is the Auto-CliniLumat™ LB 952T/16 (64), which is capable of measuring chemiluminescence at up to 250 specimens per hour. The instrument can record chemiluminescence originating from solutions, coated tubes or beads, or magnetic particles. The emitted light is measured by a high-sensitivity, low-noise photomultiplier with wavelength range of 390nm–620nm and operated with an ultrafast photon counter. The algorithm to calculate the standard curve is a smoothed spline function following a logit-log or log-log transformation of the data.

4.3. Turbidimetry and Nephelometry

These techniques measure scattered light that is the result of light interacting with suspended particles in a solution matrix. Turbidimetry is the measurement of the decrease in the intensity of the incident light owing to phenomena such as reflectance, scattering, and absorption as it passes through a suspension of particles. In turbidity, measurements of the change in light intensity are made in a manner analogous to how absorbance measurements are made; that is, at 180° from the incident beam. Because of this similarity, turbidity measurements can be made on most automated enzyme analyzers. One of the main concerns with turbidity measurements is the signal-to-noise ratio, therefore it is critical that analyzers making such measurements have photometric systems with electro-optical noise of 0.0002 absorbance units or less.

Nephelometry measures light scattered toward a detector that is not in the direct path of the transmitted light. Instruments like the Beckman ARRAY™ (65) rate nephelometer, that is, a new generation of Beckman's Immuno Chemistry System (ICS), measures the forward scatter at 70° to the incident beam to take advantage of the increased forward-scatter intensity resulting in greater analytical sensitivity. Light sources commonly used in nephelometry are xenon or quartz halogen lamps and lasers. The latter are particularly useful because of their high intensity and better-defined properties. The Behring Nephelometer Analyzer (66) uses a laser light source to measure light scattering at 13°–24° from the incident light at two time points after mixing of the assay components (67, 68).

Light-scattering measurements, whether turbidimetric or nephelometric, have been applied to immunoassays of specific proteins (69) such as immunoglobulins, transferrin, albumin, haptoglobin, C3/C4, and haptens such as theophylline, gentamicin, and tobramycin. Enzymes that have received any attention in this regard are amylase (70, 71) and lipase (72, 73, 74, 75, 76, 77). A difficulty with light-scattering assays for amylase and lipase is the restriction on the substrate concentration that must be low enough to allow some signal to reach the detector; generally, these concentrations are suboptimal for the enzyme assays (78).

4.4. Fluorescence Polarization

Fluorescence polarization (79) has been applied to the measurement of enzymes; however, these are generally mass assays rather than assays of enzyme activity.

5. NON-LIQUID REAGENT SYSTEMS

The terminology used to describe this technology is diverse enough to include such descriptions as "nonliquid" or "dry" reagent systems or "solid-phase" or "film" technology. In the context of this review, we will refer to the technology as nonliquid reagent systems. However it is termed, these systems represent one of the more innovative analytical technologies.

Nonliquid reagent systems can be regarded as complete analytical packages and typically take the form of either thin pads or films. They are complete in the sense that all the reagents required are distributed in a dry form either throughout the supporting element, as in the case of the Seralyzer (80) or compartmentalized into specific layers as in the case of the Kodak Ektachem system reagent slides.

5.1. Advantages of Nonliquid Systems

Simplicity, convenience, and the ability to use small specimen volumes, a particular advantage in pediatric and geriatric populations, are among the advantages of these systems. Because of their "dry" form, nonliquid reagent systems provide a remarkably efficient approach to stabilizing labile components, such as coupling enzymes and, because of this, these systems have prolonged storage stability, even at room temperature. Such potential for extended storage further increases the convenience of these analytical devices. For example, the spreading layer of the Kodak Ektachem test slide can not only trap cells, crystals, and other particulate matter, but it can also separate low-molecular weight compounds from potential larger-molecular weight interferents, e.g., proteins. For some tests, but not enzymes, the solution reaching the reactive underlayers resembles a protein-free filtrate. Compounds that reflect light such as TiO_2 and $BaSO_4$ are usually incorporated into the spreading layer to hide or mask color or turbidity that may be present in patients' specimens. The Reflotron (81) and the Ames Seralyzer nonliquid reagent systems use a film membrane to exclude blood cells that interfere with test procedures.

5.2. Evaluations of Nonliquid Systems

There are a many reports that evaluate the performance of the nonliquid reagent systems; however, owing to proprietary constraints, there are relatively few articles that describe the actual preparation or detail the characteristics of these analytical devices. This veil of secrecy is particularly true of the Drichem (82) and Konica nonliquid reagent systems; information on these systems is confined almost exclusively to the patent literature. Very few clinical or performance evaluations have been reported. Therefore, these systems will not be dealt with in this overview. The OPUS (83) and the Stratus II (84) Immunoassay Systems are also nonliquid reagent systems that deal with the immunoassays of therapeutic drugs, fertility hormones, thyroid function, and other metabolites. The Stratus II can also be used to measure CK-MB as the protein rather than the enzyme. An excellent review by Chan (85) describes the hardware and performance of the OPUS, Stratus, and other such systems.

The principles and applications of nonliquid reagent systems have been reviewed (86, 87, 88). Particularly useful are the reviews by Zipp and Hornby (89) and by Campbell and Price (90).

The only nonliquid reagent system that appears in the College of American Pathologists surveys of laboratories is the Kodak Ektachem (91). Because of its widespread use, this system will be described in detail as being representative of nonliquid reagent systems. The Seralyzer and the Reflotron System are also nonliquid reagent systems that measure enzymes. Reviews have been published detailing specifics regarding the Seralyzer (92, 93, 94, 95, 96, 97) and the Reflotron (98, 99, 100).

5.3. The Ektachem Systems

The Ektachem systems (101, 102, 103, 104, 105) are high-throughput, random-access analyzers. Tests are performed on a batch of specimens, and the assays are selectable by using different reagent-impregnated slides. Each specimen can be assayed for a different number of tests. It is possible to obtain only one test per slide, and the used slides are discarded. Figure 8 illustrates the integration of the various modules of an Ektachem 700 analyzer. Cartridges, usually containing 50 slides, are kept in either one of two slide supply compartments. Magnesium salt pads in one compartment maintain the relative humidity at about 33% whereas desiccant packs maintain the relative humidity in the other compartment at about 15%. Amylase and lipase, both being measured with two-point rate techniques, use slides that are stored in the higher-humidity slide compartment whereas the slides for all the other enzymes are stored in the low-humidity slide compartment. Specimens are placed in sample cups or tubes that are covered with cross-cut, flexible plastic caps to prevent evaporation. Each specimen holder or quadrant contains disposable pipets (sample probe tips) mounted with each specimen that eliminates both the contamination of one sample by another within the probe and any carryover of one specimen into the specimen in the next sample cup. The sample probe tip is mounted automatically onto a single, positive-displacement piston in the metering tower that aspirates as much specimen volume as is required. For enzyme assays, volume requirements are either 10µL or 11µL. The piston operates in such a manner that an air space is created between the piston and the sample and, therefore, the piston does not touch the specimen. Although a piston operated on the principle of positive displacement may not always deliver exactly the same amount of serum, particularly because the viscosity may vary from individual to individual, it really is not necessary to deliver exact volumes because of the spreading layer concept and the arrangement of the optics in the Ektachem system.

5.3.1. FLUID DYNAMICS ON THE EKTACHEM

To initiate the assay process, the specimen is partially expelled from the piston and the 10µL or 11µL drop is touched to the spreading layer of the slide. Figure 9 illustrates the cross-section of a typical Ektachem slide. The thickness of the spreading layer ranges from 100µm to 300µm with void volumes of 60%–90%. Dimensions such as these provide a suitable structure that allows a rapid and uniform spreading of specimen before it reaches the reagent/indicator under-

Fig. 8. Layout of the Modules of an Ektachem 700 Analyzer showing the "rate incubator" for rate-reaction measurements, the "potentiometric incubator," and "colorimetric incubator." The latter two are not used for enzyme assays.

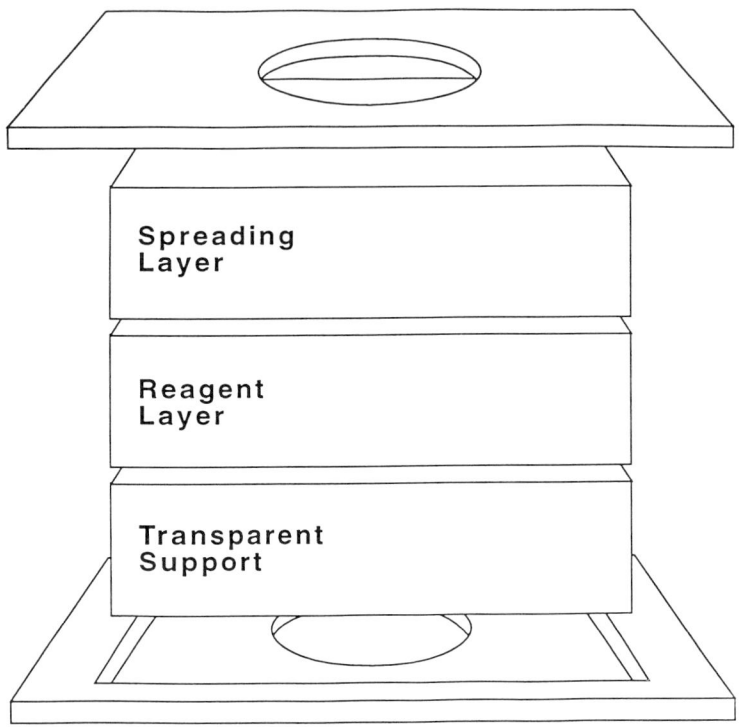

Fig. 9. Cross-section of an Ektachem test slide illustrating the major components.

layers. Once applied to the surface of the spreading layer, the sample drop diffuses into the matrix (reagent and indicator layers) where it dissolves the components of the reaction to ultimately produce a chromogenic product that is measured by reflectance spectrophotometry (106). Some specific examples of Ektachem enzyme assays are shown in Fig. 10, and Table 5 summarizes some of the properties of all the enzyme assays currently available on the Ektachem systems.

5.3.2. SPECTROPHOTOMETRY ON THE EKTACHEM

After the specimen has been applied to the slide, a distributor arm moves the slide to the proper incubator: CM for the colorimetric and two-point rate enzyme tests (acid phosphatase, amylase, and lipase), PM for the potentiometric chemistries, and RT for the rate or kinetic incubator for the multiple-point rate enzyme chemistries. Temperature control within either the CM or RT incubator is maintained at 37 ± 0.1°C by contact of the slide with the rotating thermal mass of the incubator. The products forming in the slides in either the CM or RT incubator are monitored at what are termed *read stations* by separate reflectance densitometers or reflectometers. There are, however, differences on how such measurements are made. For the enzyme slides in the CM incubator, at selected

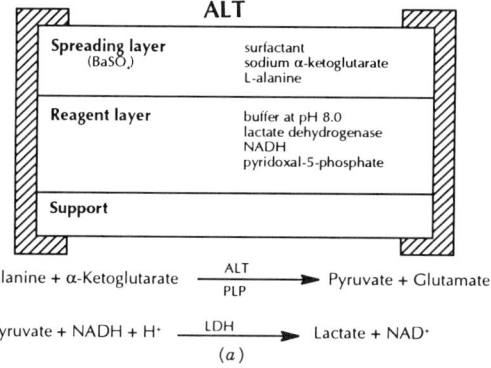

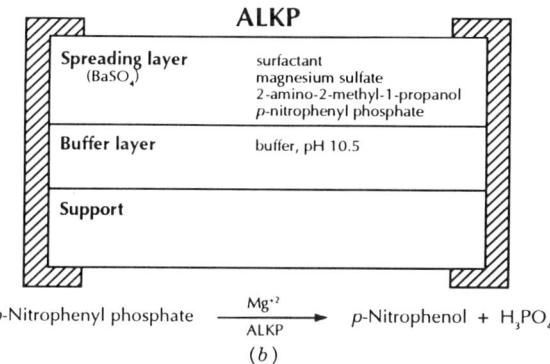

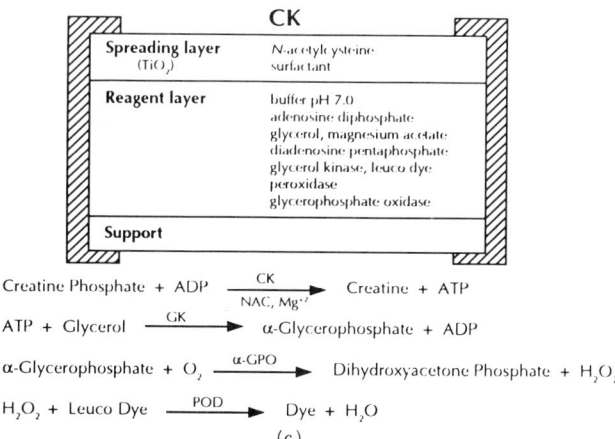

Fig. 10. Examples of enzyme assays on the Ektachem Analyzer showing the reagents in the various layers: (A) alanine aminotransferase, (B) alkaline phosphatase, and (C) creatine kinase.

TABLE 5
Summary of Enzyme Tests on the Ektachem Analyzer

Enzyme	Calibration Mode	Sample Volume (µL)	Wavelength (nm)	Substrate or Indicator
Acid phosphatase	Two-point rate	10	600	Fast red TR and azo dye
Alanine aminotransferase	Multiple-point rate	11	340	Alanine and NADH
Alkaline phosphatase	Multiple-point rate	11	400	p-Nitrophenyl phosphate/p-nitrophenol
Amylase	Two-point rate	10	540	Dyed starch and dyed saccharides
Aspartate aminotransferase	Multiple-point rate	11	340	Aspartate and NADH
Cholinesterase	Multiple-point rate	11	400	Butyrylthiocholine and ferricyanide
Creatine kinase (total and MB)	Multiple-point rate	11	670	Creatine phosphate and leuco dye
Gamma-glutamyl transferase	Multiple-point rate	11	400	L-Gamma-p-nitroanilide and p-nitroaniline
Lactate dehydrogenase	Multiple-point rate	11	340	Pyruvate and NADH
Lipase	Two-point rate	10	540	1-Oleoyl-2,3-diacetylglrlycerol and leuco dye

times, the slides are removed from the CM incubator and moved to the read station (also heated at 37°C) of the corresponding reflectometer to make the necessary reflectance reading. For the multipoint-rate enzyme chemistries (see Table 5), up to 54 readings, approximately one reading every 5-7 sec, are made at the appropriate wavelength-during the 5-min time period when the slide is still in the RT incubator. A high degree of precision in the readings is attainable; the standard deviation of the tolerance around the reflectometer is 0.0004 reflectance transmission units.

Figure 11 is a schematic of the Ektachem reflectometer; it consists of a lamp and a series of optics, and the light uniformly illuminates the slide at a 45° angle (illumination optics). The diffuse reflected light is collected (detector optics) by a fiber-optic bundle (photodiode) located at a 90° angle to the plane of the underside of the slide (Fig. 11). Because reflected light is only collected from a very small, well-defined area of about 3 mm in the center of the slide, the process is essentially volume independent; applying more specimen to the slide merely increases the size of the chromogenic product spot. A filter wheel containing narrow-band interference filters of different wavelengths is located in the path of the reflected light. It is important to note that reflectance measurements are made on the surface opposite to the specimen application side. This arrangement avoids the need to read the product of the reaction through the overlaying spreading layer (Fig. 9) that may contain potentially interfering compounds. Finally, the output of the photodiode is digitized and then passed on to the main computer for further processing.

5.3.3. PROCESSING OF ENZYME DATA ON THE EKTACHEM

The mathematical relationship of incident and reflected light is complicated by such phenomena as diffuse reflectance, ordinary transmittance, and specular reflectance of hemispherically distributed incident light. Based on an analogy to transmittance density in wet chemistry methods, a mathematical relationship that takes into account all such phenomena is as follows:

$$D_r = \log(\text{incident light}/\text{reflected light})$$
$$= \log(R_{\text{white}} - R_{\text{dark}})/(R_{\text{test}} - R_{\text{dark}})$$

where D_r is the reflective density of the slide or film element and is analogous to absorbance in transmission spectroscopy; R_{test} is the intensity of the light reflected from the test slide, R_{white} and R_{dark} refer to the reflectance of the analyzer's internal standards that span the range of potential reflectance measurements. Note that R_{white} is the intensity of reflected light read by the detector with a standard reflector (for example, a barium sulfate surface), and R_{dark} corrects for nonlinear effects caused by "flare light" and may be used to correct for other factors, if any, causing nonlinearity. The reflective density, D_r, can be transformed into transmittance density, D_t, which is linearly related to concentration or activity, using the transformation described by Williams and Clapper (107):

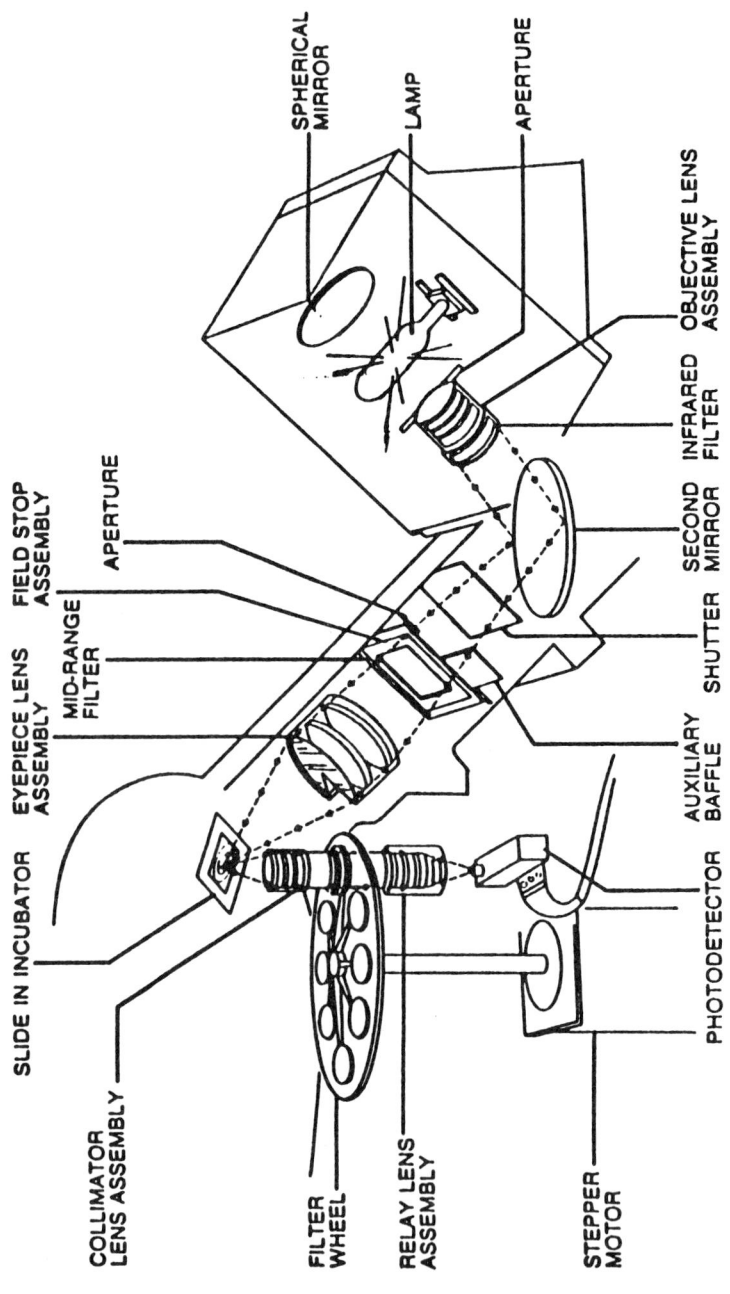

Fig. 11. Schematic of the Ektachem reflectometer. Specimen is added to the top, and reflectance measurements are made from the bottom of the slide.

$$D_t = 0.149 + 0.469D_r + (0.422/1 + 1.179e^{3.379D_r}).$$

Finally, the concentration of the analyte can be determined with:

$$C = B(D_t - D_b)$$

where C is the analyte concentration or enzyme activity, B is the reciprocal of the absorbance coefficient (a constant but analyte-dependent factor), and D_b is the blank reflectance density, that is, when $C = 0$. Curme et al. (102), Morris et al. (108), and Campbell and Price (90) provide more specific details regarding the photometric transformation of reflectance density to transmittance density.

The process whereby the analyzer uses reflectance readings to determine enzyme activity can be generalized in the following steps:

1. Accumulation of either 54 or 2 reflectance readings depending on whether the reaction is a multiple- or two-point rate (see Table 5).
2. Performing a D_r to D_t transformation to linearize the data and also remove any potential optical interferences.
3. Removing any outlier readings (with the multiple-point rate enzymes) and calculating the first derivatives of the remaining raw responses to define the linear portion of the rate curve. Early readings prior to steady-state conditions (lag time) are not used in the calculations, and later readings that are beyond certain defined tolerances of constant velocity are also not used to calculate reaction rates. Two-point enzyme rates (acid phosphatase, amylase, and lipase) do not undergo such iterations, rather, the program simply uses the two readings taken at the defined times.
4. Application of an algorithm to determine the rate of change in reflectance. A least-squares regression algorithm is used with enzymes whose response curve approximates linearity. In the case of γ-glutamyltransferase, which does not yield a sufficiently linear response, the curve is fit with a linear combination of orthogonal polynomial expressions and the maximum rate is established. Figure 12 represents these previously mentioned steps. At this point, the instrument has generated a calculated rate for each specimen that must be converted into the recognizable units of enzyme activity, U/L, and this is accomplished using a calibration curve.

The calibration curve for enzymes is defined by the relationship (calibration model):

$$C = A_0 + A_1 G(\text{rate}) + A_2(\text{rate})^k,$$

where C is the enzyme activity, A_0, A_0, and A_1 are constants that define the curve and represent, respectively, the intercept, slope, and curvature; G is the transform, and the k exponent is equal to either 0 or 2 and is method dependent. For enzymes, k is equal to 2. The calibration curve is generated by the instrument

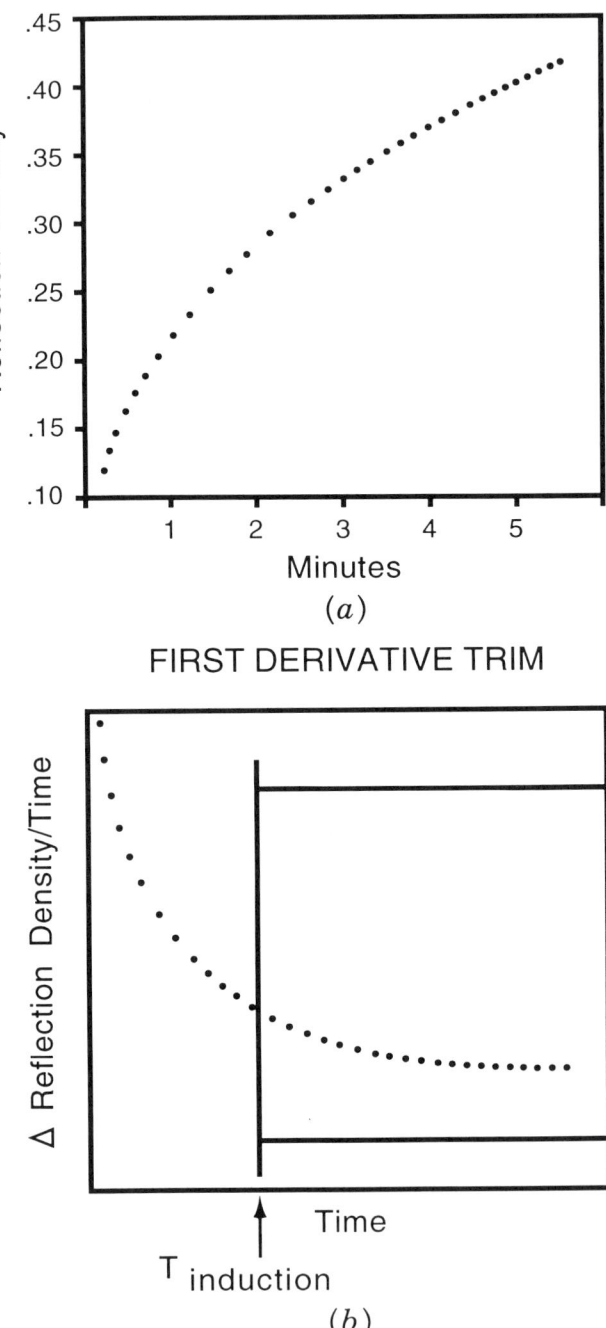

Fig. 12. Steps employed in the processing of enzyme data: (A) accumulate reflectance readings, (B) Prepare first derivative trim of the raw responses, (C) calculate rate for linear-response enzyme curves, or calculate rate for nonlinear responding tests like γ-glutamyl-transferase.

CALCULATE RATE

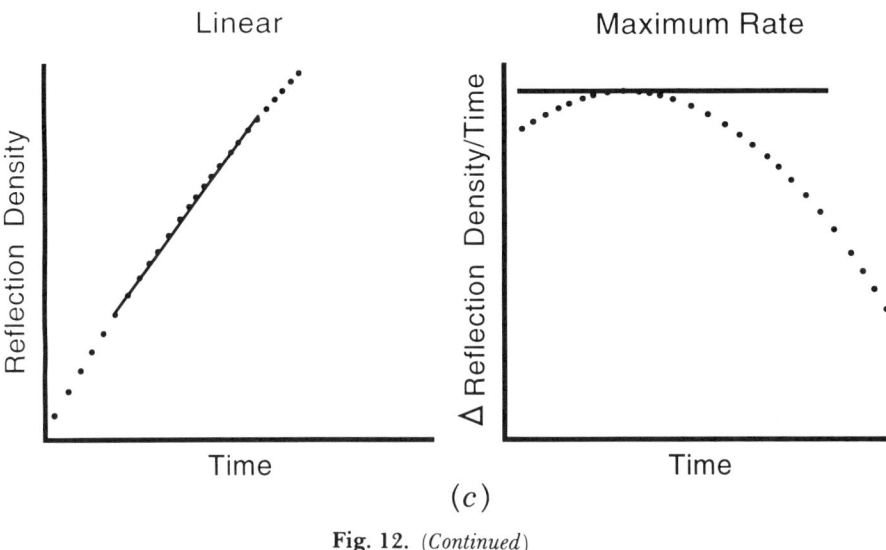

(c)

Fig. 12. (*Continued*)

manufacturer by analyses of 100 to 200 patients' specimens, usually over five days. The testing of each specimen on an Ektachem analyzer (rate analyzer response) and by the reference method is performed concurrently. Once the calibration curve or model has been defined by the analysis of patients' specimens, the enzyme activities are defined for three calibrator fluids with stabilized enzymes that were also assayed during the five-day process; this aspect is termed SAV or *Supplementary Assigned Value* assignment. Calibration data that includes SAVs, splines (transforms that linearize the calibration system), the k value (calibration model curvature term), and particularly for enzymes, information used in rate calculations (rate parameters) are provided to the user on diskettes. Subsequently, any rate arising from the assays of an unknown sample can be converted to activity using the stored calibration curve.

6. AUTOMATED ENZYME ASSAYS IN OTHER FIELDS

Enzyme assays in fields other than clinical chemistry generally are not as automated because there is typically not the same demand for throughput; there are notable exceptions in facilities that perform an extensive number of tests on animals.

6.1. Animal Testing

In veterinary medicine, enzymes used for diagnostic purposes are essentially the same as those used in the diagnosis of human diseases. There are, however, significant differences in the reference intervals among the different species and, therefore, particular attention must be paid to determining, and using, the proper reference interval in order to make a valid diagnostic assessment. In the pharmaceutical industry, because a great deal of the testing is related to toxicity studies, the most commonly measured enzymes are those that are indicators of liver, heart, and kidney damage, and these include not only the transferases and γ-glutamyltransferase but also sorbitol dehydrogenase (EC 1.1.1.14) (109). Measurement of urinary enzymes (110, 111, 112) as indicators of kidney function in toxicological studies is also common in the pharmaceutical industry.

6.2. Environmental Testing

In environmental testing, direct measurements of enzymes is not common but rather by-products of bacterial growth or contamination such as nitrate (113) and sulfite (114, 115) are measured using enzymatic procedures. Lead poisoning in smelter workers and children has become a serious societal concern and direct measurement of enzymes such as alanine dehydrogenase (EC 1.4.1.1), coproporphyrinogen oxidase (EC 1.3.3.3), and ferrochelatase (EC 4.99.1.1) has been suggested as predictors of lead toxicity (116, 117, 118, 119). Cholinesterase (EC 3.1.1.8) is valuable as a marker of hepatotoxicity and is used extensively to monitor workers involved with organophosphates (120, 121, 122). Measurement of enzymes released from lysosomes (exoenzymes) has been suggested as a measure of pollutant toxicity, i.e., organic matter degradation and nutrient regeneration in aquatic environments (123, 124). Measurement of α-amylase in certain cereals such as wheat is used to establish the fitness of the cereal for human or animal consumption (125). Increased activity of α-amylase in the cereal is indicative that the cereal has entered its "sprouting" phase and is no longer suitable for human use.

7. QUALITY CONTROL AND STANDARDIZATION OF ENZYME TESTS

7.1. Quality Control

Internal quality control is basically process control of day-to-day work. Control materials should mimic patients' specimens, and the frequency of control assays is dependent on the stability of the instrument and reagents. With contemporary instrumentation, much less frequent assaying of quality control materials suffices, e.g., once per 8-hr shift is typically quite satisfactory and results in considerable savings. Quality control is a very large topic that has been discussed extensively elsewhere, especially by Westgard and Barry (126) and Lott and Patel (127). A new issue is the use of "medical-needs criteria" in setting control limits as are typically done with Levey–Jennings charts. With today's much more precise instrumentation, e.g., the DuPont aca and Kodak Ektachem, limits based on past performance in routine testing are usually inappropriately narrow leading to unnecessary reassays, recalibration, and other steps that serve neither the patient, the physician, or the laboratory. The quality control of enzyme tests in particular must consider how the results are used. Physicians generally disregard small to moderate changes in enzyme values, particularly if the values are abnormal (128). In fact, if the values are highly abnormal, say 10-times the upper reference or normal limit, then a threefold-upper normal change in the value is generally disregarded. Such medical decision making should be translated to routine process control: Extreme precision is not needed for enzyme tests, and within-laboratory CVs of $\langle 10\%$ for the common enzyme tests is generally satisfactory.

7.2. Standardization: Now and Proposed

Currently, measurement of enzymes is based on catalytic-activity measurements, and these depend on the experimental conditions under which the enzyme is measured. It is often difficult to compare enzyme results from different laboratories or to those reported in the literature. Efforts to improve interlaboratory comparability in enzyme measurements (129, 130, 131, 132, 133, 134) have concentrated on two aspects: standardization of methodology and preparation of reference materials (135).

There have been considerable efforts over the last 10 years or so by national, particularly the Scandinavian Committee on Enzymes of the Scandinavian Society for Clinical Chemistry (136), and internationally by the IFCC to institute reference methods for many of the clinically important enzymes (137). The IFCC, through its Expert Panel on Enzymes, played a pivotal role by recommending reliable methods and clearly documenting the evaluations and recommendations being proposed. IFCC-recommended reference methods have been published for aspartate aminotransferase (138), alanine aminotransferase (139), γ-glutamyltransferase (140), alkaline phosphatase (141), and creatine kinase (142) with work just about complete on α-amylase and lactate dehydrogenase.

In many cases, standardization of enzyme methodology has significantly reduced but not eliminated interlaboratory variability (133, 134, 143). The use of recommended, reliable reference methods has hastened the elimination of unsatisfactory methods and focused the attention on the use of methods that, if not exactly like the reference method, are based on the same methodological principles. Results from external quality control schemes clearly demonstrate that it is only through the use of essentially the same recommended methods (standardization of methodology) and the availability of an enzyme reference material (ERM) that laboratories can improve their performance (133). Any system for the standardization of the catalytic activity of enzymes not only requires available reference methodology but also stable and well-characterized enzyme reference materials (135) because standardization of enzyme methodology has not been achieved, and there is some question as to whether or not it will ever be attained. National recommendations are strongly supported in some areas at the expense of international consensus. New developments in methodology and instrumentation in clinical enzymology will continue to be introduced, and small changes by the manufacturer in either the quality or concentrations of the reagents in a kit originally formulated as following the reference methodology criteria may lead to perhaps small, imperceptible changes in the reference method that may, over time, change its analytical performance. Some (135) envisioned enzyme reference materials, particularly in relation to the IFCC reference methods, as being used to: (1) maintain the IFCC methods; (2) transfer the IFCC methods to clinical laboratories; and (3) establish the traceability of regional and national enzyme methods to the IFCC reference methodologies (144). In the United States, the National Reference System for the Clinical Laboratory (NRSCL) Council of the NCCLS recognized the need for both reference methodology and materials when it instituted a formal consensus-accepted credentialing process for reference systems (135). In the case of enzymes, the credentialing process involves providing documentation to support a recognized and internationally accepted reference method, available reference materials, and documented transferability studies. To date, aspartate aminotransferase is, unfortunately, the only enzyme that has been credentialed by the NRSCL (145).

Use of the combination of reference methodology and materials to convert incompatible enzyme activity results for aspartate aminotransferase (146), alkaline phosphatase (147), and lactate dehydrogenase (148) to compatible values by use of a single scale, termed by the authors the *International Clinical Enzyme Scale* (ICES) has been suggested (149). Application of the ICES concept to the 1970 Scandinavian interlaboratory survey decreased the interlaboratory coefficient of variation from 38% to 16% for the enzymes tested. Similarly, in the 1971 New York State aspartate aminotransferase survey, the interlaboratory CV decreased from 41%-44% to 2%-5%, a major improvement. The Scandinavian Committee on Enzymes has expressed serious concerns about the philosophy of the ICES approach, and they again endorsed the widely accepted approach of ongoing development of reference methodologies and proper use of reference materials (150). Since this flurry of activity in 1984 and 1985, there has not been any further application or acceptance of the ICES concept.

7.3. Reference Materials for Enzymes

The requirements for, preparation of, and application of enzyme reference materials have been discussed extensively (134, 135, 151, 152, 153, 154, 155, 156, 157, 158). Cooperation among clinical enzymologists in Europe and the United States over the last several years has resulted in the availability of only a few enzyme reference materials. SRM 8430 from NIST is a preparation of human erythrocyte aspartate aminotransferase in a human albumin matrix (144); certified reference material (CRM) 319 from the Community Bureau of Reference (BCR) of the Commission of the European Communities is a preparation of porcine γ-glutamyltransferase in a bovine serum matrix (159). The working group of the BCR is in the process of establishing protocols and evaluating reference materials for creatine kinase, alkaline phosphatase, and lactate dehydrogenase using LD-1 (143). Although not certified by any professional group involved with standardization, there are published reports detailing the preparation and characterization of potential reference materials that apparently meet the criteria of the NCCLS (157), and these include α-amylase (160), alkaline phosphatase (161) creatine kinase (162) and prostatic acid phosphatase (163). Reference materials 8025 and 8026 from the National Institute of Public Health and Environmental Hygiene in the Netherlands (164, 165) are secondary human-serum reference materials with target values assigned for aspartate and alanine aminotransferases, creatine kinase, and lactate dehydrogenase. NIST also has available a secondary lyophilized human serum reference material, SRM 909, with target values for aspartate and alanine aminotransferases, alkaline phosphatase, creatine kinase, lactate dehydrogenase, acid phosphatase, and γ-glutamyltransferase (166). Target values in these materials have been assigned using [IFCC reference methodology in round-robin type evaluations. SRM 909a-1 and 909a-2 from NIST is a two-level reference material that will be replacing SRM 909.

7.4. Standardization of Enzymes as Proteins

Immunological methods for enzymes, more specifically isoenzymes, such as lactate dehydrogenase-1 (167, 168), mitochondrial aspartate aminotransferase (169), prostatic acid phosphatase (170, 171, 172), and creatine kinase-MB (173, 174, 175), have been in use in the clinical laboratory for 10 years. However, the use of the immunological rather than catalytic properties of enzymes has not provided the opportunities for standardization that was anticipated a number of years ago (176, 177, 178). It is only within the last year that a working group on CK-MB mass assay was formed under the auspices of the Standards Committee of the American Association for Clinical Chemistry (AACC). The objective of this working group is to prepare a reference material to calibrate methods that are based on the principle of CK-MB mass measurement.

8. THE IDEAL AUTOMATED ENZYME ANALYZER

The "ideal" analyzer described here does not exist, although various manufacturers have some of the features on their devices, the qualities described here are those currently needed in contemporary automated enzyme assays in clinical laboratories. Our aim is to set some goals for future development in this area. Owing to patients' needs and physicians' demands, clinical laboratories must always be ready to accept and analyze specimens making them somewhat different from most other analytical services, moreover, redundancy and ruggedness are needed at all times. The current needs are broadly grouped into three areas: general human requirements, mechanical/computer needs, and analytical specifications.

8.1. General Requirements

The safety of laboratory personnel has become a major focus. Infectious agents in patient-derived materials pose a significant risk for laboratory workers, particularly HIV and hepatitis B. A clear goal is to reduce the exposure to infectious materials so that any contact is unlikely or even impossible. Of course, laboratory people are well indoctrinated on the use of universal precautions in handling specimens, but we must go beyond this to improve the failsafe nature of laboratory safety. Sampling out of a "closed" tube is a primary goal. Two areas of exposure are avoided: the aerosol that occasionally forms when a tube is opened, and possible contact with droplets in and around the top of the specimen container and where it is handled. Certainly, pouring the specimen into a cup or tube should be avoided; however, the technology of direct-collection-tube sampling is not generally available today, or the existing technology needs perfecting.

An analyzer should be environmentally friendly; the waste stream, both solid and liquid, should be minimal. Exposure to reagents should be absolutely minimized, and hazardous and/or carcinogenic reagents should be eliminated. Finally, all functions where human error are possible should be eliminated. A careful analysis of all the steps in the analytical chain must be made to identify nodes of possible human blunders.

8.2. Mechanical and Computer Needs

Specimen identification should be as infallible as possible. In an ideal system, the specimen collector (e.g., a phlebotomist) carries a portable computer and label generator. This person downloads the requisitions from the central laboratory system to their portable unit prior to specimen collection. At the collection point (e.g., at the bedside), the collector, using the portable unit, wands an identifying label (e.g., a bar code on a wrist band) from the source of the specimen (e.g., the patient); a bar-coded label is generated right there and is put on the specimen. Once in the laboratory, the label is "read" by the analyzer to give unambiguous identification. In the entire chain of analytical events, the specimen is always kept in the labeled container. Extra tubes, cups, and the like are never made. The central computer system downloads the tests to be performed to the instrument for a given patient, so there are no missed tests.

The instrument does automatic, reflex reassays for out-of-range specimens. Dilutions are prepared, a less-sensitive wavelength is used, or some other technique is implemented. In short, the final result is produced without operator intervention, calculations, or other operator manipulations. A "library" of specimens is prepared by the instrument to permit retrieval of specimens with special needs such as "add-on" tests, questions from users of the data, repeats, and so on.

The reagent systems are failsafe and are in bar-code-labeled containers that identify them, the lot, expiration date, and so on. It is impossible to use the wrong reagent for a given test. The instrument monitors the inventory and quality of the reagents and alerts the operator to replenishment or maintenance needs.

Extensive safeguards exist for data maintenance including uninterruptable power supplies, redundant data storage, and retransmission capabilities to a host system where appropriate. The instrument also has extensive self-diagnostic features to monitor failures of all types including thermostatting, lamp failures, photometer misreadings of test materials, and so on.

8.3. Analytical Specifications

The specimen volume needed for analysis is minimal (e.g., 5µL), the "dead" volume in a specimen container is minimized, and the instrument has a "smart probe" as described above. Evaporation control is possible with lids of some kind, and a variety of different types of specimens, e.g., serum, plasma, urine, cerebrospinal fluid, and other body fluids can be accommodated on the same instrument. Obviously, the menu of enzyme tests must be broad enough to consolidate the testing on as few work stations as possible.

Finally, analytical accuracy, precision, and freedom from interferences must meet contemporary needs. Precision of modern instruments is largely excellent; bias from the true value is a common problem with enzyme assays, and interferences from lipemia, hemolysis, and icterus is still a problem with some instruments and methods.

9. APPENDIX 1: CHECKING ABSORBANCE ACCURACY WITH $K_2Cr_2O_7/H_2SO_4$ SOLUTIONS

Follow the procedure below:

A. Prepare 2 liters 0.005 mol/L H_2SO_4
  1. 0.05 mol/L H_2SO_4: 2.8 mL conc. H_2SO_4 (18 mol/L) Q.S. to 1 L with d. H_2O
  2. 0.005 mol/L H_2SO_4: Dilute 100 mL of 0.05 mol/L H_2SO_4 to 1 L with distilled H_2O. Stable indefinitely in glass. Label and date.
B. Dry about 15 g ACS or reagent-grade $K_2Cr_2O_7$ at 110°C for 2 hr. Cool in a desiccator.
C. Prepare a STOCK solution of 0.9350 g/L $K_2Cr_2O_7$ in 0.005 mol/L H_2SO_4. This solution has a theoretical absorbance (A) of 10.0 at 350nm. Stable indefinitely in a well-stoppered, pyrex glass container.
D. Prepare dilutions of 2.5 mL–20 mL of the STOCK $K_2Cr_2O_7$ and enough 0.005 mol/L H_2SO_4 to make 100 mL of each dilution. The theoretical absorptivity is 10.7 L/g-cm.

10. APPENDIX 2: CALCULATING UNITS OF ENZYME ACTIVITY

An enzyme assay for LD (pyruvate to lactate direction) uses 0.010 mL serum in a total volume of 1.010 mL. The change in absorbance with time (slope) in the linear region of the reaction curve is 0.022/min. Calculate the enzyme activity in U/L. The molar absorptivity or ε of NADH is 6,220 L/mol-cm, b = 1 cm.

$1 \text{ U/L} = 1 \text{ μmol/min-L}$ substrate conversion under conditions of assay.

$\Delta c / \Delta t = \text{μmol/min-L}$

$c = A/\varepsilon b$

Therefore:

$$\frac{\Delta A}{\Delta t} \cdot \frac{1}{\varepsilon b} \cdot \frac{TV}{SV} = \text{U/L}$$

$$\frac{0.022}{\text{min}} \cdot \frac{1 \text{ mol-cm}}{6220 \text{ L}} \cdot \frac{1}{1 \text{ cm}} \cdot \frac{1.01}{0.01} \cdot \frac{10^6 \text{ μmol}}{\text{mol}} = 357 \text{ U/L}.$$

11. REFERENCES

1. N. L. Alpert, *Clin. Chem.* **15**, 1198 (1969).
2. M. K. Schwartz, *Meth. Enzymol.* **21**, 5 (1971).
3. Perez-Bendito and M. Silva, *Kinetic Methods in Analytical Chemistry*, Wiley, New York, 1988, pp. 221–243; 299–302.
4. D. B. Roodyn, *Automated Enzyme Assays*, American Elsevier Publishing Co., New York, 1970, pp. 7–201.
5. T. O. Tiffany, C. A. Burtis, and N. G. Anderson, *Meth. Enzymol.* **31**, 790 (1974).
6. M. Valcárcel and M. D. Luque de Castro, *Automatic Methods off Analysis*, Elsevier, New York, 1988, pp. 428–466.
7. M. Valcárcel and M. D. Luque de Castro, *Automatic Methods of Analysis*, Elsevier, New York, 1988, p. 238.
8. J. R. Welborn and M. S. Burgess, *Clin. Chem.* **34**, 180 (1988).
9. A. Stavljenic, J. Sertic, K. Kumar, N. Vrikic, and K. Fumic, *Clin. Chem.* **34**, 1917 (1988).
10. B. G. Blijenberg, F. Braconnier, J. M. Vallez, A. Burlina, M. Plebani, M. Celadin, R. Haeckel, M. Romer, E. Hansler, and G. DeSchrijver, *J. Clin. Chem. Clin. Biochem.* **27**, 369 (1989).
11. J. A. Lott, *Clin. Chem.* **31**, 282 (1985).
12. A. T. Chow, J. E. Kegelman, C. Kohli, D. D. McCabe, and J. F. Moore, *Clin. Chem.* **36**, 1579 (1990).
13. E. Hanseler, D. Vonderschmitt, R. Haeckel, M. Romer, C. Collombel, *Eur. J. Clin. Chem. Clin. Biochem.* **29**, 81 (1991).
14. D. Knight, R. Singer, J. M. White, and C. G. Fraser, *Clin. Chem.* **34**, 1899 (1988).
15. H. Baadenhuijsen, P. M. Bayer, H. Keller, M. Knedel, N. Montalbetti, S. Brenna, L. Prencipe, and A. Vassault, *J. Clin. Chem. Clin. Biochem.* **28**, 261 (1990).
16. E. G. Lentjes, G. A. Harff-GA, and E. T. Backer, *Clin. Chem.* **33**, 2089 (1987).
17. J. Flood, R. Liedtke, H. Mattenheimer, L. Rothouse, D. Trundle, D. V. Roon, and K. Whisler, *Clin. Biochem.* **23**, 477 (1990).

18. H. Baadenhuijsen, P. M. Bayer, H. Keller, A. Vassault, L. Prencipe, N. Montalbetti, S. Brenna, and M. Knedel, *J. Clin. Chem. Clin. Biochem.* **28**, 291 (1990).
19. S. B. Schotters, J. H. McBride-JH, D. O. Rodgerson, M. H. McGinley, and M. Pisa, *J. Clin. Lab. Anal.* **4**, 157 (1990).
20. D. D. Koch, J. J. Oryall, E. F. Quam, D. H. Feldbruegge, D. E. Dowd, P. L. Barry, and J. O. Westgard, *Clin. Chem.* **36**, 230 (1990).
21. F. K. Gorus and L. De Pree, *Clin. Chem.* **36**, 685 (1990).
22. B. T. Doumas, *Clin. Chem.* **35**, 151 (1989).
23. M. G. Bissell, E. S. Pravatiner, S. S. Reddy, T. Lipscomb, and S. Hussain, *Clin. Chem.* **34**, 1511 (1988).
24. J. Boland, G. Carey, E. Krodel, and M. Kwiatkowski, *Clin. Chem.* **36**, 1598 (1990).
25. J. Kropf, A. M. Marx, J. Hildebrandt, and A. M. Gressner, *Eur. J. Chem. Clin. Biochem.* **29**, 675 (1991).
26. G. A. McClellan, H. C. Nipper, M. M. J. Horn, W. D. Burris, K. Hodges, S. Monaco, R. S. Cox, Jr., and A. Tillie, *Am. J. Clin. Pathol.* **95**, 743 (1991).
27. T. L. Isenhour, S. E. Eckert, and J. C. Marshall, *Anal. Chem.* **61**, 805A (1989).
28. College of American Pathologists, 325 Waukegan Road, Northfield, IL 60093-2750.
29. K.-H. W. Lau, T. Onishi, J. E. Wergedal, F. R. Singer, and D. J. Baylink, *Clin. Chem.* **33**, 458 (1987).
30. P. Lainé-Cessac, A. Turcant, and P. Allain, *Clin. Chem.* **35**, 77 (1989).
31. L. L. Tilzer and R. W. Jones, *Arch. Pathol. Lab. Med.* **112**, 1200 (1988).
32. J. A. Lott, *Med. Lab. Observ.* (**May**) 30 (1991).
33. R. L. Columbus and H. J. Palmer, *Clin. Chem.* **37**, 1548 (1991).
34. Roche Diagnostic Systems, Mira Instrument, 1 Sunset Avenue, Montclair, NJ 07042.
35. J. A. Lott, "Creatine Kinase in Serum," in *Selected Methods for the Small Clinical Chemistry Laboratory* (eds. W. R. Faulkner and S. Meites), American Association for Clinical Chemistry, Washington, D.C., 1982, pp. 185–190.
36. C. A. Burtis, *Clin. Chem.* **36**, 544 (1990).
37. S. P. Harrison and I. M. Barlow, *Clin. Chem.* **36**, 1520 (1990).
38. W. H. Copeland, D. A. Nealon, and R. Rej, *Clin. Chem.* **31**, 185 (1985).
39. N. W. Tietz, A. D. Rinker, C. Burtis, et al. *Clin. Chem.* **30**, 704 (1984).
40. National Institutes of Science and Technology, Gaithersburg, MD 20889.
41. J. A. Lott, *CRC Crit. Rev. Clin. Lab. Sci.* **7**, 277 (1977).
42. W. J. Korzun, W. G. Miller, *Clin. Chem.* **32**, 162 (1986).
43. M. L. Glick and K. W. Ryder, *Interferographs: Users Guide to Interferences in Clinical Chemistry Instruments*, Science Enterprises, Indianapolis, IN, 1987.
44. H. Zollner, *Handbook of Enzyme Inhibitors*, VCH Publishers, New York, 1989.
45. D. S. Young, *Effects of Drugs on Clinical Laboratory Tests*, AACC Press, 1990, pp. 3-1–3-381.
46. S. C. Kazmierczak and J. A. Lott, "Alkaline Phosphatase," in *Methods in Clinical Chemistry* (eds. A. J. Pesce and L. A. Kaplan), Mosby, St. Louis, Mo, 1987, pp. 1074–1079.
47. H.-U. Bergmeyer and K. Gawehn, *Principles of Enzymatic Analysis*, Verlag Chemie, New York, 1978, pp. 13–40.
48. H.-U. Bergmeyer, *Methods of Enzymatic Analysis*, Academic Press, New York, 1965, pp. 3–13.
49. G. Szasz, *Clin. Chem.* **22**, 650 (1976).
50. S. Udenfriend, *Fluorescence Assay in Biology and Medicine*, Academic Press, New York, 1962.
51. K. W. Pearson, R. E. Smith, A. R. Mitchell, and E. R. Bissel, *Clin. Chem.* **27**, 256 (1981).
52. M. Sergio and M. Pazzagli (eds.), *Luminescent Assays: Perspectives in Endocrinology and Clinical Chemistry*, Raven Press, New York, 1982.
53. R. L. Boeckx, *Human Pathol.* **15**, 104 (1984).

54. L. J. Kricka, P. E. Stanley, G. H. G. Thorpe, and T. P. Whitehead (eds.), *Analytical Applications of Bioluminescence and Chemiluminescence*, Academic Press, New York, 1984.
55. J. Scholmerich, R. Adreesen, A. Kapp, M. Ernst, and W. G. Woods (eds.), *Bioluminescence and Chemiluminescence: New Perspectives*, John Wiley, Chichester, 1987.
56. W. R. G. Baeyens, K. Nakashima, K. Imai, B. L. Ling, and Y. Tsukamoto, *J. Pharm. Biomed. Anal.* **7**, 407 (1989).
57. K. Van Dyke, *Luminescence Immunoassay and Molecular Applications*, CRC Press, Boca Raton, 1990.
58. L. J. Kricka, *Clin. Chem.* **37**, 1472 (1991).
59. J. G. Burr (ed.), *Chemi- and Bioluminescence*, Marcel Dekker, Inc., New York, 1985.
60. K. Van Dyke, "Commercial Instruments," in *Bioluminescence and Chemiluminescence: Instruments and Applications* (ed. K. Van Dyke), CRC Press, Boca Raton, 1985, p. 83.
61. P. E. Stanley, *J. Biolum. Chemilum.* **7**, 77 (1992).
62. K. Wulff, F. Stahler, and W. Gruber, "Standard Assay for Total Creatine Kinase and the MB-isoenzyme in Human Serum with Firefly Luciferase," in *Bioluminescence and Chemiluminescence* (eds. M. A. DeLuca and W. D. McElroy), Academic Press, New York, 1981, p. 209.
63. A. Lundin, B. Jaderlund, and T. Lovgren, *Clin. Chem.* **28**, 609 (1982).
64. Berthold Analytical Instruments, Inc., Nashua, NH.
65. Beckman Instruments, Brea, CA.
66. Behring Diagnostics, Marburg, Germany.
67. H. J. M. Salden, B. M. Bas, I. T. H. Hermans, and P. C. W. Janson, *Clin. Chem.* **34**, 1594 (1988).
68. S. B. Schotters, J. H. McBride, D. O. Rodgerson, S. Higgins, and M. Pisa, *J. Clin. Lab. Analysis* **2**, 108 (1988).
69. C. P. Price, K. Spencer, and J. Whiccher *Ann. Clin. Biochem.* **20**, 1 (1983).
70. L. Zinterhofer, S. Wardlaw, P. Jatlow, and D. Seligson, *Clin. Chim. Acta* **43**, 5 (1973).
71. J. R. Smeaton and H. F. Marquardt, *Clin. Chem.* **20**, 896 (1974).
72. W. C. Vogel and L. Zieve, *Clin. Chem.* **9**, 168 (1963).
73. J. R. Shipe and J. Savory, *Clin. Chem.* **19**, 645 (1973).
74. L. Zinterhofer, S. Wardlaw, P. Jatlow, and D. Seligson, *Clin. Chim. Acta* **44**, 173 (1973).
75. J. Ziegenhorn, U. Neumann, K. W. Knitsch, and W. Zwez, *Clin. Chem.* **25**, 1067 (1979).
76. W. Rick and M. Hockenborn, *J. Clin. Chem. Clin. Biochem.* **20**, 735 (1982).
77. U. Neumann, P. Kaspar, and J. Ziegenhorn, *Methods Enzymol. Anal.* **4**, 26 (1984).
78. J. Kusnetz and H. P. Mansberg, "Optical Considerations: Nephelometry," in *Automated Immunoanalysis* (ed. R. F. Ritchie), New York, 1978.
79. S. R. Popelka, D. M. Miller, J. T. Holen, and D. M. Kelso, *Clin. Chem.* **27**, 1198 (1981).
80. Ames Division, Miles Laboratories Ltd., Elkhart, IN.
81. Boehringer Mannheim, Corp., Indianapolis, IN.
82. Fuji Photo Film Co., Ltd., Asaka-Shi, Saitamma, Japan.
83. PB Diagnostic Systems, Inc, Westwood, MA.
84. Baxter Diagnostics Inc., Miami, FL.
85. D. W. Chan (ed.), *Immunoassay Automation: A Practical Guide*, Academic Press, New York, 1992.
86. B. Walter, *Anal. Chem.* **55**, 498A (1983).
87. R. L. Steinhausen and C. P. Price, "Principles and Practice of Dry Chemistry Systems, in *Advances in Clinical Biochemistry*, Vol. 3 (eds. C. P. Price and K. G. M. M. Alberti), Churchill Livingstone, Edinburgh, 1985, p. 27.
88. J. C. Libeer, *J. Clin Chem. Clin. Biochem.* **23**, 645 (1985).

89. A. Zipp and W. E. Hornby, *Talanta* **31**, 863 (1984).
90. R. S. Campbell and C. P. Price, *J. Inter. Fed. Clin. Chem.* **3**, 204 (1991).
91. Clinical Diagnostics Division, Eastman Kodak Company, Rochester, NY.
92. J. Greyson, *J. Auto. Chem.* **3**, 66 (1981).
93. A. Zipp, *J. Auto. Chem.* **3**, 71 (1981).
94. B. Walter, R. C. Berreth, and M. Wilcox, *Clin. Chem.* **29**, 1267 (1983).
95. L. Thomas, W. Plischke, and G. Storz, *Annu. Clin. Biochem.* **19**, 214 (1982).
96. A. Karmen and R. Lent, *J. Clin. Lab. Automat.* **2**, 284 (1982).
97. P. M. S. Clark and P. M. G. Broughton, *J. Automat. Chem.* **5**, 22 (1983).
98. F. Stahler, "Real Time Clinical Chemistry Using Dry Chemistry," in *Clinical Biochemistry, Principles and Practice* (eds. A. E. Eng and P. Garcia-Webb), 1983, p. 297.
99. P. U. Koller, *Upsala J. Med. Sci.* **91**, 135 (1986).
100. K. Roberts, *Amer. Clin. Prod.* **16** (1987).
101. R. W. Spayd, B. Bruschi, B. A. Burdick, G. M. Dappen, J. N. Eikenberry, T. W. Esders, J. Figueras, C. T. Goodhue, D. D. LaRossa, R. W. Nelson, R. N. Rand, and T.-W. Su, *Clin. Chem.* **24**, 1343 (1978).
102. H. G. Curme, R. L. Columbus, G. M. Dappen, T. W. Seder, J. E. Pinney, R. R. Rand, V. J. Sanford, T.-W. Wu, C. P. Glover, C. A. Goffe, D. E. Hill, W. H. Lawton, E. J. Muka, W. D. Fellows, and J. F. Figueras, *Clin. Chem.* **24**, 1335 (1978).
103. T. L. Shirey, *Clin. Biochem.* **16**, 147 (1983).
104. K. M. Reynolds, *Upsala J. Med. Sci.* **91**, 143 (1986).
105. T. K. Mayer and N. P. Kubasik, *Lab. Management*, April, 43 (1986).
106. W. Werner and W. Rittersdorf, "Reflectance Photometry," in *Methods of Enzymatic Analysis*, Vol. 1 (eds. J. Bergmeyer and M. Grassi), Verlag-Chemie, Weinheim, 1983, p. 305.
107. F. C. Williams and F. R. Clapper, *J. Opt. Soc. Amer.* **43**, 595 (1953).
108. D. L. Morris, P. B. Ellis, R. J. Carrico, F. M. Yeager, H. R. Schroeder, J. P. Albarella, and R. C. Gobuslaski, *Anal. Chem.* **53**, 658 (1981).
109. Y. P. Yagminas and D. C. Villenueve, *Biochem. Med.* **18**, 117 (1977).
110. R. J. Pierce, R. G. Price, A. M. Marsden, and J. S. Fowler, *Curr. Prob. Clin. Biochem.* **9**, 201 (1979).
111. R. E. Vanderlinde, *Annu. Clin. Lab. Sci.* **11**, 189 (1981).
112. K. Jung, J. Diego, V. Strobelt, D. Scholz, and G. Schreiber, *Clin. Chem.* **32**, 1807 (1986).
113. H. O. Beutler, B. Wurst, and S. Fischer, *Dtsch. Lebensm.-Rundsch.* **82**, 283 (1986).
114. H. O. Beutler and I. Schuette, *Dtsch. Lebensm.-Rundsch.* **79**, 323 (1983).
115. T. Bagaringo and R. D. Vetter, *Marine Biol. 103*, 291 (1989).
116. B. C. Campbell, M. J. Brodie, G. G. Thompson, P. A. Meredith, M. R. Moore, and A. Goldberg, *Clin. Sci. Mol. Med.,* **53**, 335 (1977).
117. G. S. Fell, *Ann. Clin. Biochem.* **21**, 453 (1984).
118. W. J. Rogan, J. R. Reigart, and B. C. Gladen, *J. Pediatr.* **109**, 60 (1986).
119. E. Rossi, K. A. Costin, and P. Garcia-Webb, *Clin. Chem.*, **36**, 1980 (1990).
120. S. S. Brown, W. Kalow, W. Pilz, M. Whittaker, and C. L. Woronick, *Adv. Clin. Chem.* **22**, 1 (1981).
121. W. J. Hayes, *The Pesticides in Man*, The Williams and Wilkins Co., Baltimore, 1982.
122. D. F. Innes, B. H. Fuller, and G. M. B. Berger, *S. African Med. J.* **78**, 581 (1990).
123. P. I. Boon, *Arch. Hydrobiol.* **115**, 339 (1989).
124. M. Tabata, Y. Kobayashi, A. Nakajima, and S. Suzuki, *Bull. Environ. Contam. Toxicol.* **45**, 31 (1990).
125. B. T. O'Connell, G. L. Rubenthaler, and N. L. Murbach, *Cereal Chem.* **57**, 411 (1980).

126. J. O. Westgard and P. L. Barry, *Cost-Effective Quality Control: Managing the Quality and Productivity of Analytical Processes*, American Association for Clinical Chemistry Press, Washington, DC, 1986.
127. J. A. Lott and S. T. Patel, "Assessment of Analytical Variability in Clinical Chemistry," in *Laboratory Quality Assurance*, (eds. P. J. Howanitz and J. H. Howanitz), McGraw-Hill, New York, 1987, pp. 185–213.
128. J. A. Lott, N. Surufka, and C. G. Massion, *Arch. Pathol. Lab. Med.* **115**, 11 (1991).
129. H. U. Bergmeyer, *Clin. Biochem.* **13**, 155 (1980).
130. J. A. Lott, D. W. Tholen, and C. G. Massion, *Arch. Pathol. Lab. Med.* **112**, 392 (1988).
131. D. Laue, "Standardization of Enzyme Assays Based on Determinations by Reference Methods," in *Selected Topics Clin. Enzymol.* (eds. M. Werner and D. M. Goldberg), Walter de Gruyter, FRG, 1984, p. 15.
132. D. Moss, *Ann. Clin. Biochem.* **25**, 1 (1988).
133. R. T. P. Jansen and A. P. Jansen, *Ann. Clin. Biochem.* **20**, 52 (1983).
134. D. G. Bullock, D. W. Moss, and T. P. Whitehead, *Ann. Clin. Biochem.* **23**, 577 (1986).
135. G. N. Bowers, Jr., G. C. Edwards, and R. N. Rand (eds.), *A Reference System for Clinical Enzymology: Proceedings of the Workshop*, The National Committee for Clinical Laboratory Standards, Villanova, 1986.
136. Scandinavian Committee on Enzymes of the Scandinavian Society for Clinical Chemistry, *Scand. J. Clin. Lab. Invest.* **33**, 291 (1974); and loc cit. **41**, 107 (1981).
137. M. Horder, "Standardization of Enzyme Analysis: Findings of the Expert Panel on Enzymes of the IFCC," in *Clinical and Analytical Concepts in Enzymology* (ed. H. A. Homburger), College of American Pathologists, Skokie, 1983, p. 267.
138. H.-U. Bergmeyer, M. Horder, and R. Rej, *J. Clin. Chem. Clin. Biochem.* **24**, 497 (1986).
139. H.-U. Bergmeyer, M. Horder, and R. Rej, *J. Clin. Chem. Clin. Biochem.* **24**, 481 (1986).
140. L. M. Shaw, J. H. Stromme, J. L. London, and L. Theodorsen, *J. Clin. Chem. Clin. Biochem.* **21**, 633 (1983).
141. N. W. Tietz, A. D. Rinker, and L. M. Shaw, *J. Clin. Chem. Clin. Biochem.* **21**, 731 (1983).
142. M. Horder, R. C. Elser, W. Gerhardt. M. Mathieu, and E. J. Sampson, *J. Inter. Fed. Clin. Chem.* **130** (1989).
143. E. Colinet, G. Siest, and D. W. Moss, *Ann. Clin. Biochem.* **23**, 361 (1986).
144. G. N. Bowers Jr., R. B. McComb, D. Syed, G. C. Edwards, R. C. Paule, N. Greenberg, L. A. Malick, R. J. Ferris, and M. Horder, *Clin. Chem.* **34**, 450 (1988).
145. National Reference System for the Clinical Laboratory, *The National Reference System for Aspartate Aminotransferase*, NCCLS Document RS-2 Proposed, National Committee for Clinical Laboratory Standards, Villanova, 1987.
146. G. N. Bowers, Jr. and R. B. McComb, *Clin. Chem.* **30**, 1128 (1984).
147. R. B. McComb and G. N. Bowers Jr., *Amer. J. Clin. Pathol.* **84**, 67 (1985).
148. G. N. Bowers, Jr., W. C. Bowers, and R. B. McComb, *Clin. Chem.* **31**, 1005 (1985).
149. R. B. McComb, "A Universal Reference System for Clinical Enzymology," in *A Reference System for Clinical Enzymology: Proceedings of the Workshop* (eds. G. N. Bowers, Jr., G. C. Edwards, and R. N. Rand), National Committee for Clinical Laboratory Standards, Villanova, 1986, p. 73.
150. Scandinavian Committee on Enzymes of the Scandinavian Society for Clinical Chemistry, *Clin. Chem.* **31**, 342 (1985).
151. C. F. Fasce, R. Rej, W. H. Copeland, and R. Vanderlinde, *Clin. Chem.* **19**, 5 (1973).
152. R. Rej and R. Vanderlinde, "Use of Purified Enzyme Reference Materials in Enzyme Activity Measurements," in *Proc. Second International Symposium on Clinical Enzymology* (eds. N. W. Tietz and A. Weinstock), American Association for Clinical Chemistry, Washington, DC, 1976, p. 249.

153. J. A. Lott, "Standard Materials," in J. H. Boutwell (ed.), A National Understanding for the Development of Reference Materials and Methods for Clinical Chemistry: Proceedings of a Conference, *American Association for Clinical Chemistry*, Washington, DC, 1978, p. 107.
154. J.-P. Bretaudiere, G. Dumont, R. Rej, and M. Bailly, *Clin. Chem.* **27**, 798 (1981).
155. J. L. Beecham, K. B. Whitaker, and D. W. Moss, "The Use of Calibration Materials in Clinical Enzymology," in *Biologie Prospective-5e Colloque International de Pont-a-Mousson* (eds. M. M. Galteau, G. Siest, and J. Henny), Masson et Cie, Paris, 1983, p. 331.
156. M. Horder and R. Rej, "Requirements and Functions of Reference Materials for Enzymes," in *Progress in Clinical Enzymology*, Vol. 2 (eds. D. M. Goldberg and M. Werner), Masson Publishing USA, New York, 1983, p. 29.
157. National Committee for Clinical Laboratory Standards, NRSCL 3-T (National Reference System for the Clinical Laboratory): Guidelines for the Development of Certified Reference Materials for Use in Clinical Chemistry, National Committee for Clinical Laboratory Standards, Villanova, 1984.
158. R. Rej, R. W. Jenny, and J.-P. Bretaudiere, *Talanta* **31**, 851, (1984).
159. F. Schiele, J. Muller, E. Colinet, and G. Siest, *Clin. Chem.* **33**, 1971 and 1978 (1987).
160. E. J. Sampson, P. H. Duncan, D. M. Fast, V. S. Whitner, S. S. McKneally, M. A. Baird, M. L. MacNeil, and D. D. Bayse, *Clin. Chem.* **27**, 714 (1981).
161. P. H. Duncan, S. S. McKneally, M. L. McNeil, D. M. Fast, and D. D. Bayse, *Clin. Chem.* **30**, 93 (1984).
162. V. S. Whitner, L. G. Morin, S. S. McKneally, and E. J. Sampson, *Clin. Chem.* **28**, 41 (1982).
163. P. H. Duncan, R. L. Von Etten, M. L. McNeil, and L. M. Shaw, *Clin. Chem.* **30**, 1327 (1984).
164. J. C. Koedam, G. M. Steentjes, S. Buitenhuis, E. Schmidt, and R. Klauke, *Clin. Chem.* **32**, 1901 (1986).
165. J. C. Koedam, H. J. van Dreumel, W. Ham, G. M. Steentjes, and J. B. A. Terlingen, *Clin. Chem.* **32**, 1906 (1986).
166. G. N. Bowers, Jr., R. Alvarez, J. P. Cali, K. R. Eberhardt, et al., *National Bureau of Standards (NBS/NIST)*, Special Publication **260-83**, US Government Printing Office, Washington, DC, 1983.
167. M. Usategui-Gomez, R. W. Wicks, and M. Warshaw, *Clin. Chem.* **25**, 729 (1979).
168. R. Rej, *Clin. Biochem.* **16**, 17 (1983).
169. R. Rej, *Clin. Chem.* **26**, 1695 (1980).
170. S. N. Davies and J. C. Griffiths, *Clin. Chim. Acta* **122**, 29 (1982).
171. R. Bais and J. B. Edwards, *Crit. Rev. Clin. Lab. Sci.* **18**, 291 (1982).
172. M. F. Lin, C.-L. Lee, and T. M. Chu, *Clin. Chim. Acta* **130**, 263 (1983).
173. U. Wuerzburg, "Measurement of Creatine Kinase Isoenzyme Activity by Immunological Methods," in *Creatine Kinase Isoenzymes: Pathophysiology and Clinical Application* (ed. H. Lang), Springer-Verlag, New York, 1981, p. 49.
174. D. Neumeier, "Measurement of Creatine Kinase Isoenzyme Concentration by Immunoassay," in *Creatine Kinase Isoenzymes: Pathophysiology and Clinical Application* (ed. H. Lang), Springer-Verlag, New York, 1981, p. 75.
175. H. C. Vaidya, Y. Maynard, D. N. Dietzler, and J. H. Ladenson, *Clin. Chem.* **32**, 657 (1986).
176. J. Bohner and W. Stein, *J. Clin. Chem. Clin. Biochem.* **22**, 943 (1984).
177. M. G. Scott, *Trends Biotechnol.* **3**, 170 (1985).
178. R. E. Vanderlinde, "Enzyme Reference Materials- Early Activities," in *A Reference System for Clinical Enzymology: Proceedings of the Workshop* (ed. G. N. Bowers, Jr., G. C. Edwards, and R. N. Rand), The National Committee for Clinical Laboratory Standards, Villanova, 1986, p. 97.
179. R. N. Rand, *Clin. Chem.* **15**, 839 (1969).

BIOANALYTICAL INSTRUMENTATION   VOLUME 37

Rapid-Scanning Stopped-Flow Spectrophotometry

PETER S. BRZOVIĆ AND MICHAEL F. DUNN

Department of Biochemistry, University of California, Riverside, Riverside, California

1. Introduction
2. Applications of Rapid-Scanning, Stopped-Flow (RSSF) UV-Visible Spectroscopy to the Study of Biological Systems
    2.1. Artificial Chromophoric Substrates, Coenzymes and Cofactors
        2.1.1. α-Chymotrypsin (α-Ct) Catalysis
        2.1.2. Horse Liver Alcohol Dehydrogenase (LADH) Catalysis
        2.1.3. Conformational Transitions in the Insulin Hexamer
        2.1.4. Sheep Liver Aldehyde Dehydrogenase Catalysis
    2.2. Systems with Intrinsic Chromophores
        2.2.1. RSSF Investigations of Pyridoxyl Phosphate (PLP) Enzymes
        2.2.2. Tryptophan Synthase
        2.2.3. Reaction of Tryptophan Synthase with L-Serine
        2.2.4. Reaction of the Tryptophan Synthase α-Aminoacrylate Intermediate with Indole
        2.2.5. Observation of Allosteric Interactions in the Tryptophan Synthase Bienzyme Complex
        2.2.6. Tryptophanase
        2.2.7. Investigation of Substrate Structural Elements on Tryptophanase Reactivity
        2.2.8. Cystathionine γ-Synthase Catalysis
    2.3. Investigation of Enzyme Structure–Function Relationships by Site-Directed Mutagenesis and RSSF Spectroscopy
        2.3.1. Mutations in the β-Subunit of Tryptophan Synthase
        2.3.2. Mutations in the α-Subunit of Tryptophan Synthase
        2.3.3. Tryptophanase
    2.4. RSSF Studies of Heme Containing Proteins and Enzymes
        2.4.1. Hemoglobin and Myoglobin
        2.4.2. Electron Transport Proteins
        2.4.3. RSSF Investigation of Hydroperoxidases
3. Basic Principles and Instrumentation
    3.1. Time Domains, Wavelength Domains, and Biological Processes
        3.1.1. Time Domains
        3.1.2. Wavelength Domains

Bioanalytical Instrumentation, Volume 37, Edited by Clarence H. Suelter.
ISBN 0-471-58260-3 © 1993 John Wiley & Sons, Inc.

3.2. Detection Strategies: Phototubes versus Solid-State Detectors
   3.2.1. Single Element Detectors
   3.2.2. Array Detectors
   3.2.3. SPD and MCP-SPD Array Detectors
   3.2.4. Charge Injection Devices
   3.2.5. CCD and LCT-CCD Arrays
3.3. Rapid-Mixing, Rapid-Scanning Stopped-Flow Systems
   3.3.1. Single Element Detector Systems
   3.3.2. Array Detector Systems
4. Data Analysis Strategies
   4.1. Qualitative and Semiquantitative Analyses
   4.2. Quantitative Analysis of RSSF Data
5. Concluding Remarks
6. References

1. INTRODUCTION

Within the field of rapid-mixing, stopped-flow kinetics, work on biological systems is dominated by UV-visible absorbance spectrophotometry. Detection of fluorescence emission is a distant second, and use of other forms of spectroscopy (CD, Resonance Raman) or other regions of the electromagnetic spectrum are rather uncommon. Despite early promise (1), rapid-scanning, rapid-mixing applications to the study of biological systems have been restricted almost exclusively to the measurement of UV-visible absorbance spectra. This is in marked contrast to certain areas of photobiology where, in addition to measurement of UV-visible absorbance spectra, time-resolved FT-IR, CD, Raman, and Resonance Raman spectra have been employed to examine events in the photocycles of plant and bacterial reaction centers, the bacteriorhodopsins, and animal rhodopsins. These time-resolved rapid-scanning experiments are conducted on ps to ms time scales and are initiated by excitation of the primary electron donor (or dark adapted chromophore) by excitation flashes of brief duration. Especially for FT-IR and Raman studies, each spectrum of the time-resolved set usually is obtained by signal averaging of data acquired at a defined time interval following each of a large number of excitation flashes. By varying the time interval between the excitation flash and data acquisition, a time-resolved set of spectra is obtained. For reversible systems, the measurements are made on a static system. For irreversible systems, the sample is flowed through an appropriate flow cell system, which allows the excitation and interrogation beams to be continuously supplied with fresh sample during data acquisition.

Due to the time-resolution limitations imposed by mixing dead-times in the 1-ms range, the technology of rapid mixing has not been very relevant to those areas of photobiology. The need for significant signal-averaging usually encountered with FT-IR, Raman and Resonance Raman spectroscopies has discouraged the application of rapid-mixing, rapid-scanning technologies incorporating these modes of detection to biological problems.

The development of single-wavelength, stopped-flow (SWSF), UV-visible absorbance spectroscopy as a tool for investigating the dynamic behavior of biomacromolecules made it obvious early on that the measurement of UV-visible spectra on fast time scales could give important information that would assist the assignment of structures to transient intermediates. Early efforts to measure the spectra of transient species resorted to the reconstruction of spectra from SWSF measurements as a function of wavelength. In this way, Chance (2) in 1949 was able to determine the spectra of compound I in the reaction of catalase with hydrogen peroxide. Because the reconstruction of spectra from SWSF data is sample intensive and laborious, this approach has not been much used to determine the spectra of transient species. Early attempts to construct rapid-scanning, stopped-flow (RSSF) instrumentation were frustrated by the limitations imposed both by the mechanical mechanisms employed to achieve scanning of the dispersed spectrum and by the lack of adequate data acquisition systems for collecting, storing, and analyzing the large data sets, which are generated in the RSSF experiment.

The limited availability of affordable commercial RSSF instruments has been an important factor that has prevented the widespread application of RSSF spectroscopy to the study of biological systems. However, in the past year, a significant change in the availability of commercial instrumentation has come about. There currently are at least five manufacturers of computerized rapid-scanning detector systems. The choices in commercial instrumentation range from a mechanically scanned system with a single photomultiplier detector to photodiode array detector systems. This review includes descriptions of the currently available commercial systems. Because the authors' experience in the field of RSSF spectroscopy is limited to the use of diode array detector systems and because most of the commercial instruments have appeared on the market just within the past 12 months, it has not been possible to make detailed performance evaluations and comparisons of the new commercial systems.

2. APPLICATIONS OF RAPID-SCANNING, STOPPED-FLOW (RSSF) UV-VISIBLE SPECTROSCOPY TO THE STUDY OF BIOLOGICAL SYSTEMS

The examples in this section have been chosen to provide an in-depth presentation showing how RSSF currently has been applied to the study of biological systems. These applications include the study of isotope effects on enzyme-catalyzed reactions, the investigation of substrate-metal ion interactions in metalloenzymes, the search for and identification of covalent intermediates in enzyme-catalyzed processes, the analysis of the effects of site-directed mutations on enzyme catalytic mechanism, and the exploitation of natural and artificial chromophores as probes of allosteric processes.

Meaningful RSSF UV-visible spectroscopy requires a chromophoric probe that is responsive to the physical–chemical events under study. The examples described in this section have been chosen to illustrate the wide variety of probes, which have proven suitable. It will become obvious that the choice of systems for study need not be restricted to those with intrinsic chromophores. Through innovation in the design and selection of cofactor or substrate analogues, suitable probes can be found for many "colorless" systems (in the following paragraphs, two such examples are α-chymotrypsin and the insulin hexamer). Indeed, for the liver alcohol dehydrogenase system, it has been possible to supplement the intrinsic chromophoric properties of NADH with chromophoric signals derived from the active site metal ion (3), the aldehyde/alcohol substrate (4, 5), and the uptake/release of H^+ (6, 7) during catalysis by judicious choice of chromophoric probes. While the use of analogues in the study of enzyme catalysis introduces the possibility that the analogue undergoes reaction by an altered catalytic mechanism, it is usually the case that analogues follow a reaction pathway that is like that of the natural substrates, but with altered free energies of the ground states and activated complexes. This can have the added advantage that certain chemical intermediates along the pathway, which do not accumulate with the natural substrate, become detectable species in the analogue reaction. In some instances, the analogue follows a reaction pathway, which (due to its intrinsic chemical properties) branches away from that of the natural substrate at some point along the path, see Koerber et al. (8) and Cochran et al. (9). Such deviations often are detected by virtue of the generation of an unexpected reaction product, covalent modification of a catalytic residue, or by the accumulation of an unexpected chemical intermediate. Sometimes altered pathways can provide valuable new insights about the properties of the catalytic residues involved in the catalysis of bond scission/bond-forming processes even though the reaction catalyzed is different.

The RSSF work published in the area of enzyme catalysis is centered on the search for transient chemical intermediates and their chemical identification, and much of this work has been qualitative in form. Strategies for quantitative analysis are discussed in Section 4.2.

2.1 Artificial Chromophoric Substrates, Coenzymes, and Cofactors

Workers engaged in bioorganic mechanistic studies and enzyme mechanism work have long appreciated the utility of UV-visible spectroscopy for the study of biological systems. Aromatic and/or conjugated chromophores generally exhibit large extinction coefficients and long wavelength electronic transitions with bandwidths, band heights, and energies that are sensitive to the immediate chemical surroundings. Photon detectors (phototubes, diode arrays, and various other solid-state devices) have fast response times and high sensitivity, therefore, detection of UV-visible changes generally meets the demands for sensitivity and time resolution. Unfortunately, the majority of proteins do not exhibit a useful UV-visible signature of biological function. This limitation has been circumvented in a wide variety of systems by substitution of an artificial chromophoric analogue for a natural substrate, coenzyme, or cofactor. This section presents a variety of examples, which illustrate how artificial chromophores have been used successfully with RSSF technology to introduce a UV-visible probe that signals the essential chemical bonding interactions that occur in a biological process. Most of the available examples are from the enzymology literature. The two examples involving the insulin hexamer have been included to illustrate how different aspects of the dynamic behavior of a nonenzymic metalloprotein system can be probed by artificial chromophoric probes and RSSF spectroscopy.

2.1.1. α-CHYMOTRYPSIN (α-CT) CATALYSIS

The Acyl-E hypothesis is well-accepted dogma for α-Ct and other serine proteases. However, the chemical bonding events, which bring about substrate activation, have continued to be a lively topic of discussion (10-12). Bernhard and Lau (13) and others (14-17) have investigated the transient time course under single turnover conditions for the reaction of α-Ct with N-furylacryloyl L-tryptophan methyl ester (FATME) and the ethyl ester. This artificial substrate analogue has a chromophoric moiety (the N-Furylacryloyl group) attached via an amide linkage (Equation 1) to the α-amino N of L-Trp methyl ester. The sissile bond in this specific substrate is the ethyl ester linkage (Equation 1). Neverthe-

less, during a single turnover the spectrum of the chromophoric group shifts first to the red as covalent bonding interactions at the estercarbonyl occur (13) (Bernhard and Lau, 1971). Deacylation and release of the carboxylate ion product is accompanied by a shift back to a spectrum that is similar to that of the substrate. While it has been made obvious from detailed, single-wavelength, stopped-flow rate measurements that the spectrum of the chromophore is red-shifted as covalent bonding occurs to form the first tetrahedral intermediate and/or the acyl-enzyme (13-15), the precise nature of the spectral shift has not been well described.

The RSSF data presented in Figure 1A compare the time-resolved spectral changes in the N-Furylacryloyl chromophore that occur under single turnover conditions. These spectra (from which the enzyme spectrum has been subtracted) are characterized by a transient shift to longer wavelengths (spectra 1-4) followed by a shift to shorter wavelengths (spectra 5 and 6).

The dye, proflavin (Str 1) (λ_{max} H_2O = 444 nm), binds rapidly and reversibly to the substrate specificity pocket of α-Ct (13, 16, 18, 19).

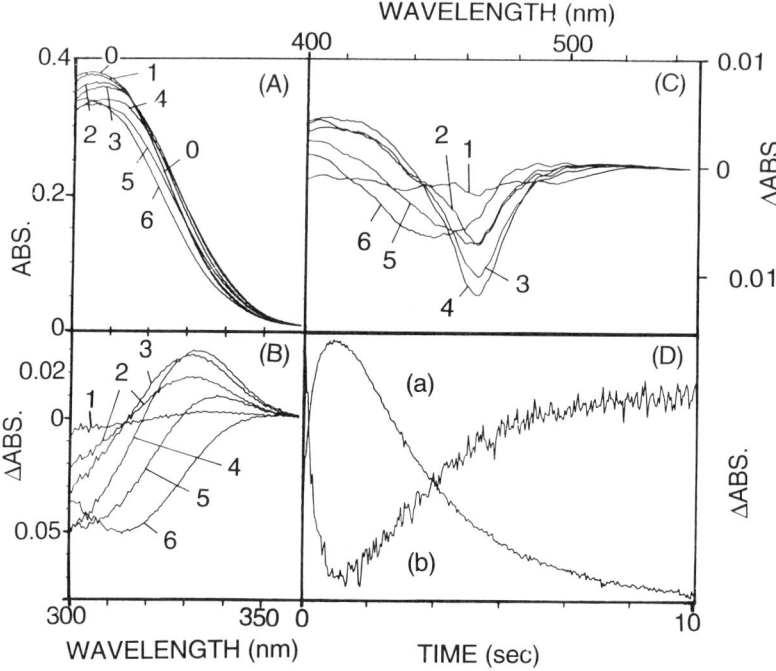

Fig. 1. Rapid-scanning stopped-flow (RSSF) study of the reaction of N-furylacryloyl L-tryptophan methyl ester (FATME) with α-chymotrypsin (α-Ct) at pH 5.0 in the absence and presence of proflavin. (A) RSSF difference spectra for the reaction of 19 μM α-Ct with 7.5 μM FATME in 0.1 M pH 5.0 sodium acetate buffer at 25°. Spectrum 0 is 7.5 μM free FATME, spectra 1–5 are difference spectra measured during reaction wherein the spectrum of α-Ct has been subtracted from the set. Spectrum 6 is the spectrum of the hydrolysis product furylacryloyl L-tryptophan with the spectrum of α-Ct removed by subtraction. Spectra were measured at the following time intervals after flow had stopped: (1) 8.54, (2) 162.3, (3) 341.6 (4) 1409.1 and (5) 3074.4 ms. Spectrum 6, t = ∞.

Proflavin

Upon binding, the spectrum of proflavin is red-shifted by 14 nm (18); consequently, proflavin is a convenient chromophoric indicator that can be used to monitor changes in the concentration of *free* enzyme resulting from the binding and reaction of α-Ct with substrate. Figures 1B and C compare the time-resolved difference spectra for the FATME spectral changes measured in the absence of proflavin with the difference spectra that result from displacement of proflavin by binding and reaction of FATME.

These RSSF data are fully consistent with a reaction mechanism (Equation 2) in which covaent bonding occurs to form covalent species along the reaction path (presumably the first tetrahedral intermediate, T_1, and/or the acyl enzyme intermediate, acyl-E, with release of methanol, P_1), followed by deacylation and formation of the carboxylate ion product (P_2).

$$E + FATME \underset{k_{-1}}{\overset{k_1}{\rightleftharpoons}} E(FATME) \rightleftharpoons E\text{-}T_1 \rightleftharpoons Acyl\text{-}E \rightleftharpoons \cdots \rightleftharpoons E + P_2 \quad (2)$$
$$+ P_1$$

Under the conditions of the experiments shown in Figure 1, the amount of the Michaelis complex, E(FATME), formed and/or the absorption changes for this process are too small to give a detectable signal (13). Thus, in Figure 1A, spectrum 0 is the spectrum of free FATME before reaction, the red-shifted spectrum (spectrum 4) is predominantly that of the covalent species (T_1 and/or Acyl-E), and spectrum 6 is the spectrum of the final carboxylate product. These assignments are supported by the accompanying changes in the amount of bound proflavin (Fig. 1B). The time-resolved difference spectra in B and C compare the FATME spectral changes (300–400 nm) with the proflavin spectral changes (400–500 nm) in an experiment where the E(proflavin) complex is mixed with FATME. The pattern of FATME spectral changes (data not shown) are the same as in A. The proflavin changes indicate a rapid displacement of the dye occurs, which corresponds to the red-shifting of the FATME spectrum. The rate of this process (Fig. 1D) is identical (within experimental error) to the rate at which the FATME spectrum is red-shifted and therefore corresponds to the accumulation of covalently bound enzyme-substrate species (Fig. 1C, spectrum 4). Finally, as the red-shifted spectrum of the N-Furylacryloyl intermediate is converted to the spectrum of the free carboxylate ion product (P_2), the proflavin spectrum is red-shifted as proflavin rebinds to the regenerated enzyme (compare Figs. 1B, C, and D).

2.2.1 HORSE LIVER ALCOHOL DEHYDROGENASE (LADH) CATALYSIS

The catalytic mechanisms of NAD(P)/NADH(P) requiring dehydrogenases, such as horse LADH, have been extensively studied in large part because the 1,4-dihydronicotinamide ring of the reduced coenzyme provides a chromophoric signal that is directly coupled to the essential chemical process, hydride transfer, catalyzed by this class of oxo-reductases. Although much can be learned from the spectral changes that accompany the binding and reaction of the reduced coenzyme chromophore, work using chromophoric artificial substrate analogues and cofactors has added greatly to the current level of understanding of the LADH catalytic mechanism, and RSSF spectroscopy has played a significant role in bringing this about.

Because LADH exhibits a very broad specificity for substrates, it has been possible to substitute a wide variety of chromophoric aldehydes/alcohols for the presumed natural substrates (acetaldehyde and ethanol) (4, 5, 8, 20). These analogues have revealed the nature of the bonding interactions between the reacting substrate, the coenzyme and the active site metal ion, and they have provided new insight about the nature of the electrostatic fields within the active site (20–23). LADH is a Zn(II)-requiring enzyme, and the active site Zn(II) is directly involved in catalysis. Substitution of chromophoric transition metal ions, such as Co(II), Ni(II), or Cu(II), for the spectroscopically silent Zn(II) ion gives catalytically functional enzyme derivatives and introduces sensitive UV-visible probes of the bonding interactions between the active site metal ion and the reacting substrate (3, 24–26). Because these systems include complex overlapping chromophoric signals derived from the coenzyme, the active site metal ion and/or the substrate, use of RSSF spectroscopy to obtain time-resolved spectra has been essential for understanding the bonding interactions between substrate, coenzyme, and metal ion during catalysis.

Early work (5, 27) established that the active site zinc ion plays a Lewis acid catalytic role in activating aldehyde substrates for reduction via the transfer of hydride ion (or its equivalent of a hydrogen nucleus and two electrons). Using RSSF UV-visible spectroscopy and the chromophoric aldehyde, 4-*trans*-N,N-dimethylaminocinnamaldehyde (DACA), Dahl and Dunn (1984a, b) were able to show that the electrostatic fields at the active site of LADH are strongly modulated by the interconversion of NADH and NAD$^+$. The reaction of DACA (λ_{max} 398 nm) with the E(NADH) complex gives a rapidly formed chromophoric intermediate wherein the DACA spectrum is red-shifted to 464 nm (a shift of 66 nm) via innersphere coordination to the active site zinc ion (5, 25, 27–31). The RSSF studies of Dahl and Dunn (22), viz. Fig. 2, show that the reaction of DACA with the E(NAD$^+$) complex gives a rapidly formed species with λ_{max} = 495 nm. This 97-nm shift arises from the combined effects of innersphere coordination to the active site zinc ion and the positive charge of the nicotinamide ring of NAD$^+$. This species is slowly converted to a mixture of the E(NADH,DACA) complex (λ_{max} 464 nm) and the E(NAD$^+$,DACA) complex (λ_{max} 495 nm). The NADH complex was concluded to arise from traces of an alcohol impurity, which, in the presence of excess NAD$^+$ and enzyme, is oxidized. In the LADH system, trace amounts of alcohol contaminants are a well-known nuisance. In the study of Dahl and Dunn (22), acquisition of time-resolved spectra proved to be an important step in determining the true spectrum of the E(NAD$^+$,DACA) complex. Prior studies (32) failed to detect the 495-nm (red-shifted) spectrum of the E(NAD$^+$,DACA) complex probably because these workers lacked the time resolution capabilities necessary to resolve the spectra of the two ternary complexes. Dahl and Dunn (23) extended these studies to the investigation of LADH carboxymethylated at Cys-46 (CME). This Cys residue is one of the active site metal ion ligands. Carboxymethylation places the negatively charged carboxymethyl carboxylate in close proximity to the reacting atoms of the substrate, the active site zinc and the 4-position of the nicotinamide ring. The consequence is an altered electrostatic field which perturbs the spectrum of innersphere-coordinated DACA in both the NADH and NAD$^+$ complexes; the CME(NADH,

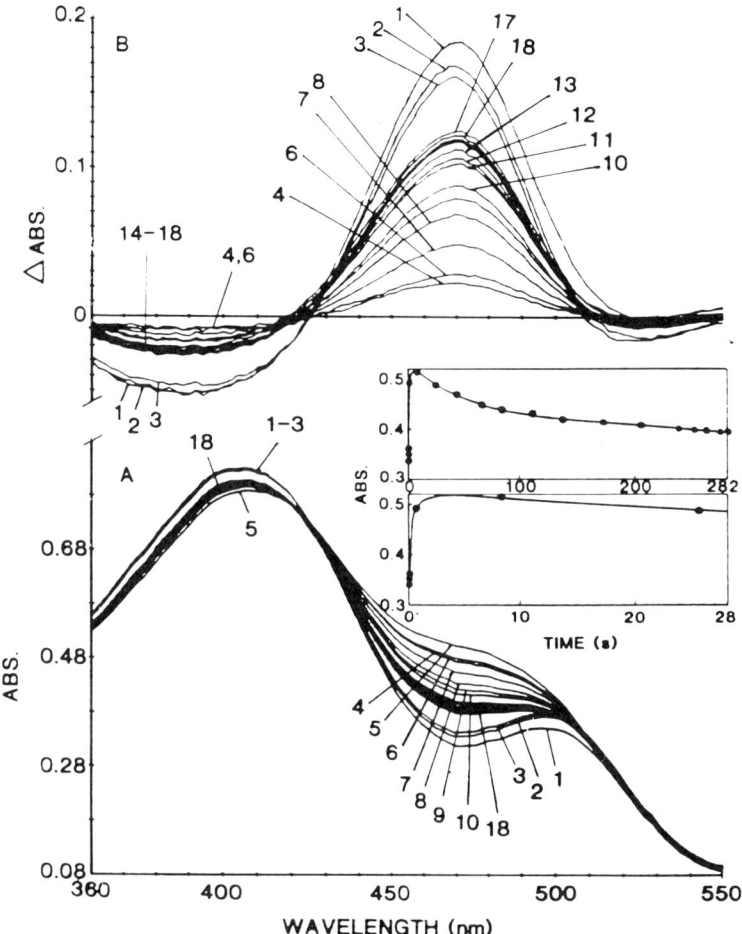

Fig. 2. Rapid-scanning stopped-flow spectra (A), difference spectra (B), and the single-wavelength time course at 464 nm (inset to panel A) for the reaction of the E(NAD$^+$) complex with DACA. (A) The scan acquisition time was 8.605 ms/scan with delays introduced between scans to give the pattern of scans indicated in the inset. (Note that two time scales are shown, 0-28 s and 0-282 s.) Collection of the first scan was initiated approximately 10 ms after flow stopped. The conditions after mixing were the following: syringe A, 30 µN LADH and 1.0 mM NAD$^+$; syringe B, 20 µM DACA. Both syringes contained 0.05 M TES/Na$^+$, pH 7.0, 0.1 mg/mL LDH, and 1.0 mM pyruvate. (B) The family of time-resolved difference spectra were calculated from the data in (A) by subtracting spectra 1-4 and 6-18 from spectrum 4 and are numbered in chronological order. [Taken from Dahl and Dunn (22) with permission.]

DACA) complex, with λ_{max} 436 nm, is red-shifted by only 38 nm, while the CME(NAD$^+$,DACA) complex, with λ_{max} 458 nm, is red-shifted by only 60 nm. Together, the RSSF and static UV-visible studies of Dahl and Dunn (22, 23) show that the active site of the E(NADH) complex is highly electrophilic, and conversion to the E(NAD$^+$) complex considerably enhances this electrophilicity as a consequence of the additional positive charge carried by the nicotinamide ring of NAD$^+$. The electrophilic interaction arising from the Coulombic charge–charge and charge–dipole interactions between coenzyme, metal ion, and substrate constitute an electronic strain-distortion effect, which distorts the aldehyde carbonyl toward the structure of the activated complex for hydride transfer. In the direction of alcohol oxidation, these positive electrostatic fields combine to increase the concentration of the reactive species, alkoxide ion, by lowering the pK_a of the innersphere coordinated alcohol. The negative charge of the carboxymethyl carboxylate of the CME decreases the electrophilic character of the site and thereby decreases the magnitude of the red-shifts of the DACA complexes. Thus, it appears that the vectoral force fields at the site provide electrostatic interactions with the bound substrate species, which specifically stabilize the electronic configuration of the activated complex (20, 21, 23).

Substitution of the active site Zn(II) of LADH by Co(II), Ni(II), Cu(II), or Fe(II) introduces spectroscopic probes, which are sensitive to the coordination geometry, the electrostatic environment, and the protein conformation. UV-visible absorbance signatures consisting of d→d and ligand-to-metal charge transfer (LMCT) transitions have proven to be useful probes of the site environment (3, 24–26, 29, 34, 35). The Co(II) and Ni(II) derivatives exhibit catalytic parameters (i.e., hydride transfer rates, K_m and k_{cat} values) that are similar to those of the native Zn(II)-enzyme (25). The Cu(II) and Fe(II) enzymes show reduced reactivities and catalytic parameters, which indicate these derivatives are of limited use as analogues of the native enzyme (24).

Because catalysis proceeds via a mechanism which involves the obligatory innersphere coordination of substrate (5), the d→d, and LMCT electronic transitions of the tetrahedrally coordinated Co(II)- and Ni(II)-enzymes provide UV-visible absorbancies that are sensitive to the ligand substitution and covalent chemical bonding changes that occur during catalysis (3). The binding of coenzyme has been shown to trigger a change in the structure of LADH from an "open" conformation to a "closed" conformation (36). The d→d and LMCT transitions of the Co(II)- and Ni(II)-enzymes also appear to be sensitive to this conformational transition (26).

The RSSF data presented in Fig. 3 show the spectral changes which occur when the Co(II)-E(NADH) complex reacts with p-nitrobenzaldehyde (3). The time-resolved spectra presented in Fig. 3A show the changes that take place in the 300–450-nm region. These changes consist of the oxidation of enzyme-bound NADH and alterations in the Co(II) LMCT bands, which accompany the oxidation of NADH, the reduction of aldehyde, and the formation and decay of inner-sphere-coordinated alkoxide ion. The complete set of spectra (a) are also presented as subsets (b) and (c), which show the changes that occur, respectively,

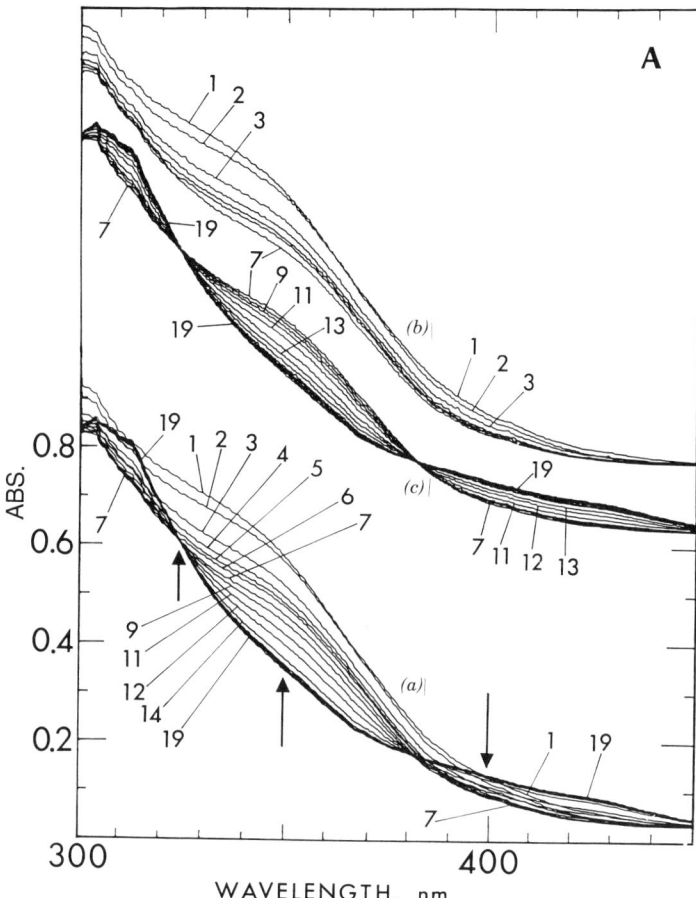

Fig. 3. Time-resolved UV-visible spectra for the reaction of the Co(II)E(NADH) complex with p-nitrobenzaldehyde (NBZA). Panel A shows the wavelength region 300–450 nm. Panel B shows the region 440–740 nm. To aid in visualizing the complex spectral changes, part (a) of each panel presents the combined set of spectra for both phases of the reaction; parts (b) and (c) present the subsets of spectra for the fast phase (spectra 1–7) and for the slow phase (spectra 7–19) offset from each other. For clarity, the 9th, 10th, 13th, and 15th spectra have been omitted from part (a) of panel B. Reaction is limited to a single turnover by the inclusion of the potent inhibitor pyrazole in the reaction mixture. The first scan in each set was initiated approximately 5 ms after flow had stopped in the rapid-mixing, stopped-flow apparatus. The repetitive scan rate was 16.48 ms/scan with additional delays introduced at longer times to space the 19 acquired scans over the 47.5 s time course of the biphasic reaction. Conditions after mixing: (syringe 1), [Co(II)E] = 27.5 μN (79% active site substitution); (syringe 2), [NADH] = 46.8 μM, [NBZA] = 85.8 μM, [pyr] = 19 mM, 0.1 M sodium pyrophosphate-50 mM TES buffer containing 38 mM NaCl, final pH 8.36 and 25°C. The absorbance values refer to a 2-cm light path. The data have not been smoothed. [Taken from Koerber et al. (3) with permission.]

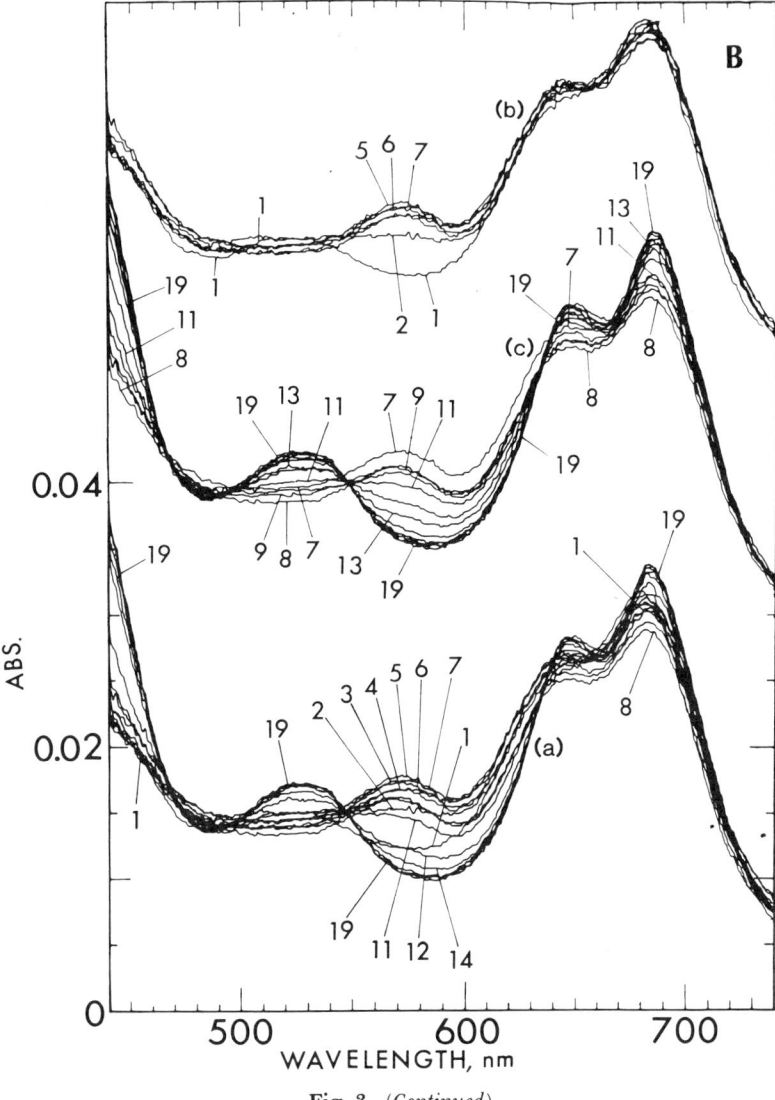

Fig. 3. (*Continued*)

during intermediate formation (b) and intermediate decay (c). Figure 3b presents the spectral changes in the 450–750-nm region due to the Co(II) d→d transitions which take place during the same reaction. Again, the complete set is shown in (a), subset (b) corresponds to intermediate formation and subset (c) corresponds to intermediate decay. The single-wavelength time courses measured at 324 (A), 350 (B), 398 (C), and 575 nm (D) are shown in Fig. 4A–D. Comparison of these RSSF and SWSF data establish that the changes in the d→d transitions occur concomitant with the changes in the spectrum of NADH and the changes in the LMCT transitions. Thus, the rapid oxidation of NADH that occurs during inter-

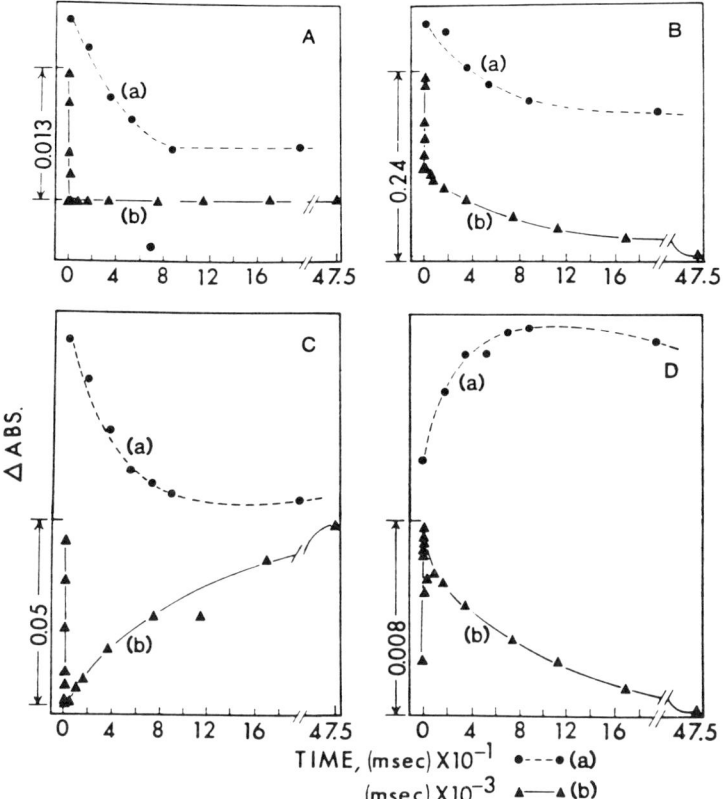

Fig. 4. Single-wavelength reaction time courses reconstructed from the rapid-scanning spectra presented in Fig. 3. The "time slices" are for (A) 324, (B) 350, (C) 398, and (D) 575 nm, respectively. Note that two time scales are shown in each panel: 0–210 ms (●), and 0–21 s (▲). The final absorbance value (▲) taken from the 19th spectrum at t = 47.5 s is shown at the extreme right margin of each panel. [Taken from Koerber et al. (3) with permission.]

mediate formation is accompanied by changes in the d→d transitions, most notably the appearance of a new band at ~575 nm [compare subset (a) of Fig. 4A with subset (a) of Fig. 4B]. As the intermediate decays, the changes in the LMCD transitions observed between 300 and 450 nm [viz. subset (b) of Figure 4A] occur concomitant with the disappearance of the 575-nm band and the appearance of the product complex [with d→d bands at 525, 650, and 685 nm, viz. subset (b) of Fig. 4B]. Because the formation of the intermediate coincides with the oxidation of NADH, the intermediate must be some form of bound alcohol. Since innersphere-coordinated anions give Co(II)-E(NAD$^+$,anion) ternary complexes with 575-nm bands similar to the transient 575-nm band of the intermediate, it was concluded that the intermediate almost certainly must be the innersphere-coordinated alkoxide ion in a ternary complex with NAD$^+$ (3, 26).

From inspection of Figs. 3a and b, it is obvious that the complexity of the spectral changes makes the analysis by RSSF essential. The RSSF data identify the temporal relationships between critically important spectral changes, and point out special wavelengths (e.g., the apparent isosbestic points located at 324, 380, 467, 495, 547, and 632 nm) for single wavelength measurements. With the nature of the spectral changes defined by RSSF spectroscopy, the quantitative analysis of the system using SWSF and relaxation kinetics becomes a much more straightforward endeavor.

A more extensive examination of the Co(II)-E by RSSF spectroscopy has shown that the spectral changes, which occur during the reaction of p-nitrobenzaldehyde with the Co(II)-E(NADH) complex (Fig. 3A, B) are characteristic of a wide variety of aldehydes (26). When reaction is studied in the direction of alcohol oxidation (26), again, RSSF studies show that the 575-nm intermediate accumulates. However, as might be anticipated, the appearance of this species precedes the appearance of NADH, a finding that is fully consistent with the assignment of the intermediate as the ternary Co(II)-E(NAD$^+$, alkoxide ion) complex wherein the alkoxide ion is directly coordinated to the active site metal ion (26).

The substitution of Ni(II) for Zn(II) at the LADH active site gives an enzyme derivative with kinetic properties nearly indistinguishable from the native enzyme (25, 37). Like the Co(II)-substituted enzyme, the d–d and LMCT transitions of the Ni(II) enzyme also provide spectroscopic signatures, which are sensitive to the bonding interactions that occur during catalysis. The RSSF spectra shown in Fig. 5 compare the time-resolved spectral changes for the reaction of benzaldehyde with the Ni(II)-E(NADH) complex under single-turnover conditions (38). These data show that upon mixing, the spectrum of the enzyme (spectrum 1) is rapidly transformed to the spectrum of a transient intermediate (spectrum 3) as NADH is oxidized. Then, in a much slower step, the intermediate decays to the spectrum of the Ni(II)-E(NAD-Pyrazole) adduct. As in the case of the Co(II)-E (viz. Fig 3A, B), these spectral changes are most simply interpreted as reflecting a mechanism in which the transient intermediate, which accumulates, is a Ni(II)-E ternary complex involving NAD$^+$ and the innersphere coordinated alkoxide ion.

A variety of chromophoric aldehydes and aldehyde analogues have been investigated as substrates for horse liver alcohol dehydrogenase. Notable among these are 4-*trans*-N,N-dimethylaminocinnamaldehyde (DACA), 3-hydroxy-4-nitrobenzaldehyde (20), and the aryl nitroso compounds p-nitroso-N,N-dimethylaniline (NDMA) and p-nitroso-N-phenyniline (NPA) (8, 9). The two nitroso compounds are intense chromophores ($\varepsilon_{max} \geqslant 3 \times 10^4$ M^{-1} cm^{-1}) that undergo novel, LADH-catalyzed, redox elimination reactions (8) wherein the nitroso functionality is reduced to the hydroxylamine via reaction with enzyme-bound NADH. Then the hydroxylamine eliminates hydroxide ion to form, in the case of NDMA, the benzoquinonediiminium ion and, in the case of NPA, the corresponding benzoquinonediimine (8). The reaction of NDMA with the E(NADH) complex was one of the first enzyme-catalyzed reactions to be examined by RSSF spectroscopy (9).

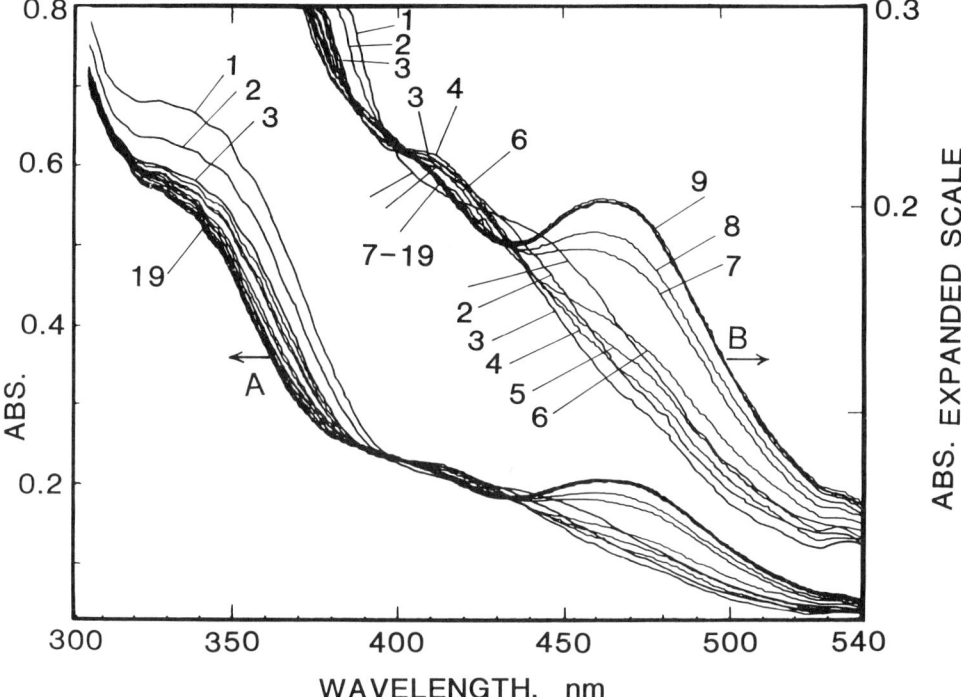

Fig. 5. Rapid-scanning, stopped-flow UV-visible kinetic study of the Ni(II)-E catalyzed reduction of benzaldehyde by NADH at 25°C (38). Concentrations after mixing: syringe 1, $[\text{Ni(II)-E}]_o = 21.8$ μN; syringe 2, $[\text{NADH}]_o = 40.6$ μM; $[\text{benzaldehyde}]_o = 0.476$ mM $[\text{KCl}]_o = 12.5$ mM; $[\text{pyrazole}]_o = 12.2$ mM; 25 mM potassium-TES, 25 mM potassium-TAPS buffer, pH 8.75. The repetitive scan rate was 8.605 ms/scan. Scans were collected at the following intervals after flow stopped: 8.6, 17.2, 25.8, 34.4, 43.0, 51.6, 103.3, 129.0, 516.3, 929.3, 1979.2, 4130.4, 6282, 9263, 11875, 13596, 17038, 19619, 24782 ms. (A) wavelength region 300–540 nm. (B) Wavelength region 380–540 nm shown on an expanded scale (right ordinate). Unpublished results of M. Gerber, M. Zeppezauer and M. F. Dunn.

Although the chemistry that occurs in the LADH-catalyzed reactions of these nitroso compounds diverges from the course of the physiological reaction, the discovery of an LADH-mediated elimination reaction has provided new insights into the Lewis acid character of the active site zinc ion. The mechanism proposed for the reactions of the nitroso compounds invokes a Lewis acid catalytic role for the active-site zinc in which coordination of the nitroso oxygen activates the nitroso nitrogen for hydride attack. The resulting zinc-coordinated hydroxylamine then undergoes a rapid elimination of hydroxide ion to form the benzoquinonediimine product.

RSSF studies of the LADH-catalyzed interconversion of 3-hydroxy-4-nitrobenzaldehyde and the corresponding alcohol have provided additional insights into the electrostatic properties of the LADH catalytic site and substrate binding cleft (20, 21, 33). The chromophoric properties of the 3-hydroxy-4-nitrobenzyl moiety make this system a useful spectroscopic probe for the study of active site environments. Ionization of the phenolic hydroxyl shifts the aldehyde spectrum from 360 to 433 nm (pK_a = 6.0), while the alcohol spectrum shifts from 350 to 417 nm (pK_a = 6.9). At pH = 8.75 and under conditions where the reaction from either direction is limited to a single turnover of enzyme sites, the transient kinetic time course is characterized by the formation and decay of an enzyme-bound chemical intermediate. The time-resolved spectra and difference spectra for the reaction of the aldehyde with the E(NADH) complex are shown in Figure 6A, B. These data establish that the intermediate has spectral properties distinctly different from either the aldehyde substrate or the alcohol product at this pH. The spectra show the presence of two apparent isoabsorbance points located at 371 and 441 nm that occur during the slow phase of the biphasic reaction. The single-wavelength insets compare the time course at 320 nm (dominated by the disappearance of NADH) with the time courses at 371 nm and 428 nm.

Careful kinetic analysis of the disappearance of aldehyde and NADH, the formation and decay of the intermediate, the binding of substrate and ^2H isotope effects established that NADH oxidation takes place at the rate of the slower relaxation, and that the events that occur in the slower relaxation actually precede those that occur in the faster relaxation. This is an uncommon kinetic circumstance where intermediate formation is slower than intermediate decay, but the rates are close enough to allow detectable amounts of intermediate to accumulate.

Intermediate formation is characterized by spectral changes (viz., Fig. 6A, B) that indicate the intermediate has a much lower extinction coefficient at 428 nm than either the substrate or product. The long wavelength spectral bands of substrate and product at pH 8.75 are due to the o-nitrophenoxide ion chromophore. Because no new red-shifted peak appears in the RSSF data set, it was concluded that the spectrum of the intermediate must be strongly blue-shifted and, therefore, the most likely intermediate is a reduced species in which the phenolic group is neutral (viz. structure 2).

2

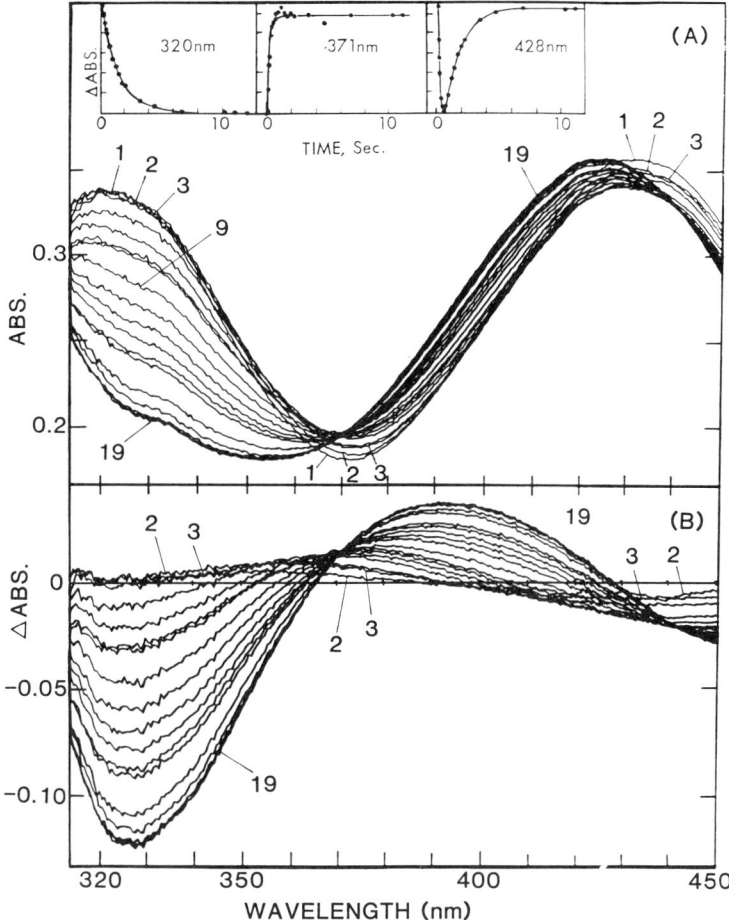

Fig. 6. Time-resolved, rapid-scanning, stopped-flow (RSSF) spectra (A) and difference spectra (B) for reaction of the liver alcohol dehydrogenase-NADH complex with 3-hydroxy-4-nitrobenzaldehyde (HOnPhCHO) under single-turnover conditions at pH 8.75 and 25 ± 1°C. The insets present the reaction time courses at (a) 320 nm, (b) at the 371-nm isoabsorbance point, and (c) at 428 nm (where substrate and product have identical extinction coefficients). Spectra were collected with a repetitive scanning rate of 8.605 ms/scan. At the longer times, delays of variable length were introduced between the scans to give the timing pattern indicated by the data points in the insets. Spectra are numbered consecutively. The difference spectra (B) were calculated by subtracting scan 1 from the other scans in (A). Conditions after mixing: [E] = 20 μN; [NADH] = 25 μM; [HOnPhCHO] = 40 μM; [pyrazole] = 20 mM; 10 mM sodium pyrophosphate buffer, pH 8.75. By inclusion of the potent inhibitor, pyrazole, reaction is limited to essentially a single turnover of enzyme sites. [Taken from MacGibbon et al. (20) with permission.]

If this interpretation is correct, then the pK_a of the phenolic hydroxyl of the intermediate must be perturbed to a higher value by at least 2 pka units (i.e., a value $\geqslant 8.75$), while the pKa of the $-CH_2OH$ group is perturbed to a value <9 in this ternary complex. The blue-shifted spectrum of the intermediate is consistent with structure 2. Studies with other substrates (5, 26, 27, 39) strongly support a catalytic mechanism in which the hydride transfer step involves the interconversion of innersphere-coordinated aldehyde and innersphere-coordinated alkoxide ion. The data of Kvassman et al. (39) and Sartorius et al. (26) indicate a pK_a for the coordinated alcohol of 6 or 7 is not unreasonable (21, 40).

At the LADH catalytic site, the pK_a values of the phenolic hydroxyl and the $-CH_2OH$ group are strongly perturbed in opposite directions. The lowering of the $-CH_2OH$ ionization constant by 9 to 10 orders of magnitude is believed to be largely a consequence of two electrostatic fields, one from the active site zinc ion and one from the positively charged nicotinamide of NAD^+ [see also the preceding discussion of the DACA complexes with E(NADH) and E(NAD^+)]. The increased pK_a of the phenolic hydroxyl is proposed to result from the interaction of this hydroxyl with a nonpolar region of the substrate binding cleft. The X-ray structures of LADH-ternary complexes indicate that when the substrate $-CH_2OH$ group is coordinated to the active site zinc, the phenolic hydroxyl would reside in a hydrophobic region of the binding site. This picture of the active site and binding site regions predicts a steep electrostatic field gradient, which changes over a few angstroms distance from an environment more polar than water at the zinc ion to a hydrocarbon-like environment for the substrate binding cleft.

2.1.3. CONFORMATIONAL TRANSITIONS IN THE INSULIN HEXAMER

Crystallographic structure determinations (41–45) have established that the subunits of the zinc-insulin hexamer can assume two quite different conformations. Solution studies (46–53) have established that ligand binding interactions can drive the interconversion of these two conformations. Kaarsholm et al. (1989) proposed that the transition is allosteric and designated the various forms as T_6, T_3R_3, and R_6. In the T_6 conformation, residues 1–8 of each insulin B-chain assume an extended conformation. The two identical zinc ions per hexamer reside in distorted octahedral ligand fields (three water molecules and three His-B10 imidazole rings) located 16 Å apart on the hexamer three-fold symmetry axis (Fig. 7A). In the R_6 conformation, residues 1–8 of each B-chain take up an α-helical conformation (Fig. 7B). In this state, the zinc ions reside in distorted tetrahedral ligand fields [one exogenous ligand and the three His-B10 imidazole rings (Fig. 7B)]. The transition from the extended chain to the α-helical conformation causes each PheB 1 residue to move by more than 30 Å, and this rearrangement of the B chain exposes a cryptic cavity in each subunit which binds phenol and structurally similar compounds [(Fig. 7b (c)] (43, 50). Consequently, these phenolic compounds are positive effectors of the T to R allosteric transition. The work of Brader et al. (50), Brader and Dunn (53), and Choi et al. (54) has established that heterotropic interactions between the phenolic pockets and the fourth coordination position of the R-state tetrahedral metal sites strongly shift this allosteric transition in favor of the R-state. Under certain conditions, the binding of anions induce crystallization of a T_3R_3 hexamer (41). It is likely that significant amounts of T_3R_3 are formed in solution under certain conditions (47).

Substitution of Co(II) for Zn(II) gives an insulin hexamer, which undergoes the T- to R-conformational transition (48). This substitution introduces a sensitive UV-visible spectroscopic probe of the allosteric process. In the T_6 state, the Co(II) centers give a broad, diffuse envelope of d-d transitions typical of an octahedral coordination geometry (λ_{max}, 490 nm; $\varepsilon \simeq 90$ M^{-1} cm^{-1}). Conversion to the R_6 state shifts the d-d envelope to the red and gives the relatively intense, narrow bands typical of tetrahedral coordination (λ_{max} 530–620 nm, depending on the exogenous anion, and $\varepsilon \simeq 500$ M^{-1} cm^{-1}). As part of an effort to investigate the mechanism of this allosteric transition, Gross and Dunn (55) carried out a series of RSSF studies to characterize the time-course for the phenol-induced transition of Co(II)-T_6 to Co(II)-R_6 in the presence of different exogenous anions. Figure 8 compares the sets of time-resolved spectra and difference spectra obtained when the exogenous anion is either phenoxide ion (A and C) or Cl^- (B and D). From inspection of the data set in A, it is obvious that a

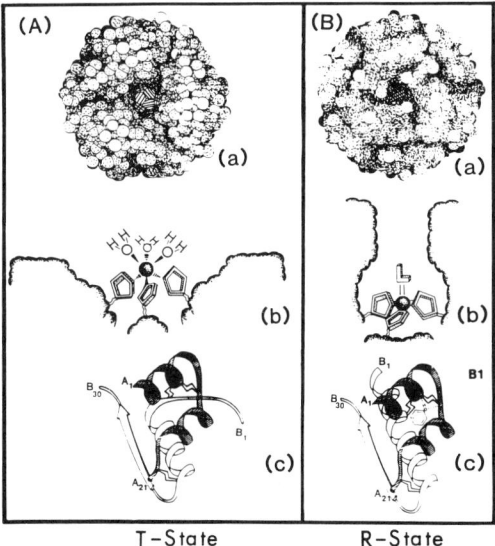

Fig. 7. (a) Space-filling models viewed down the 3-fold symmetry axes. The three water molecules coordinated to Zn(II) in the T-state are shown as striped balls. The exchangeable ligand coordinated to Zn(II) in the R-state is shown as a cross-hatched ball at the center. (b) Cartoons depicting the metal cavities and coordination geometries. (c) Cartoons depicting the extended (T-state) and helical (R-state) conformations of the B-chain residues 1–9. [Redrawn from Smith et al. (41) and Brader et al. (50) with permission.]

spectroscopically distinct transient species is formed in the reaction. In B, the method of scaled subtractions, has been used to remove the spectrum of the reactant, Co(II)-T_6, and the spectrum of the final species, Co(II)-R_6, from the spectrum of the intermediate in each trace. This decomposition reveals that the intermediate has a d–d envelope with $\lambda_{max} \simeq 560$ nm. While the transient appearance of an intermediate is not as obvious when the exogenous ligand is Cl^- (viz. Fig. 8C), the scaled subtractions in D show that an intermediate with a highly similar (perhaps identical) spectrum is formed. From these and other RSSF studies, Gross and Dunn (55) conclude that the intermediate is a Co(II)-R state species with at least one tetrahedral metal center, and the fourth coordination position is occupied by either a water molecule or hydroxide ion [ligand L in Fig. 7B (b)]. Conversion to the final Co(II)R_6 complex is a process which includes exchange of the water molecule (or hydroxide ion) for the dominant exogenous ligand from solution [e.g., phenoxide ion in Figure 8A, or Cl^- in Fig. 8(B)].

The crystalline zinc-insulin hexamer is the form in which insulin is stored in the mature secretory granules of the β-cells in the pancreas. During exocytosis, these vesicles fuse with the plasma membrane and eject the microcrystals into the intercellular space. Within a few seconds, the crystalline insulin hexamers dissolve and dissociate to the biologically active insulin monomer. Consequently, the dynamics of the dissociation of the T_6 and R_6 forms of the insulin hexamer have been a subject of some interest. The unassisted dissociation of Zn(II) from the Zn(II)-T_6 hexamer is a relatively slow process, the dissociation of the Zn(II)-R_6 complex is orders of magnitude slower (47, 56–58). In contrast to the slow rates of the unassisted dissociations, the chelator-assisted sequestering and removal of Zn(II) from the T_6 complex is a relatively rapid process (47, 56, 58). The two Zn(II) ions of the Zn(II)-T_6 hexamer are located at the bottom of shallow, dish-shaped clefts on opposite faces of the cylindrically shaped hexamer (Fig. 7). Each zinc ion is coordinated by three His residues and by three water molecules. The coordinated waters face out into the shallow cleft and appear accessible for exchange with other ligands. The studies of Dunn et al. (56), Kaarsholm and Dunn (58), and Kaarsholm et al. (47) provide strong evidence indicating that the facilitated removal of Zn(II) from these sites by suitable chelators occurs via a mechanism in which: (1) the chelator first forms a complex with the protein-bound metal ion by displacement of the three water molecules, (2) then, in a second, slower step, the chelator—metal ion mono complex dissociates from the protein, and (3) in the presence of excess chelator, a second molecule of chelator combines with the mono complex to give the bis complex.

The RSSF and SWSF data shown in Fig. 9A–D compare the time-resolved spectra and difference spectra for the reaction of 1-(2-pyridylazo)-2-naphthol (PAN, Str 3),with the Zn(II)-T_6 hexamer (59).

PAN

Structure 3

PAN forms chromophoric mono and bis complexes with Zn(II) and many other divalent metal ions. Spectrum 1 in Fig. 9A is the first spectrum measured 16 ms after mixing PAN with Zn(II)-T_6 and is dominated by free PAN. Spectrum 19 is the spectrum measured after completion of reaction and corresponds to the Zn(II) bis complex with PAN. Panels B and C are difference spectra calculated by subtraction of Spectrum 1 from every other spectrum in the set. Panel D shows the SWSF time course for intermediate formation and decay at 500 nm and the time course for formation of the *bis* PAN-Zn(II) product complex at 560 nm. The wavelength and absorbance scales in C are expanded to show details of the difference spectra in the vicinity of 500 nm. These details show that there is not a true isosbestic point, but rather, two apparent isoabsorbance points, one at ~494 nm and one at ~506 nm. Thus, reaction appears to involve formation and decay of an intermediate. This conclusion is reinforced by the single wavelength time courses measured at 500 nm and 560 nm (Panel D). Both time-courses are biphasic, consistent with the formation and decay of an intermediate. Since the spectrum of the intermediate must be similar to that of the product, the simplest interpretation is that the intermediate is a ternary complex formed between PAN and the protein-bound zinc ion. Therefore, it appears that the chelator-assisted dissociation of the Zn(II)-T_6 hexamer proceeds via the rapid formation of an intermediate in which PAN replaces the coordinated water molecules. This sequestering of the metal ion by the chelator weakens the coordination to the HisB10 residues, thereby facilitating the loss of zinc. Since zinc coordination to the HisB10 residues stabilizes the hexamer, the sequestering and removal of zinc triggers the dissociation of the hexamer to monomer and dimer (60, 61).

2.1.4. SHEEP LIVER ALDEHYDE DEHYDROGENASE CATALYSIS

The NAD^+-requiring aldehyde dehydrogenases catalyze the quasi-irreversible oxidation of aldehyde to the corresponding carboxylates. Like the liver alcohol

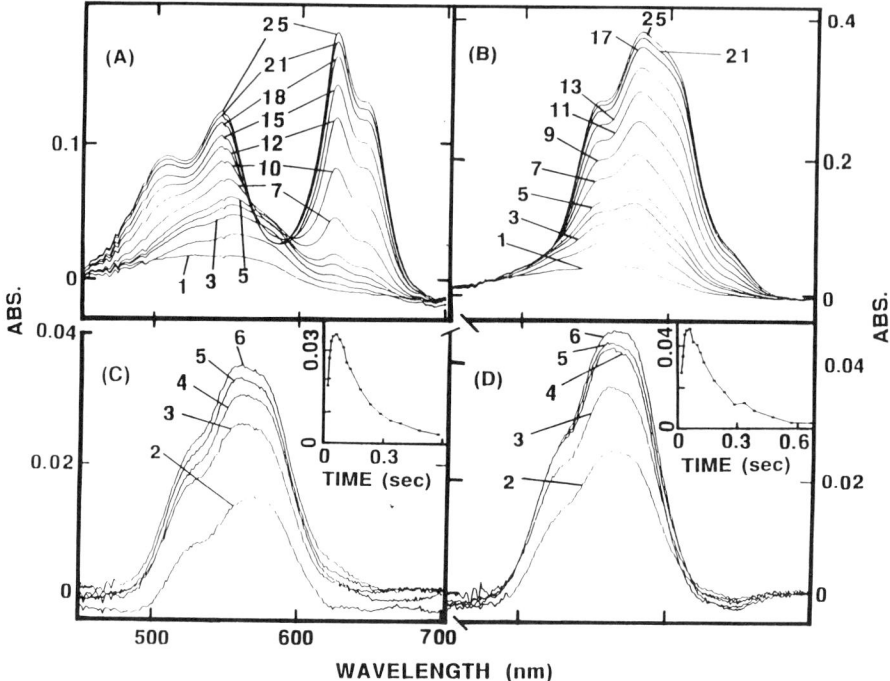

Fig. 8. The rapid-scanning spectroscopic time courses for the reaction of 0.17 mM Co(II)-T_6 with 100 mM phenol at pH 8.0 in the absence (A, C) and presence (B, D) of 100 mM chloride ion are shown. (A) Reaction in the absence of chloride ion. The time interval between scans is 8.54 ms for the first five spectra, followed by spectra at successively longer intervals afterward (see insets in C and D). The total acquisition time was 1.71 s for the 25 spectra collected; only spectra numbers 1–5, 7, 10, 12, 15, 18, 21, and 25 are shown. (B) Reaction in the presence of 100 mM chloride ion. The timing sequence of the spectra is the same as that used in (A). For clarity, spectra 6, 8, 10, 12, 14–16, 18–20, and 22–24 have been omitted. (C) The scaled, subtracted spectra, calculated from the second to the sixth spectrum of part A, correspond to the time-course for intermediate formation. The time course plotted in the inset shows the absorbance change at 560 nm for the complete set of scaled, subtracted spectra as a function of time. (D) Scaled, subtraction spectra numbers 2 to 6, as in part C, for the data part B, with chloride ion present. The inset plot also shows the time course at 560 nm obtained from the complete set. [Taken from Gross and Dunn (55) with permission.]

dehydrogenases, the liver aldehyde dehydrogenase from sheep is characterized by a fairly high coenzyme specificity and a very broad substrate specificity. Consequently, the sheep liver enzyme catalyzes the oxidation of simple aliphatic and aromatic aldehydes with similar efficacy. This happenstance has made it possible to investigate the catalytic mechanism of the enzyme using RSSF spectroscopy by comparing the spectral changes that take place in the spectrum of the coenzyme with the spectral changes that accompany the oxidation of chromophoric aldehydes.

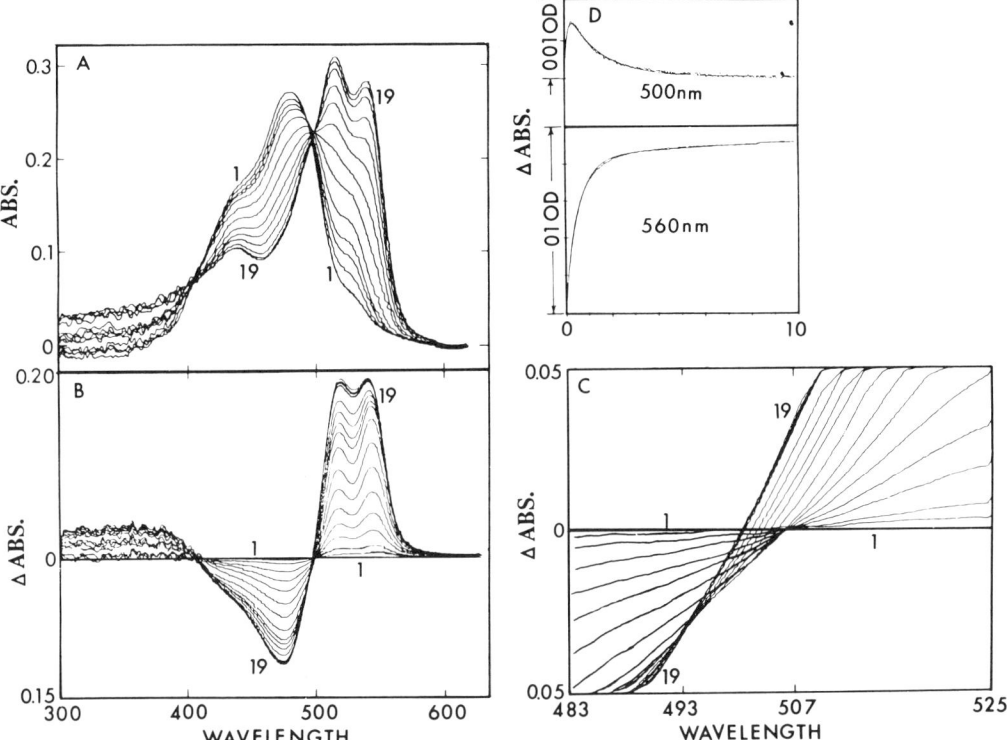

Fig. 9. Time-resolved RSSF spectra (A), difference spectra (B, C) and single-wavelength time courses (D) for the reaction of 1-(2-pyridylazo)-2-naphthol (PAN) with the Zn(II)-T_6 insulin hexamer (59). The expanded difference spectra shown in (C) and the single-wavelength time course measured at 500 nm in (D) establish that reaction occurs via the formation and decay of an intermediate. Concentrations after mixing in (A-C): [Zn(II)-T_6] = 2.5 μM, [PAN] = 15 μM. In (D), [Zn(II)-T_6] = 2.5 μM, [PAN] = 15 μM. A 50 mM pH 8.0 Tris-HCl buffer at 25°C was used in both experiments. Unpublished work of G. G. Gould and M. F. Dunn.

The early SWSF studies of this enzyme indicated that the reaction occurs via a chemical intermediate. Due to the apparent similarity of the chemistry catalyzed by D-glyceraldehyde-3-phosphate dehydrogenase (GPDH) and by aldehyde dehydrogenase, it was speculated that the chemical intermediate formed in this reaction might be an acyl-enzyme (viz., Equation 3) analogous to the GPDH acylenzyme intermediate

$$E(NAD^+) + RCHO \rightleftharpoons E(NAD^+,RCHO) \rightleftharpoons T_1 \rightleftharpoons \overset{H_2O}{Acyl-E(NADH)} \rightleftharpoons T_2 \rightleftharpoons \cdots \rightleftharpoons RCO_2^- + H^+ + E + NADH \quad (3)$$

(62). In Equation 3, T_1 is the tetrahedral intermediate that undergoes oxidation by enzyme-bound NAD^+, while T_2 is a tetrahedral adduct with water (or hydroxide ion). However, identification of the intermediate as an acyl-enzyme species was only achieved via use of chromophoric subtrates, a chromophoric NAD$^+$ analogue (ThioNAD$^+$), and RSSF spectroscopy (63, 64).

Buckley and Dunn (63) used RSSF spectroscopy to investigate the reactions of several chromophoric aldehydes with the E(NAD$^+$) complex. Figure 10 presents the time-resolved spectral changes for the reaction of 4-*trans*-N,N-dimethylaminocinnamaldehyde (DACA) under single turnover conditions where [E(NAD$^+$)] >> [DACA]. These data establish that a new species with λ_{max} ~463 nm forms rapidly and then decays to a mixture of enzyme, NADH and 4-*trans*-N,N-dimethylaminocinnamate ion (λ_{max} = 323 nm). Similar results were obtained with several other chromophoric aldehydes. Because the wavelength region below 370 nm is complicated by strong overlapping absorbancies from DACA, NADH, and enzyme, it was not possible to directly compare the time course for intermediate formation and decay with the time course for NAD$^+$ reduction. To circumvent this problem, NAD$^+$ was replaced with ThioNAD$^+$ so that the rate of ThioNADH (λ_{max} 395 nm) could be compared with the rate of appearance of the intermediate. As is true for most NAD$^+$-requiring dehydrogenases, ThioNAD$^+$ is a functional analogue of NAD$^+$ in the aldehyde dehydrogenase system. Close inspection of the RSSF data in Fig. 10A reveals an apparent isoabsorption point at 422.8 nm during the rapid phase for the interconversion of DACA and the intermediate when NAD$^+$ is the coenzyme. By substituting ThioNAD$^+$ for NAD$^+$ and monitoring the absorbance changes at 422.8 nm, a wavelength where ThioNADH appearance can be measured without complications from other chromophores in the system, Buckley and Dunn (63) were able to show that, within the limits of experimental error, the rate of appearance of ThioNADH and the rate of appearance of the intermediate are identical (Fig. 10B). Therefore, the intermediate almost certainly must be an oxidized derivative of DACA, presumably an Acyl-enzyme species (both tetrahedral species shown in Equation 3 would give blue-shifted spectra). The red-shifted spectrum of the intermediate must reflect a fully conjugated π-system with sp^2 centers at the carbonyl of the chromophore. It is noteworthy that the position of the λ_{max} at 463 nm is significantly red-shifted in comparison to the spectrum of any reasonable model acyl compound yet synthesized (64). The list of model acyl compounds tried includes the oxy and thioesters, the amide and the N-acyl imidazole derived from 4-*trans*-N,N-dimethylamino cinnamic acid. While it is likely that the Acyl-enzyme is a thioester, this has not yet been firmly established. Like the serine proteases (65) and GPDH (62, 66), Dunn and Buckley (64) conclude that the Acyl-enzyme derived from DACA resides in a special microenvironment provided by the enzyme catalytic site. Interaction of a strong electrophile ($y^{\delta+}$) with the carbonyl oxygen of the 4-*trans*-N,N-dimethylaminocinnamyl moiety would give a red-shifted spectrum to the Acyl-enzyme (structure 4). Such an interaction would activate the acyl group for deacylation.

[Structure 4 diagram]

Structure 4

2.2. Systems with Intrinsic Chromophores

2.2.1. RSSF INVESTIGATIONS OF PYRIDOXYL PHOSPHATE ENZYMES

Pyridoxal 5′-phosphate dependent enzymes constitute an important class of proteins involved predominately in amino acid metabolism. The PLP-cofactor is capable of catalyzing a variety of reactions at the α-, β-, and/or γ-carbons of amino acid substrates. These reactions include tranamination, racemization, decarboxylation, and aldoyltic cleavage reactions at the α-carbon and elimination/substitution reactions at either the β-, or γ-position of the amino acid substrate (67–74) The chemical properties of the cofactor (67–71) are responsible for the great diversity of reactions catalyzed by PLP, while reaction specificity is ultimately determined by the active site environment imposed by the surrounding apo-protein to which the cofactor is covalently bound (69).

Such catalytic versatility requires that PLP must serve the dual purpose of generating and stabilizing both carbanionic and electrophilic intermediates. This is accomplished by establishing a highly conjugated π-system between the cofactor and amino acid substrate. Through a series of chemical transformations and/or isomerizations, this π-system may be extended to the α-, β-, and even the γ-carbons of the substrate (68). PLP-dependent catalysis necessitates the need for multi-intermediate reaction mechanisms in which reactive species differ both in structure and in the extent of the conjugated π-system. Thus, the intrinsic chromophoric properties of the PLP-cofactor make both UV-visible spectroscopy and rapid-kinetic techniques particularly useful for the direct detection and characterization of catalytic intermediates, information that is crucial in the investigation of mechanism and structure–function relationships within this important class of enzymes.

The interpretation of UV-visible spectra is greatly facilitated by the large amount of information available in the literature regarding the electronic spectra of both PLP-model compounds in solution and from stable enzyme-substrate complexes (71, 75–80). Most PLP-species display spectral bands that absorb between 300 nm and 550 nm, wavelengths well separated from the intense protein absorption bands that arise from aromatic amino acid residues. Secondly, catalytic turnover rates, especially for enzymes catalyzing either β- or γ-elimination/replacement reactions, are generally in a range suitable for study by rapid mixing experiments. The combination of these factors make many PLP-dependent enzymes ideally suited for study by RSSF spectroscopy.

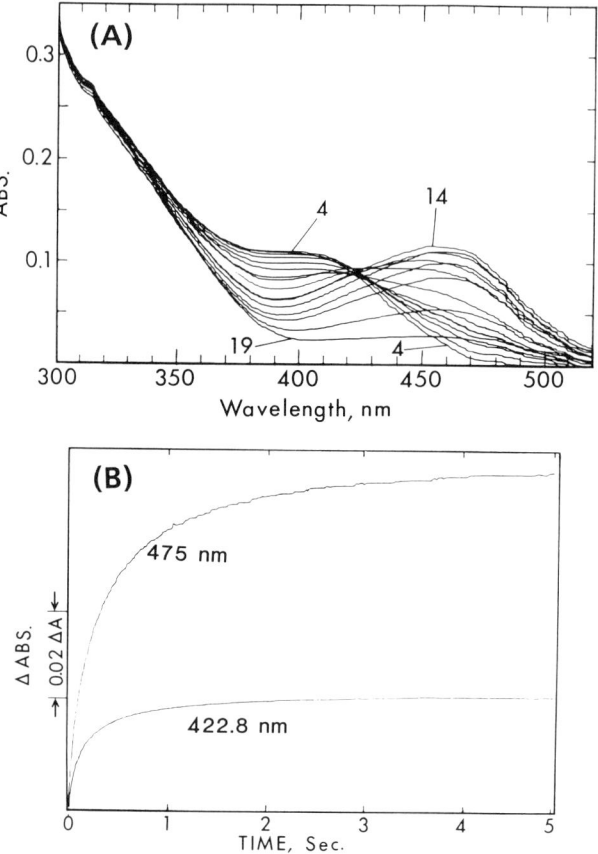

Fig. 10. A. Time-resolved UV-visible spectra for the reaction of the aldehyde dehydrogenase-NAD$^+$ complex with DACA. The first scan in each set was initiated approximately 5 msec after flow had stopped. The repetitive scan rate was 8.605 msec/scan with additional delays introduced at longer times to space the 19 scans over 81 sec in order to follow both the fast build up of intermediate at 460 nm and its subsequent slow decay. Little change occurred during the first three scans, which are omitted. Conditions before mixing: (syringe 1) [aldehyde dehydrogenase] = 42.7 μM; [NAD$^+$] = 3.18 mM; (syringe 2) [DACA] = 5.6 μM; both in 35 mM phosphate buffer (pH 7.6) at 25°C. (Taken from Buckley and Dunn (63) with permission.) B. Data obtained from the single-wavelength, stopped-flow spectrophotometer measured at 475 nm and 422.8 nm on mixing the aldehyde dehydrogenase-thio-NAD$^+$ complex with DACA. Conditions before mixing (syringe 1) [aldehyde dehydrogenase] = 64 μM; [thio-NAD$^+$] = 3.09 mM; (syringe 2) [DACA] = 8.9 μM; both in 35 mM phosphate buffer (pH 7.6) at 25°C. The best fit assuming two consecutive first-order processes: (time course at 475 nm) k_1, 5.0 s^{-1}, amplitude, 0.04 AU; k_2, 0.8 s^{-1}, amplitude, 0.04 AU; (time course at 422.8 nm) k_1, 6.4 s^{-1}, amplitude, 0.02 AU; k_2, 1.0 s^{-1}, amplitude, 0.01 AU. [Taken from Buckley and Dunn (63) with permission.]

2.2.2. TRYPTOPHAN SYNTHASE

The tryptophan synthase bienzyme complex from enteric bacteria provides an important example wherein RSSF has been used to good advantage for the study of both enzyme mechanism and protein structure–function relationships. This enzyme complex is composed of heterologous α- and β_2-subunits arranged in a nearly linear α-β-β-α array (81). The α-subunit catalyzes the aldolytic cleavage of IGP to indole and G3P, while the β-subunit catalyzes the PLP-dependent condensation of L-Ser and indole to yield L-Trp. The $\alpha\beta$-reaction is essentially the sum of the individual α- and β-reactions (scheme I). Indole, the common intermediate produced at the α-site, is directly channeled to the β-active site via a tunnel located in the interior of the protein complex which directly interconnects the α- and β-catalytic centers (81–84). Although the individual subunits may be isolated and are functional, formation of the bienzyme complex not only increases the catalytic activities of the separate subunits by nearly 100-fold, but also alters the thermodynamic stability of β-site reaction intermediates and introduces heterotropic allosteric interactions between sites.

The mechanism of the PLP-dependent β-reaction involves a number of different chemical transformations (scheme 1B). The reaction requires the formation/scission of C–C, C–O, C–N, C–H, N–H, and O–H bonds and the pathway for the synthesis of L-Trp from L-Ser and indole involves a minimum of at least eight distinct PLP-intermediates. RSSF spectroscopy allows direct detection and spectral characterization of the various catalytic intermediates, which accumulate during the course of the reaction (85, 86). Information from RSSF spectroscopic investigations is greatly enhanced by the use of both isotopically labeled substrates (85) and substrate analogs (82), which alter the accumulation of intermediates during the presteady state phase of the reaction. Direct comparison of RSSF spectra for deuterium labeled substrates with the isotopically normal compounds is a powerful tool for the identification and assignment of chromophoric reaction intermediates (85). Finally, structure–function relationships within the bienzyme complex may be addressed by careful comparison of the time-resolved RSSF spectra for reactions of native and mutant enzyme species (87–89).

2.2.3. REACTION OF TRYPTOPHAN SYNTHASE WITH L-SERINE

The PLP-dependent β-reaction may be broadly divided into two stages or half-reactions (scheme 1B); stage I involves the formation of the quasi-stable electrophilic α-aminoacrylate complex [E(A-A)] from L-Ser. In the absence of a reactive nucleophile, the E(A-A) complex slowly decomposes to yield pyruvate and ammonium ion. Stage II of the β-reaction involves the formation of a covalent C–C bond between the nucleophilic C-3 carbon of indole and the electrophilic β-carbon of the E(A-A), ultimately resulting in the synthesis of L-Trp. Drewe and Dunn (85, 86) have investigated the presteady state RSSF spectral changes that occur during both stage I and stage II of the β-reaction catalyzed by the $\alpha_2\beta_2$ complex from *E. coli*. The spectral changes for the reaction of L-Ser with

(α -REACTION)

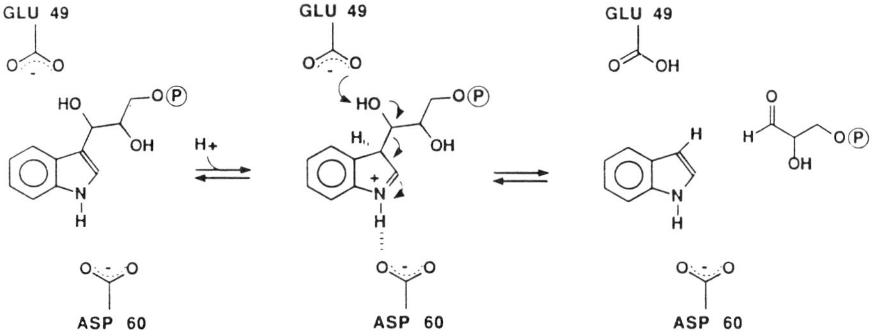

GLU 49
HO
O(P)
OH
H+
N
H
ASP 60
H
OH

(β -REACTION)

Stage I

HO
CO_2^-
$^+NH_3$
ENZ
H+
E(Ain)
410nm
E(GD)
310-340nm
E(Aex1)
422nm
E(Q1)
460nm?
-OH
E(A-A)
350nm
shoulder 380-530nm

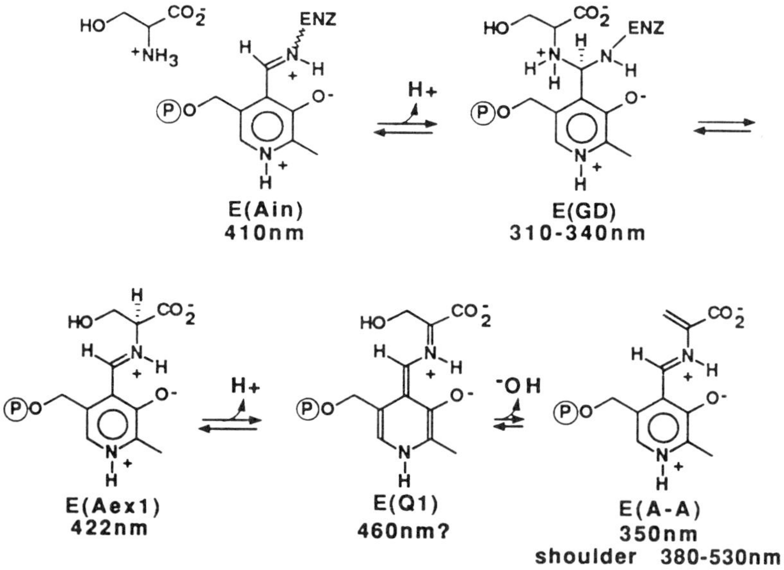

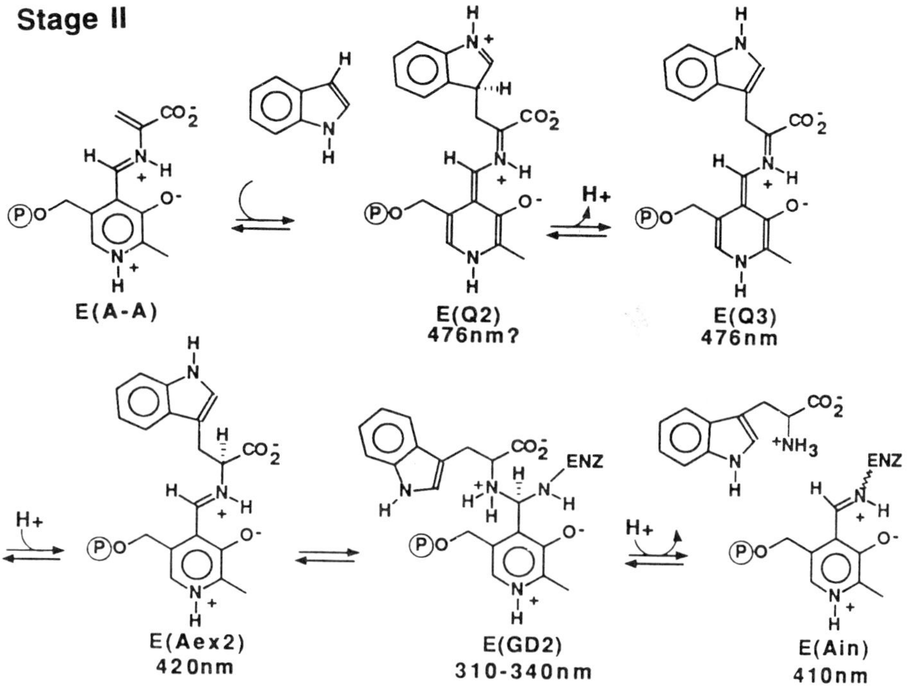

Stage II

E(A-A)   E(Q2) 476nm?   E(Q3) 476nm

E(Aex2) 420nm   E(GD2) 310-340nm   E(Ain) 410nm

Scheme I. Summary of the reactions catalyzed by the tryptophan synthase bienzyme complexes from bacteria. The α-subunit (α-reaction) catalyzes the reversible cleavage of 3-indole-D-glycerol 3′-phosphate to give indole and D-glyceraldehyde 3-phosphate. (The postulated catalytic roles of αE49 and αD60 are indicated.) The β-subunit (β-reaction) catalyzes a PLP-dependent β-replacement reaction wherein the β-hydroxyl of L-Ser is replaced by indole to form L-Trp and a water molecule. UV-visible spectral band assignments are given for the various intermediates. Note that the β-reaction is subdivided into two stages; Stage I is the synthesis of the α-aminoacrylate species, E(A-A), and Stage II is the conversion of E(A-A), via reaction with indole, to L-Trp. [Redrawn from Brzović et al. (88) with permission.]

the native enzyme are shown in Fig. 11. The spectrum of the native enzyme is characterized by an absorption band centered at 412 nm with a shoulder at shorter wavelengths (spectrum 0, Figure 11A). The PLP-cofactor is covalently linked via a Schiff's base to the ε-amino group of Lys 87 at the β-active site. When the enzyme is rapidly mixed with L-Ser, a new spectral band centered at 422 nm accumulates ($1/\tau_1$, spectrum 1). This band subsequently decays in a biphasic process ($1/\tau_2 > 1/\tau_3$) that is accompanied by increases in absorbance both in the 300- to 385-nm and 470- to 530-nm regions of the UV-visible spectrum. The final E(A-A) spectrum is characterized by an absorbance maximum at 350 nm with a broad shoulder between 385 nm and 530 nm (spectrum 18). The spectral changes observed during $1/\tau_2$ and $1/\tau_3$ are nearly identical, although the relative amplitudes for each process ($1/\tau_2 > 1/\tau_3$) are different.

The RSSF data presented in Fig. 11B show that deuterium substitution at the α-carbon of L-Ser results in the increased accumulation of the 422-nm species in the $\alpha_2\beta_2$ catalyzed reaction, a narrowing of the spectral bandwidth, and a reduction in the rate of decay of the 422-nm species to the E(A-A). These results show that removal of the α-proton is partially rate-determining in the reaction to form E(A-A), and the 422 nm species represents the accumulation of an intermediate prior to α-proton abstraction. Comparison of the spectral changes which occur during the reaction of L-Ser with the isolated β_2-dimer with the $\alpha_2\beta_2$-catalyzed reaction identifies the 422 nm species as the external aldimine, $E(Aex_1)$, formed between L-Ser and the PLP-cofactor. A species exhibiting a 420-nm spectral band, which corresponds to $E(Aex_1)$, also accumulates in the reaction of L-Ser with the β_2-subunit; this species is strikingly similar to the transient 422-nm species observed in the reaction of the $\alpha_2\beta_2$ complex with L-Ser (Fig. 12). In addition to similar λ_{max} and ε values, these two species are also highly fluorescent. However, no other subsequent intermediates are observed in the reaction catalyzed by the isolated β_2-subunit. Clearly, formation of the $\alpha_2\beta_2$ complex from the isolated subunits alters the relative stability of covalent intermediates in reactions catalyzed by tryptophan synthase.

The increased bandwidth observed in the transient accumulation of $E(A_{ex1})$ during the $\alpha_2\beta_2$-catalyzed reaction provides evidence for the presence of an additional reaction intermediate. Direct comparison of the spectrum of the transient 422-nm species with the 420-nm spectral band observed in the β_2-reaction with L-Ser shows that the new intermediate absorbs in the 460-nm region of the spectrum and likely has a very sharp spectral band shape (Fig. 12). These spectral characteristics are highly indicative of a quinonoidal species [$E(Q_1)$], formed upon removal of the α-proton. Since α-proton removal is completely rate determining in reactions catalyzed by the isolated β_2-subunit (90), this intermediate does not accumulate in the β_2-catalyzed reaction. Similar analysis of the $\alpha_2\beta_2$ reactions shows that the 460-nm species accumulates to a much lesser extent if α-^{2}H-Ser is the substrate. Thus, deuterium substitution at the α-carbon alters the distribution of intermediates formed during $1/\tau_1$. This spectral analysis provides direct evidence that $E(Q_1)$ is rapidly formed during $1/\tau_1$ of the $\alpha_2\beta_2$-catalyzed reaction and is likely in rapid-equilibrium with the $E(A_{ex1})$ intermediate.

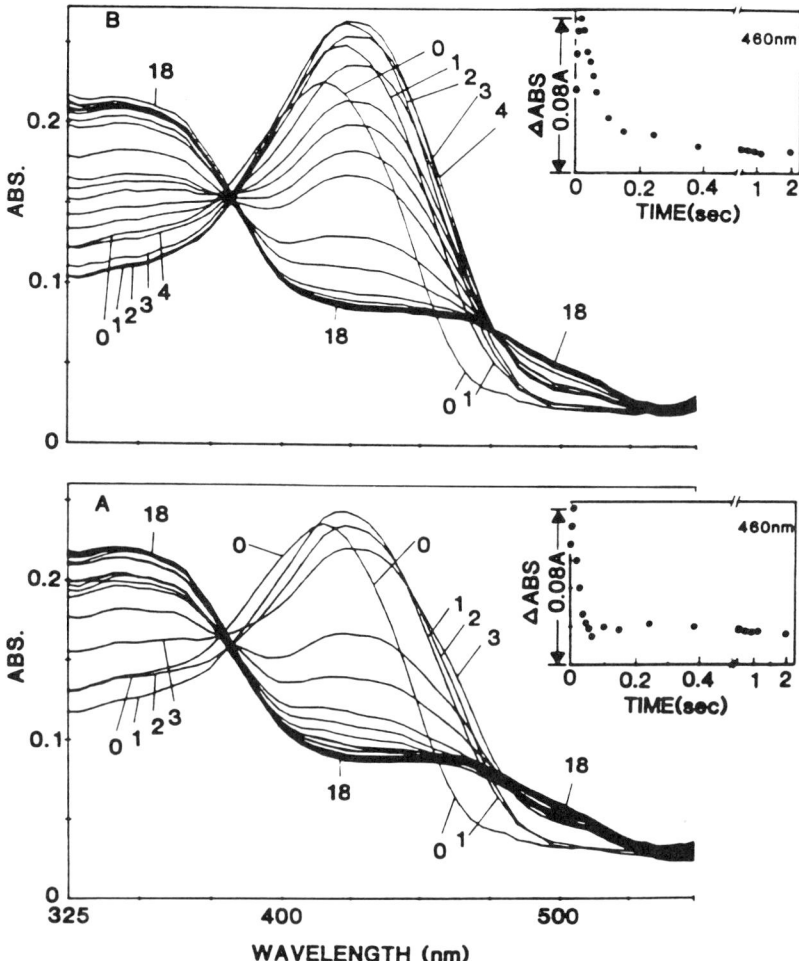

Fig. 11. Rapid-scanning, stopped-flow data comparing the accumulation of transient intermediates during the reaction of 8 mM DL-[α-^{1}H]-serine (A) and 8 mM DL-[α-^{2}H]-serine (B) with 13.3 μM $α_2β_2$ from *E. coli*. The trace designated 0 is the reconstructed spectrum of the reactants before mixing. The insets to both (A) and (B) are 460-nm reaction time courses reconstructed from the RSSF data. Each experiment was conducted using 0.1 M potassium phosphate and 1 mM EDTA buffer at pH 7.80 and 25°C. All conditions refer to concentrations immediately after mixing. The initiation of scanning in both (A) and (B) occurred 2 ms after flow stopped. Scans 2 through 19 were collected at 4.7, 9.3, 14.0, 18.7, 23.4, 28.0, 32.7, 42.0, 51.4, 60.7, 70.1, 107, 154, 247, 387, 761, 1135, and 1980 ms after the first scan, respectively, with a repetitive scan rate of 4.7 ms/scan. [Taken from Drewe and Dunn (85) with permission.]

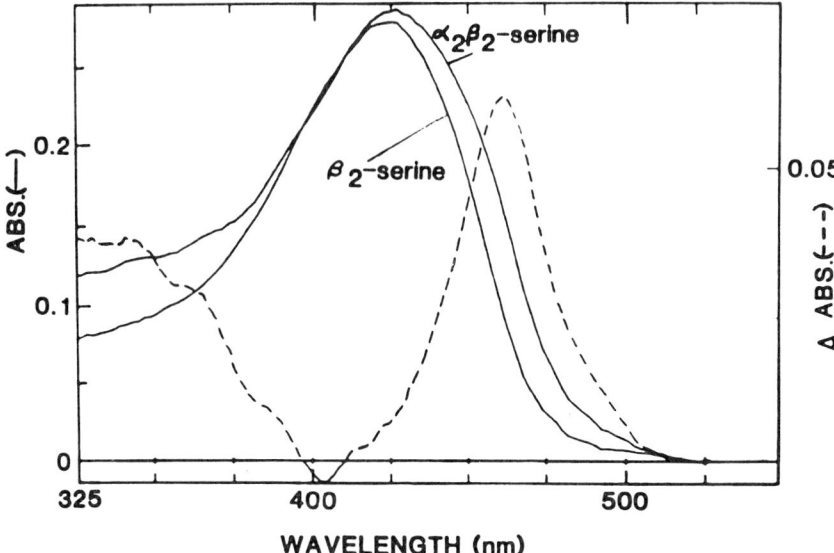

Fig. 12. Normalized spectra (—) and difference spectrum (--) compmaring the 425 nm $\alpha_2\beta_2$ transient species with the quasi-stable 420 nm β_2 species formed in the L-serine reactions. The spectrum labeled β_2-serine is the final spectrum at 1970 ms derived from RSSF data set for the reaction of the isolated β_2-subunit with L-serine (data not shown). The spectrum labeled $\alpha_2\beta_2$-serine is the first scan derived from a RSSF data set for the reaction of 40 mM L-serine with the $\alpha_2\beta_2$-bienzyme complex. These spectra were baseline zeroed and normalized by adjusting each spectrum to the same amplitude. The difference spectrum was generated by subtracting the normalized β_2-serine final spectrum from the normalized $\alpha_2\beta_2$-serine spectrum. [Taken from Drewe and Dunn (85) with permission.]

Because quinonoidal species generally exhibit large extinction coefficients [20,000–50,000 $M^{-1}cm^{-1}$; (80, 91)] the amount of $E(Q_1)$ detected represents only a small fraction of the total enzyme active sites.

O-acetyl-L-serine sulfhydrylase (OASS) is another PLP-dependent enzyme which catalyzes the synthesis of L-Cys via a β-replacement reaction in which the O-acetyl group of O-acetyl-L-Ser is replaced by HS^-. This enzyme also reacts with cognate substrate to form an E(A-A) intermediate. The OASS E(A-A) spectrum is characterized as a broad spectral band centered at 470 nm with another band in the 350-nm region of the spectrum (92). The spectrum assigned to the tryptophan synthase E(A-A) species is qualitatively similar. Consistent with the observed sequence of spectral changes, the increase in absorbance at 480-nm is due to the formation of the E(A-A) species. The large 350-nm band may result from a different tautomer, protonation state, or conformational isomer of the E(A-A) species. Because the binding of L-Ser to the β-subunit is reversible, the final reaction spectrum is composed of an equilibrium mixture of the various stage I intermediates.

Another interesting feature of the reaction of L-Ser with the $\alpha_2\beta_2$ complex is that the spectral bands observed during $1/\tau_2$ and $1/\tau_3$ are nearly identical throughout the entire spectral region (Fig. 11A). If $1/\tau_2$ and $1/\tau_3$ represented the formation and decay of another reaction intermediate, then the spectral changes during $1/\tau_2$ and $1/\tau_3$ would be very different. On the other hand, if formation and decay of a covalent intermediate is ruled out, then the most likely remaining explanations are (a) the existence of two slowly interconverting conformations of the bienzyme complex, which react to form E(A-A) at different rates, or (b) a conformational isomerization, which occurs after the initial formation of E(A-A) [E(A-A) $\rightleftharpoons$ E*(A-A)] (85).

2.2.4. REACTION OF THE TRYPTOPHAN SYNTHASE α-AMINOACRYLATE INTERMEDIATE WITH INDOLE

When L-Ser and indole are mixed simultaneously with the $\alpha_2\beta_2$ complex, the initial spectrum observed after mixing ($1/\tau_1$) is identical to that formed in the reaction of L-Ser alone and is subject to the same isotope effects described above when ^{2}H-L-Ser is substituted for isotopically normal L-Ser [spectrum 1, Fig. 13; (86)]. As the reaction progresses toward the steady state, the RSSF spectra show the accumulation of a new intermediate with λ_{max} at 476 nm (spectrum 18, Fig. 13). This species has also been observed in steady-state spectra of the β-reaction and in the binding of the product, L-Trp, to the β-active site of the $\alpha_2\beta_2$ complex (93). This intermediate is the L-Trp quinonoidal species, $E(Q_3)$. Rapid kinetic experiments with ^{2}H-C_3 indole as the reactive nucleophile failed to detect any isotope effect, indicating that $E(Q_2)$ likely does not accumulate to any considerable extent (94). The time-course for the accumulation of quinonoid in this experiment is biphasic ($1/\tau_2 > 1/\tau_3$), and the observed rates of quinonoid formation and the relative amplitudes of the processes closely parallel the decay of the 422-nm Schiff base observed in the reaction of $\alpha_2\beta_2$ with L-Ser alone. These findings clearly illustrate the order of catalytic events at the β-site (scheme I). As expected, the transimination reaction to form $E(A_{ex1})$ and α-H abstraction are obligatory events that must take place before the condensation of indole with E(A-A) can occur. Under these experimental conditions, the rate-limiting step for L-Trp quinonoid formation is the production of the electrophilic E(A-A) intermediate. The presence of indole in the reaction mixture does not appear to significantly affect the rate of E(A-A) formation. Equilibrium dialysis experiments failed to detect any specific binding of indole to the β-site of the free $\alpha_2\beta_2$ complex (95). Therefore, it is likely that indole can only bind upon formation of E(A-A) at the β-site.

The RSSF experiment may be performed in a slightly different manner. Tryptophan synthase may be pre-equilibrated with L-Ser to form the E(A-A) complex [Fig. 14, spectrum 0; (86)]. Rapid mixing of E(A-A) with indole results in RSSF spectral changes that are very different from those previously described (compare Fig. 13 and Fig. 14). Under these reaction conditions, four relaxation processes ($1/\tau_1 > 1/\tau_2 > 1/\tau_3 > 1/\tau_4$) are detected. Within the time required to

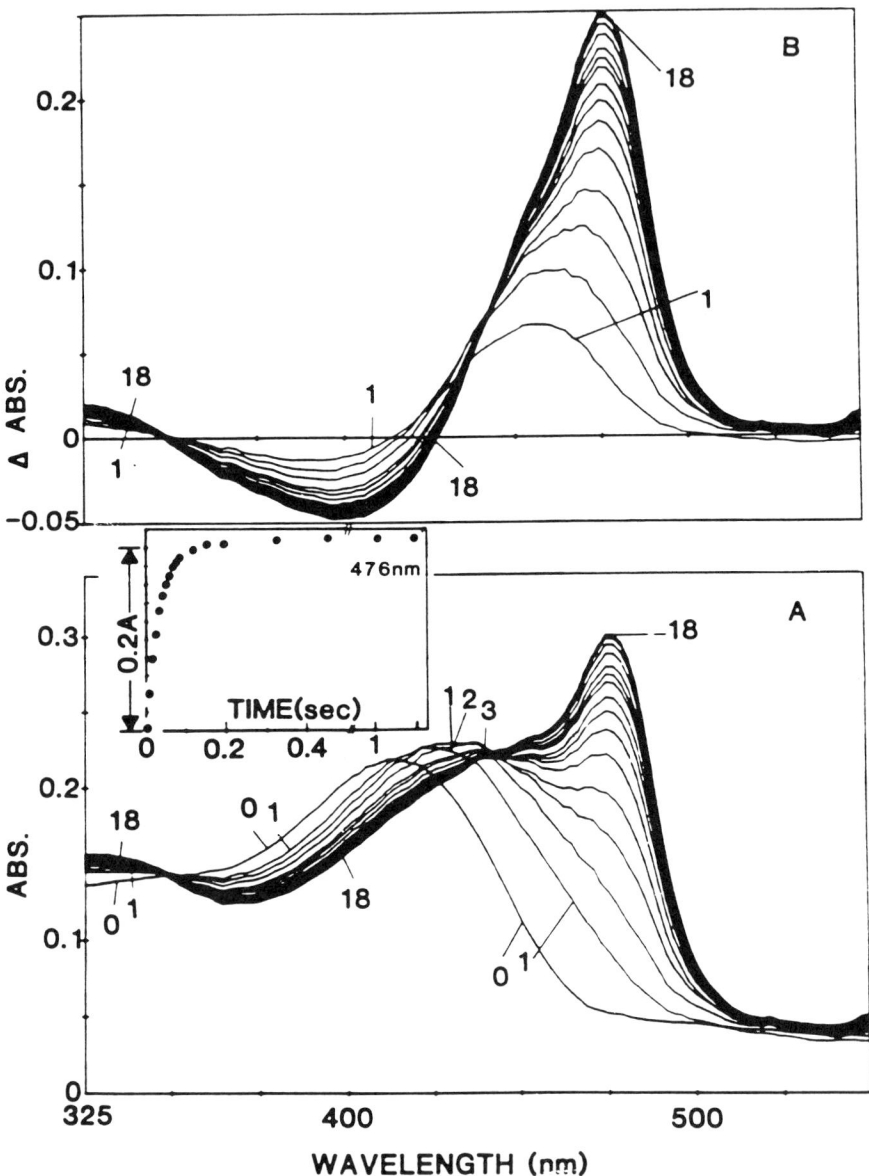

Fig. 13. Rapid-scanning, stopped-flow spectra and difference spectra showing the pre-steady-state spectral changes for the $\alpha_2\beta_2$-catalyzed condensation of indole and L-serine. The spectra (A) and difference spectra (B) were measured at high concentrations of serine and indole. The enzyme contained in one syringe was mixed with indole and L-serine premixed in the second syringe. In (A), the trace designated 0 is the reconstructed spectrum of the reactants before mixing. The first scan in (A) was initiated 4 ms after flow stopped. The inset in the 476-nm reaction time course is reconstructed from the RSSF data. Difference spectra (B) were computed as $(scan)_t - (scan)_o$ from the data presented in (A). Conditions after mixing: [L-Ser] = 40 mM, [indole] = 5 mM, $[\alpha_2\beta_2]$ = 13.3 μM, 0.1 mM potassium phosphate, 1 mM EDTA, pH 7.80, and 25°C. Spectra were collected at 4.0, 12.6, 29.8, 38.4, 47.0, 55.6, 64.2, 72.8, 81.4, 90.1, 98.7, 133, 168, 211, 340, 469, 1114, and 1974 ms after flow stopped. [Taken from Drewe and Dunn (86) with permission.

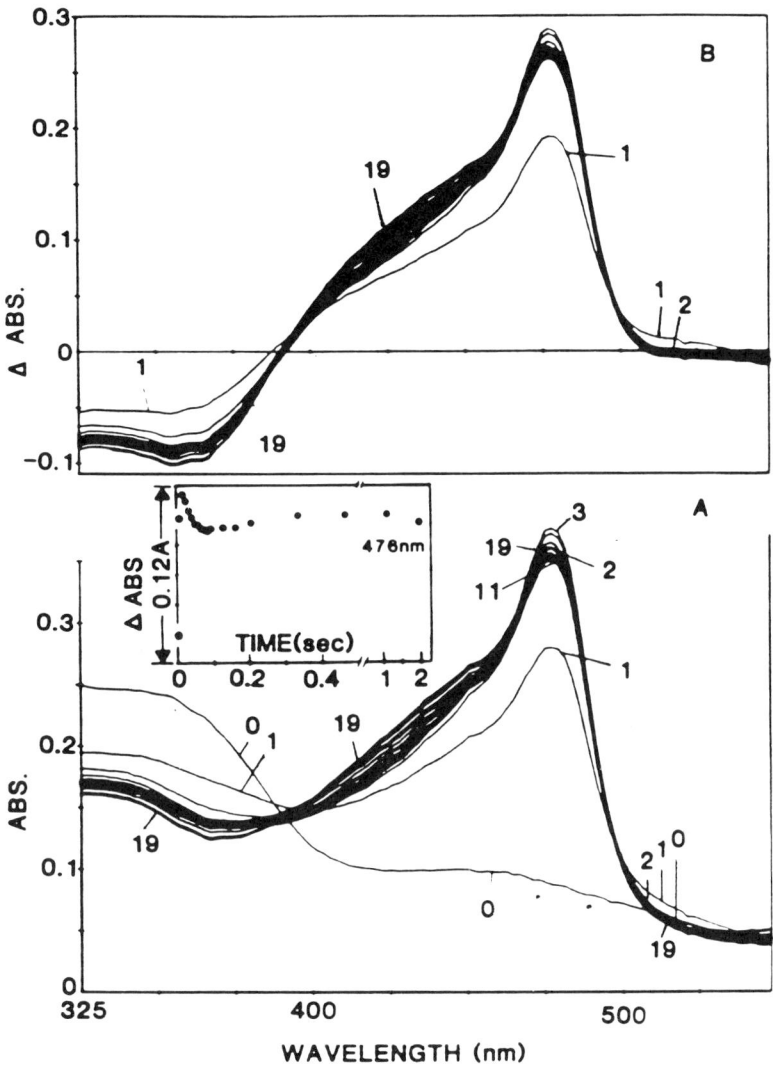

Fig. 14. Rapid-scanning, stopped-flow spectra and difference spectra showing the pre-steady-state spectral changes for the reaction of the E(A-A) intermediate (spectrum 0) with indole measured at high substrate concentrations. Both the enzyme and L-Ser were premixed in one syringe while indole was present in the second syringe. The initiation of scanning occurred 1 ms after flow stopped. Difference spectra (B) were computed as $(scan)_t-(scan)_o$ from the data presented in (A). The 476-nm time course reconstructed from the RSSF data is shown in the inset to (A). Conditions after mixing: [L-Ser] = 40 mM, [indole] = 5 mM, $[\alpha_2\beta_2]$ = 13.3 µM, 0.1 M potassium phosphate, 1 mM EDTA, pH 7.8 and 25°C. Spectra were collected at 1.0, 9.6, 18.2, 26.8, 35.4, 44.0, 52.6, 61.2, 69.8, 78.4, 87.1, 95.7, 130, 165, 208, 337, 466, 1111, and 1971 ms after flow stopped. [Taken from Drewe and Dunn (86) with permission.]

measure the first spectrum, there is a rapid but small increase in absorbance between 500 and 540 nm ($1/\tau_1$; spectrum 1 of Fig. 14), which rapidly disappears during $1/\tau_2$. The spectral changes during $1/\tau_1$ (previously undetected in earlier single-wavelength rapid kinetic experiments) have been attributed to a perturbation in the equilibrium distribution of stage I intermediates caused by the binding of indole to the $\alpha_2\beta_2$ complex prior to the reaction with E(A-A).

Formation of a carbon–carbon bond between indole and E(A-A) results in the rapid accumulation during $1/\tau_2$ of an intense spectral band centered at 476 nm with a shoulder at approximately 440 nm (Fig. 14, spectrum 3). The spectral characteristics of this intermediate are consistent with the assignment of this species to the L-Trp quinonoid. Accumulation of the L-Trp quinonoidal species is followed by a further increase in absorbance ($1/\tau_3$) in the 420-nm region of the spectrum (Fig. 14, spectrum 19). If the reaction is carried out in the absence of phosphate ion (an allosteric effector of reactions at the β-site), then the spectral changes during $1/\tau_3$ are even more pronounced and a distinct spectral band at 420 nm is evident (data not shown). The shape and position of the 420-nm band, in conjunction with the observed sequence of spectral events, is consistent with the assignment of this band to the Schiff base species formed between the cofactor and L-Trp, the product amino acid, E(Aex$_2$) (86, 87).

In the case of tryptophan synthase, qualitative examination of the RSSF spectra has resulted in the direct detection of most of the expected reaction intermediates, and in the elucidation of the sequence of catalytic events, information crucial to the determination of the reaction mechanism. The RSSF data also provide a rational approach both for the selection of wavelengths for the detailed analysis of the dependence of relaxation rates on substrate concentrations by SWSF and for the accurate determination of isoabsorptive points by single-wavelength methods (85, 86). The presence of apparent isoabsorptive points during one or more phases of a multistep reaction simplifies greatly the interpretation of physical events observed during either RSSF or SWSF rapid-kinetic studies.

2.2.5. OBSERVATION OF ALLOSTERIC INTERACTIONS IN THE TRYPTOPHAN SYNTHASE BIENZYME COMPLEX

Allosteric interactions between the heterologous α and β-subunits of the bienzyme complex serve to coordinate catalytic interactions between the two active sites. Binding of an allosteric ligand at the α-active site changes both the affinity of substrates and the subsequent distribution of intermediates at the β-site (96–98). In the tryptophan synthase system, a thorough understanding of how allosteric effectors influence catalysis requires a knowledge of which steps in the catalytic sequence change in response to effectors. Since RSSF spectroscopy allows for the simultaneous detection of a number of separate catalytic intermediates, changes in the rates of formation and deecay of multiple catalytic intermediates induced by the binding of allosteric ligands may be directly monitored.

Binding of L-Ser to the $\alpha_2\beta_2$ complex from *Salmonella typhimurium* results in spectral changes that are very similar to those observed for the closely related enzyme from *E. coli* [Fig. 15A; (87–89, 98)]. Rapid formation of the external aldimine ($1/\tau_1$) is followed by a biphasic decay to the E(A-A) complex ($1/\tau_2 > 1/\tau_3$). The rate of $1/\tau_2$ in this experiment saturates at a rate of nearly $10\ s^{-1}$. Once again, the presence of indole in the reaction mixture does not significantly perturb the rate of $1/\tau_2$ (Fig. 15B), though the intermediates, which are observed to accumulate, are consistent with the formation of the L-Trp quinonoid and the L-Trp E(Aex_2) external aldimine forms of the cofactor. However, if L-Ser and IGP (the substrate of the α-subunit) are simultaneously mixed with the enzyme complex (Fig. 15C), then the decay of the E(A_{ex1}) to form E(A-A) is increased threefold. Preincubation of the bienzyme complex with the nonreactive IGP analog, GP, results in an eight-fold increase in the rate of E(A-A) formation from E(A_{ex1}). It is clear from the RSSF and SWSF data that binding of a ligand to the α-site affects the reactivity at the β-site nearly 30 Å away. In a reciprocal fashion, several groups have shown that the formation of E(A-A) at the β-site increases the reactivity of the α-subunit by a factor of 20- to 30-fold (83, 98–100). Therefore, both the binding of substrate at the α-site, and the formation of a discrete covalent intermediate at the β-site results in protein conformational changes that communicate allosteric information within the bienzyme complex. The reciprocal allosteric interactions between subunits serves to synchronize the catalytic activities at the α- and β-active centers in order to ensure the efficient synthesis of L-Trp without the loss of indole from the enzyme complex (97, 98).

2.2.6. TRYPTOPHANASE

The approaches that have been described in some detail for tryptophan synthase may be applied to other PLP-dependent enzymes. Tryptophan indole lyase, or tryptophanase, catalyzes the PLP-dependent β-elimination of indole from tryptophan to yield indole, pyruvate, and NH_4^+ (Equation 4).

$$\text{L-Trp} + H_2O \rightleftharpoons \text{Pyr} + \text{Indole} + NH_4 \qquad (4)$$

This enzyme represents an interesting contrast to tryptophan synthase, which catalyzes the essentially irreversible formation of L-Trp. The spectrum of the native enzyme, which is highly pH dependent, is characterized by two absorbance bands centered at 420 nm and 337 nm. Early RSSF investigations utilizing rapid incremental jumps in pH showed that the two spectral bands arise from different protonation states of the covalently bound internal aldimine, E(Ain), form of the cofactor (101). Studies with a variety of amino acid inhibitors of tryptophanase (amino acids, which react reversibly with the enzyme to form covalent PLP-intermediates, but cannot complete the β-elimination reaction to form products), showed that the 420-nm species is the reactive form of the cofactor. The 337-nm species must be converted to the 420-nm species before reaction with the amino group of the substrate will occur. The 420-nm species represents a ketoenamine form of the cofactor in which the iminium nitrogen of the Schiff's base is protonated (102).

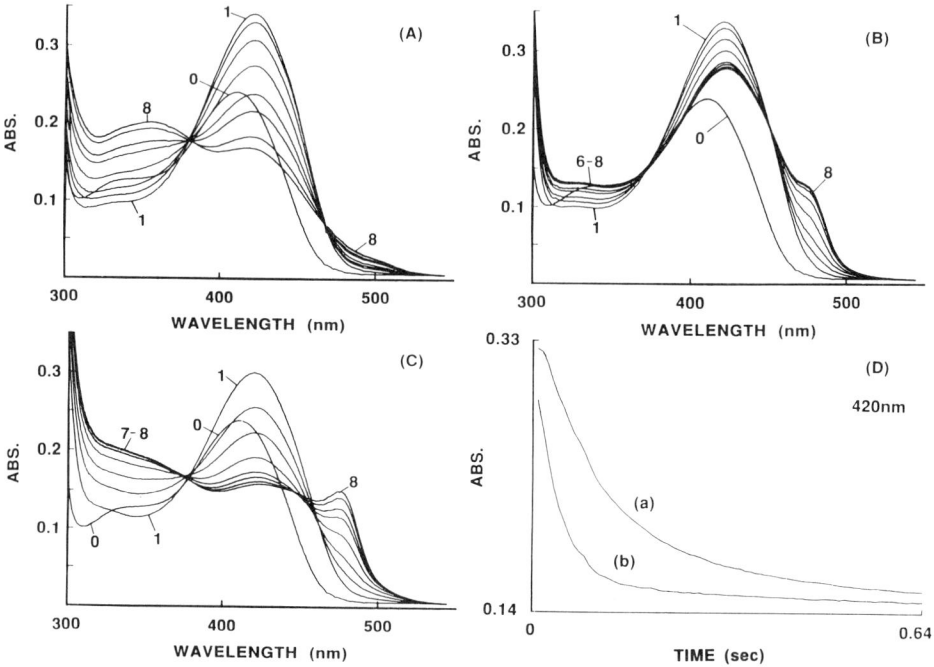

Fig. 15. Time-resolved RSSF spectra for the reaction of 10 μM $\alpha_2\beta_2$ from *S. typhimurium* with (A) 40 mM L-Ser, (B) 40 mM L-Ser and 0.5 mM indole, and (C) 40 mM L-Ser and 0.22 mM IGP. The native enzyme was in one syringe, the substrates were in the other. The spectra shown were recorded at 8.54, 25.63, 42.72, 76.9, 128.2, 179.4, 358.9, and 640.1 ms after flow stopped. Spectrum 0 represents the spectrum of the enzyme in the absence of substrates. (D) Single-wavelength, stopped-flow time courses at 420 nm are depicted for the reaction shown in panel A (trace a) and panel C (trace b). Time courses were derived from 74 successive RSSF spectra for a total acquisition time of 0.64 s with 8.54 ms between each data point. All concentrations refer to conditions immediately after mixing. [Taken from Brzović et al. (98) with permission.]

In equilibrium binding experiments, Kazarinoff and Snell (103) found that indole binds to complexes of amino acids with short side chains, but could detect no binding to the free enzyme. Phillips (104) used RSSF and SWSF methods to study the binding of both indole and the isoelectronic indole analog benzimidazole (BZ) to the complex of L-alanine with tryptophanase. Before mixing, the initial equilibrium mixture is dominated by a quinonoidal species with λ_{max} = 501 nm (Fig. 16A, spectrum 0). Rapid mixing with indole results in a rapid decrease at 501 nm and an increase between 400 and 480 nm, which is complete within the instrument mixing time (Fig. 16A, spectrum 1). Following this rapid process, there is a slower phase in which the intensity of the quinonoid peak at 501 nm decreases with a concomitant increase in absorbance at 420 nm (Fig. 16A, spectrum 12). An isoabsorptive point at 446 nm is observed during this

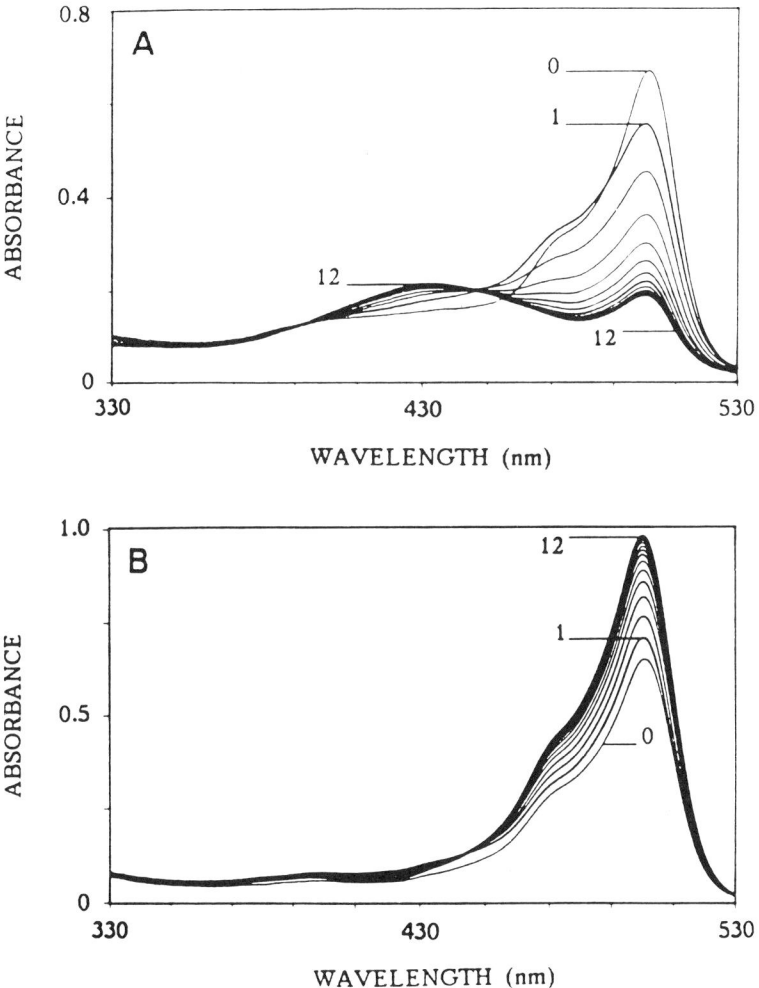

Fig. 16. Rapid-scanning data for the reactions of 5 mM indole (A) and 5 mM benzimidazole (B) with 17.2 μM tryptophan indole-lyase (tryptophanase) that has been preequilibrated with 0.25mM L-alanine. L-Alanine was premixed in both syringes to prevent unwanted concentration changes. In panel A, scans were collected at 15, 92.5, 170, 247.5, 325, 402.5, 480, 557.5, 635, 712.5, 790, 867.5, and 945 ms after flow stopped. Spectrum 0 represents the spectrum of the enzyme-alanine complex before mixing with indole. In panel B, scans were collected at 15, 390, 765, 1140, 1515, 1890, 2265, 2640, 3015, 3390, 3765, and 4140 ms after mixing. Spectrum 0 represents the enzyme-alanine complex before mixing with benzimidazole. [Taken from Phillips (104) with permission.]

phase of the reaction, indicating that the slow process corresponds primarily to a redistribution between two different intermediates. Examination of the dependence of the observed rate at 501 nm on the indole concentration (data not shown) yields data that is consistent with a mechanism in which there is a rapid binding of indole to the tryptophanase L-Ala complex, which is followed by a slow isomerization between quinonoid and $E(A_{ex})$ forms of the enzyme. Indole apparently binds preferentially to the $E(A_{ex1})$ form of the enzyme complex, which absorbs at 420 nm.

In contrast to these results, the RSSF spectral changes, which occur upon mixing BZ with the L-Ala complex, show a large increase in the amount of the L-Ala quinonoid present (Fig. 16B). The rate dependence of the spectral changes at 501 nm with respect to BZ concentration show that the rate decreases with increasing ligand. These data suggest a mechanism in which a slow conformational change of the enzyme complex occurs before BZ binds to the enzyme. Since the amount of the L-Ala quinonoid increases significantly, BZ binding must stabilize this conformation of the enzyme complex.

S-alkyl-L-cysteines are good substrates for the β-elimination reaction catalyzed by tryptophanase. RSSF spectra for the reaction of tryptophanase with S-benzyl-L-cysteine (SBC) show a rapid formation of an intense quinonoid band centered at 512 nm as the reaction approaches the steady state (Fig. 17A). Whereas mixing with indole does not noticeably perturb the steady-state spectrum, simultaneous mixing of SBC and BZ with tryptophanase show the rapid formation of the quinonoid band (Fig. 17B, spectrum 4) is followed by a slower phase that is characterized by a decrease at 512 nm and the formation of a new spectral band centered at 345 nm (Fig. 17B, spectrum 13). There is an apparent isoabsorptive point at 362 nm. Qualitatively similar spectral changes were observed for the reaction of the physiological substrate L-Trp in the presence of BZ. The accumulation of the 345-nm band occurs at a rate that is faster than the turnover rate of the enzyme, suggesting that the 345-nm species is kinetically competent to be a reaction intermediate. This previously uncharacterized species also accumulates in the reverse reaction from a mixture of pyruvate and ammonium ion if BZ is present. When this mixture is rapidly mixed with indole, a 505-nm quinonoidal species is observed to accumulate rapidly (data not shown), suggesting that the 345-nm species is capable of either reacting directly with indole or rapidly forming a reactive intermediate.

Based upon the observed sequence of catalytic events described above, Phillips (104) has proposed that the 345-nm band is derived from an E(A-A) species similar to that observed in the reaction between L-Ser and tryptophan synthase. Because the extended conjugation of the E(A-A) π-system is expected to give a species absorbing at longer wavelengths, Phillips has proposed that the 345-nm species corresponds either to a different tautomeric state of the E(A-A) species, or to an intermediate with tetrahedral geometry at the C-4' position of the cofactor, perhaps a gem-diamine intermediate formed between the α-aminoacrylate and the ε-amino group of the active site lysine residue, Lys-270. It is possible that the α-amino acrylate gem-diamine species represents a labile intermediate formed during transamination between Lys-270 and the E(A-A) Schiff's base (structure 5A).

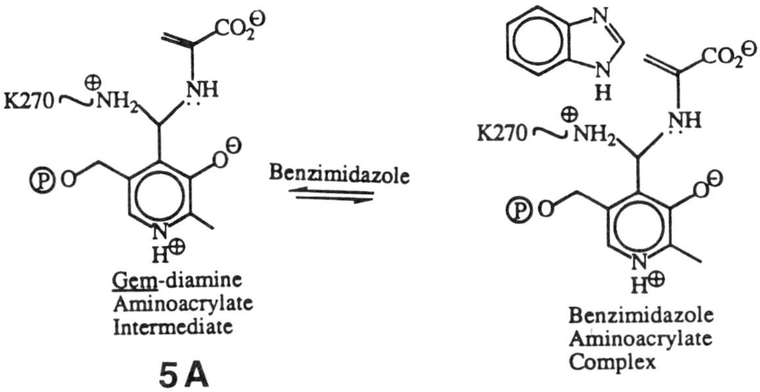

5 A Gem-diamine Aminoacrylate Intermediate ⇌ Benzimidazole ⇌ **5 B** Benzimidazole Aminoacrylate Complex

Protonation at the β-carbon to yield a protonated imine that is subsequently hydrolyzed to pyruvate and ammonia provides a reasonable mechanism for the final steps in the β-elimination reaction catalyzed by tryptophanase. BZ may bind preferentially to this gem-diamine species thereby stabilizing a labile reaction intermediate (structure 5B).

2.2.7. INVESTIGATION OF SUBSTRATE STRUCTURAL ELEMENTS ON TRYPTOPHANASE REACTIVITY

As previously mentioned, tryptophanase is inhibited by a variety of different amino acids, which react with the PLP-cofactor to form covalent intermediates (102), but the structure of the substrate prevents completion of the reaction pathway. At equilibrium, these quasisubstrates generally form intense absorption bands in the 500-nm region of the spectrum, which result from the accumulation of a stable quinonoidal species. Phillips et al. (105) utilized RSSF in conjunction with SWSF studies to investigate the mechanisms of reaction for various aminoacid analogs of L-Trp in order to determine substrate structural elements important both for substrate binding and reactivity with the enzyme (structures 6–9).

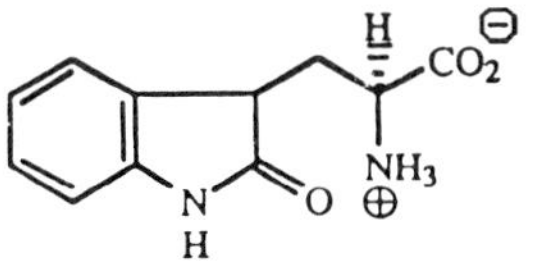

6 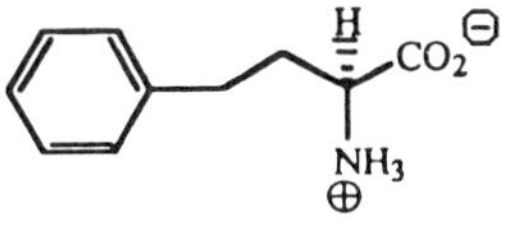   **7**

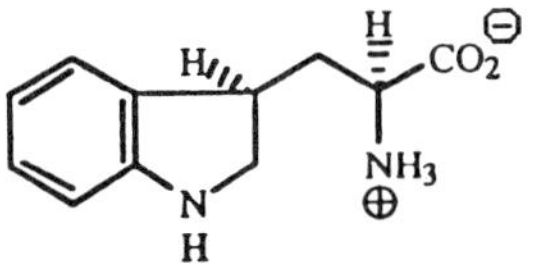

8 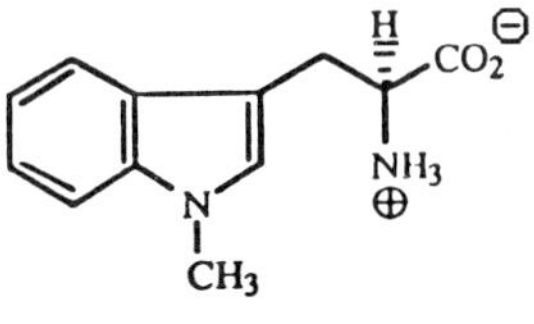   **9**

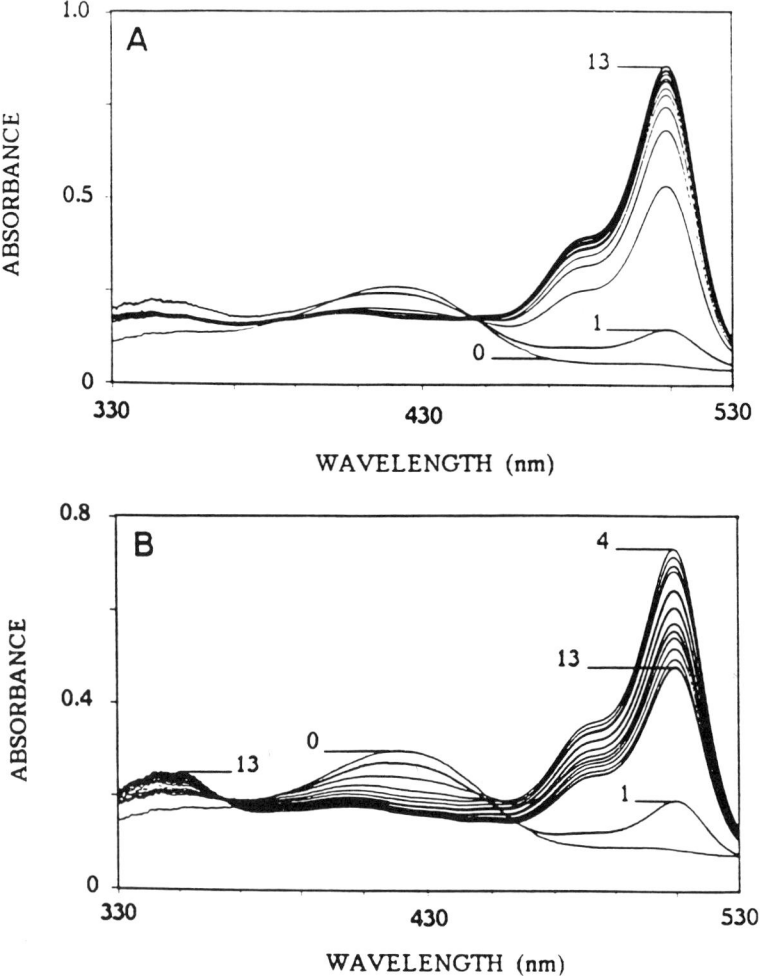

Fig. 17. Rapid-scanning data for the reaction of 20.9 μM tryptophanase with 20 mM S-ethyl-L-cysteine in the absence (A) and presence (B) of 5 mM benzimidazole. Spectra shown in both (A) and (B) were collected at 10, 31, 52, 73, 94, 115, 136, 157, 178, 199, 220, 241, and 262 ms after flow stopped. Curve 0 is the spectrum of the native enzyme in the absence of substrates. [Taken from Phillips (104) with permission.]

The mechanism of the β-elimination reaction catalyzed by tryptophanase is thought to require tautomerization of the indole ring to an activated indolenine intermediate with tetrahedral geometry at the C-3 carbon of the indole ring. Therefore, interactions between active-site residues and the ring nitrogen are likely to be very important for binding and catalysis.

Comparisons of the RSSF spectral changes observed with these analogs show large differences among the various substrates both in the rates of formation and accumulation of reaction intermediates and in the concentration of intermediate species present at equilibrium. As an example, the set of spectral changes for the reaction of 2 mM L-homophenylalanine with tryptophanase are shown in Fig. 18. These data show a rapid loss of the E(Ain) absorbance band at 420 nm with a concomitant accumulation of a previously undetected species absorbing at 340 nm. This initial rapid phase is followed by a decay of the 340-nm species and the formation of a quinonoid band at 508 nm (Fig. 18, spectrum 13). The sequence of catalytic events and the position and shape of the 340-nm band suggest that this intermediate corresponds either to a gem-diamine or enolimine Schiff base form of the substrate-cofactor complex (structures 10A and 10B).

10 A $\lambda_{max} \sim 330$ nm
External aldimine—Enolimine

10 B $\lambda_{max} \sim 325$ nm
Gem-diamine

Since ketoenamine Schiff bases are usually the reactive form of the cofactor (102), the 340-nm species likely arises from accumulation of a gem-diamine intermediate.

This order of events was confirmed by a study of the dependence of the observed rate at 340, 420, and 508 nm on the concentration of L-homophenylalanine in a series of SWSF experiments. These studies showed that in the concentration range between 0 and 1.2 mM L-homophenylalanine, the fast processes at 420 nm and 340 nm exhibit a linear dependence of rate on concentration, giving an apparent bimolecular rate constant of 2.2×10^4 $M^{-1}s^{-1}$ with an off-rate of nearly 15 s^{-1}. Quinonoid formation at 508 nm exhibited a hyperbolic dependence of rate on L-homophenylalanine concentration, suggesting that quinonoid formation is closely coupled to a bimolecular binding step. These data were used to determine a K_1 for L-homophenylalanine of 110 µM, in good agreement with the K_1 determined by steady-state methods.

Similar experiments with other L-Trp substrate analogs revealed large changes in both the rate of deprotonation to form the quinonoid and the equilibrium distribution of species. Substrates with a heterocyclic NH functional group in the ring, such as L-Trp and L-oxyindole alanine, react much more readily to form quinonoidal species than do "acyclic" analogs such as L-homophenylalanine. Therefore, the structure of the aromatic ring, which is removed from the point of covalent bond transformation at the α-carbon, plays an essential role in the formation and accumulation of various reaction intermediates via the formation of specific interactions at the indole subsite. These interactions likely involve both van der Waals contacts and hydrogen bonding interactions with the heterocyclic

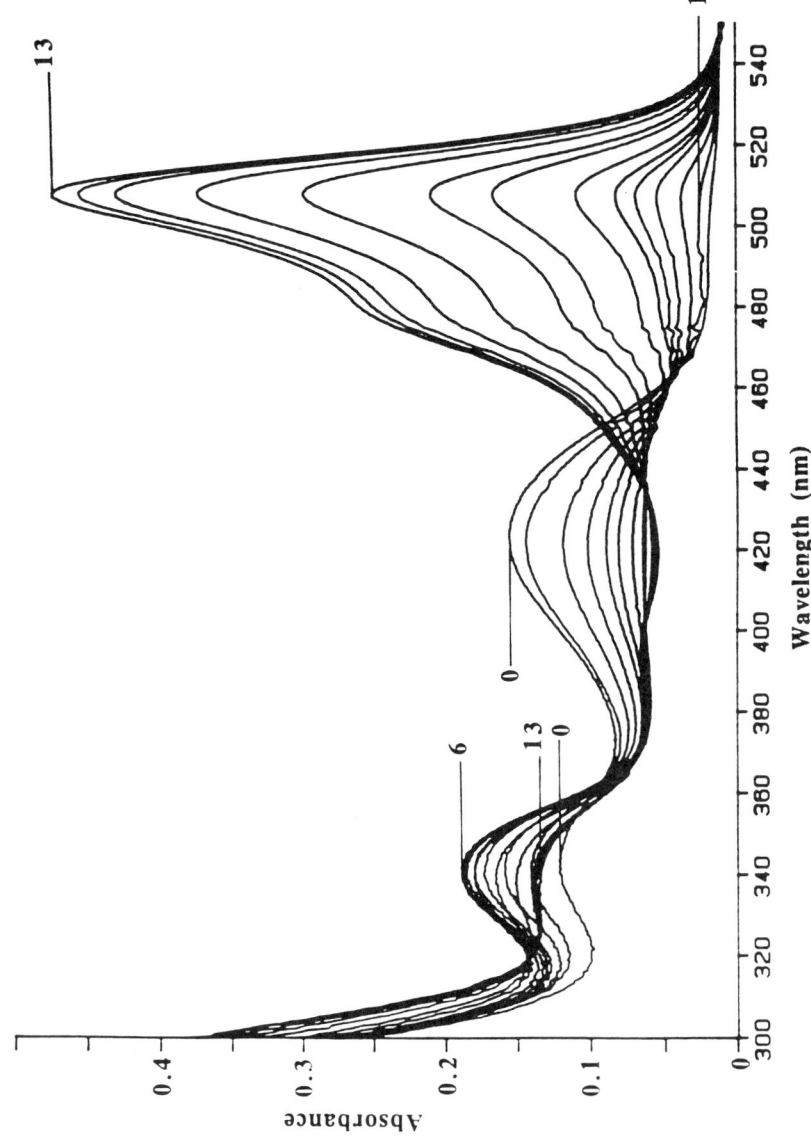

Fig. 18. Rapid-scanning, stopped-flow spectra for the reaction of 20 μM tryptophanase with 2 mM L-homophenylalanine in 0.02 M potassium phosphate buffer, pH 8.0, containing 0.16 M KCl and 5 mM 2-mercaptoethanol. Scans were collected at 8.54, 17.08, 25.62, 34.16, 42.70, 59.78, 93.94, 128.1, 230.6, 384.3, 606.1, 1025, and 1708 after flow stopped. Curve 0 is the spectrum of the native enzyme in the absence of substrates. [Taken from Phillips et al. (106) with permission.]

NH of the aromatic ring. Removal of the α-proton changes the hybridization of the α-carbon from an sp^3 to an sp^2 center. This change in geometry around the α-carbon atom would involve obligatory motions not only in the position of the β-carbon but of the aromatic ring as well. Thus, the conformation of the active site, and perhaps the rest of the protein, must adjust in order to accommodate the various configurations of the different structural intermediates. Alterations in substrate structure likely effect the ability of ligand–protein interactions to elicit the necessary protein conformational changes required for catalysis in tryptophanase.

2.2.8. CYSTATIONINE γ-SYNTHASE CATALYSIS

Cystathionine γ-synthase (CGS) is a rather unique PLP-enzyme that catalyzes a transsulfuration reaction important in microbial methionine biosynthesis. It is the only known enzyme whose function is the catalysis of a PLP-dependent replacement reaction at the γ-carbon of the amino acid substrate; the succinyl moiety of O-succinyl-L-homoserine is replaced by L-Cys to give the thioether linkage of L,L-cystathionine (scheme II). In the absence of L-Cys, the enzyme catalyzes a net γ-elimination reaction from OSHS (scheme II). Because both reactions require the elimination of succinate, the catalytic pathways must diverge from a common reaction intermediate. It was originally hypothesized that a vinylglycine quinonoidal intermediate (structure 11)

Structure 11

was the key partitioning intermediate (68). However, in order to partition between γ-elimination and replacement pathways, this single species would be obliged to display both electrophilic and nucleophilic properties at the reactive γ-carbon of the substrate, a chemically implausible behavior.

Model studies aimed at producing the vinylglycine quinonoid in solution indicate that this species should absorb in the 500-nm region of the spectrum (79). If this species is the key partitioning intermediate, then it is reasonable to expect the accumulation of an absorbance band in this region of the spectrum.

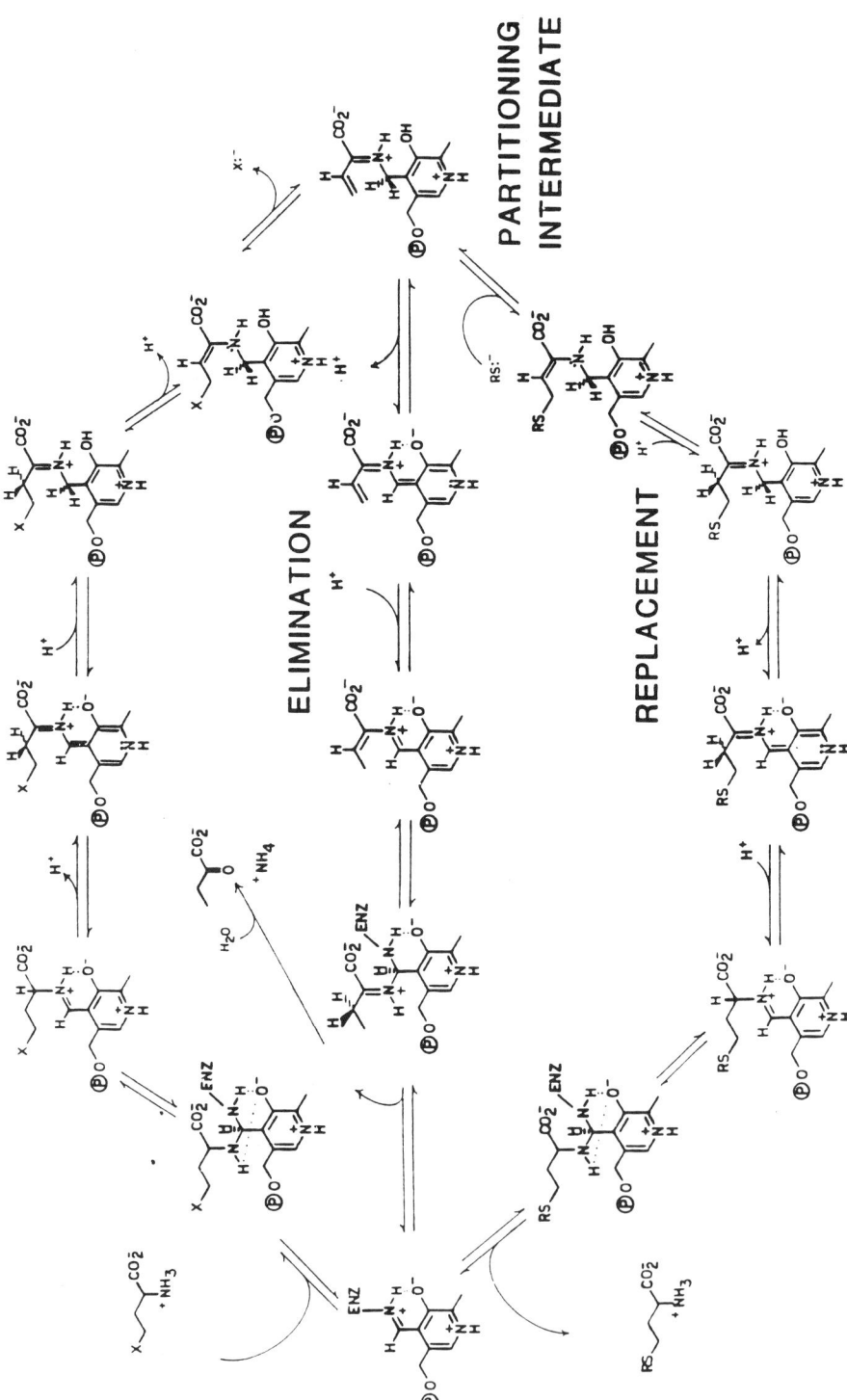

Scheme II. Reaction mechanism for the PLP dependent cystathionine γ-synthase γ-elimination and γ-replacement reactions proposed by Brzović et al. (107). [Redrawn from (107) with permission.]

Figure 19 shows the observed spectral changes, which occur during the γ-replacement reaction catalyzed by CGS (107). The native enzyme exhibits a spectral band centered at 422 nm, which arises from the E(Ain) form the cofactor that is covalently bound to the enzyme (Fig. 19A, spectrum 0). Upon mixing with both L-Cys and OSHS, the RSSF spectra show there is an initial rapid shift ($1/\tau_1$) in the absorbance maximum at 422 nm to 425 nm within the instrument mixing time (Fig. 19a, spectrum 1). This shift is consistent with the formation of a Schiff's base between the substrate OSHS and the cofactor. The initial rapid process is followed by a decrease in absorbance at 425 nm ($1/\tau_2$). Difference spectra, calculated by substraction of the native enzyme spectrum (spectrum 0) from the subsequent RSSF spectra (spectra 1–8), establish the concomitant accumulation of a new, previously undetected, intermediate absorbing at 300 nm (Fig. 19B). An apparent isoabsorptive point exists at 340 nm during this phase of the reaction. It is clear from the RSSF spectra that no transient intermediates, which absorb in the 500-nm region, accumulate during the presteady-state phase of the γ-replacement reaction.

If L-Cys is excluded from the reaction mixture, the presteady-state spectral changes for the net γ-elimination reaction are very different. Figure 20A reveals that there is a rapid loss of the 422 nm E(Ain) spectral band ($1/\tau_1$). The calculated difference spectra derived from the RSSF data (Fig. 20B) show the concomitant accumulation of the 300-nm species during $1/\tau_1$ with an apparent isoabsorptive point at 335 nm. After this process is complete, there is a biphasic accumulation of a new species, which absorbs at 485 nm. This intermediate is unique to the γ-elimination pathway. Comparison of the spectral changes during both the γ-elimination and replacement reactions show that the 300-nm species is common to both pathways. Titration of CGS with a substrate analog, L-allyl glycine, also results in the loss of absorbance at 422 nm and a corresponding increase in absorbance at 300 nm (data not shown). No intermediates absorbing at wavelengths above 422 nm could be detected. NMR studies in D_2O show that protons at both the α- and β-positions of the substrate are exchanged upon incubation with CGS, but with L-allylglycine no reaction products consistent with a net γ-elimination reaction were found. However, rapid mixing of the L-allylglycine-enzyme complex with L-Cys results in the rapid loss of the 300-nm species (Fig. 21). The product of this reaction was isolated and identified as γ-methyl-L,L-cystathionine.

These results establish that there is a partitioning between the γ-elimination and replacement pathways that occurs from the 300-nm intermediate, a species that is common to both reaction pathways. Based upon the spectra of PLP-compounds in solution, this intermediate is almost certainly a pyridoxamine form of the cofactor-substrate adduct. Removal of the α-proton and subsequent protonation at the C-4' position of the cofactor would effectively remove electron density from the α-carbon (scheme II). The resulting protonated imine would effectively increase the acidity of the β-carbon protons. The β-carbanion produced by abstraction of a β-proton may then lend anchimeric assistance toward elimination of the γ-substituent and formation of a β-γ unsaturated

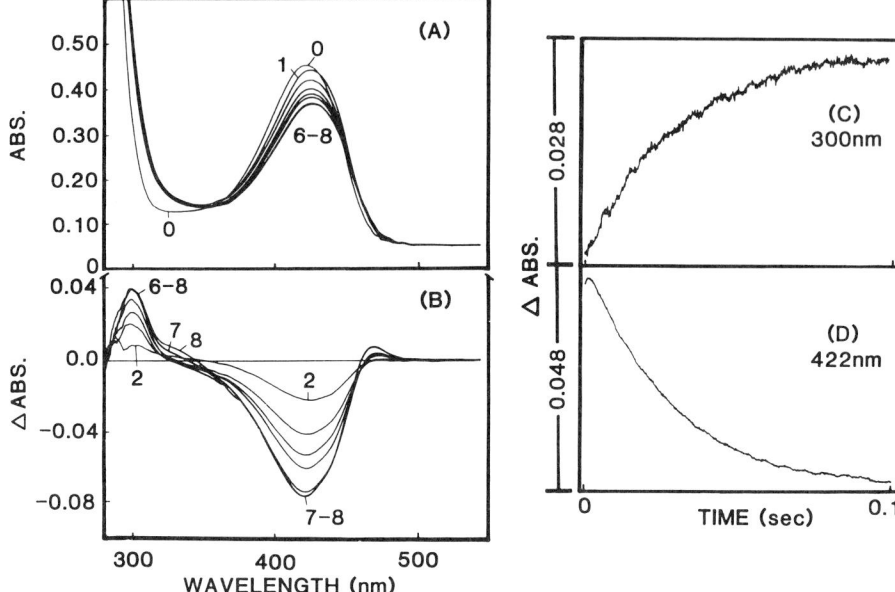

Fig. 19. Rapid-scanning, stopped-flow spectra (A) and difference spectra (B) of the γ-replacement reaction with OSHS and L-cysteine. Before mixing, enzyme was contained in one syringe while OSHS and L-Cys were premixed in the other. All concentrations refer to those immediately after mixing: [OSHS] = 10 mM, [L-Cys] = 10 mM, [CGS] = 6.25 µM, 0.1 M potassium phosphate, 5 mM DTE, 1 mM EDTA, pH 7.2 at 25°C. (A) RSSF data were obtained by collection of 79 sequential scans with a repetitive scan time of 8.9 ms. The scans shown are a representative subset of the 79 scans collected for each experiment. The trace designated 0 is a spectrum of the enzyme in the absence of substrates. The initiation of scanning occurred at 1.3 ms after flow stopped. The spectra shown were collected at 1.3, 10.2, 19.1, 28.0, 26.9, 72.5, 250.5, and 695.5 ms after flow stopped. (B) Difference spectra were computed as $(scan)_t - (scan)_o$ from the data presented in (A). (C, D) Time courses at 300 nm (C) and 422 nm (D) collected in conventional single-wavelength stopped-flow experiments under conditions identical with the RSSF experiments. [Taken from Brzović et al. (107) with permission.]

ketimine. This species (scheme II, structure XI) is susceptible toward Michael addition of a nucleophile at the γ-carbon, ultimately resulting in a γ-replacement reaction. In the absence of an appropriate nucleophile, a net tautomerization occurs to form an aminocrotonate species absorbing at 485 nm, which is committed along the γ-elimination pathway. A similar 485-nm species has also been observed in the reaction of CGS with β-halo amino butyrate substrates (108). These compounds cannot participate in γ-replacement reactions and are committed along the γ-elimination pathway. Thus, this mechanism removes the ambiguity of requiring the L-vinylglycine quinonoidal species to display both electrophilic and nucleophilic character at the γ-carbon and provides a mechanism for the effective labilization of protons at the β-position of amino acid substrates. It is probable that all PLP-dependent enzymes, which catalyze reactions that require labilization of protons at the β-portion, employ a similar mechanism.

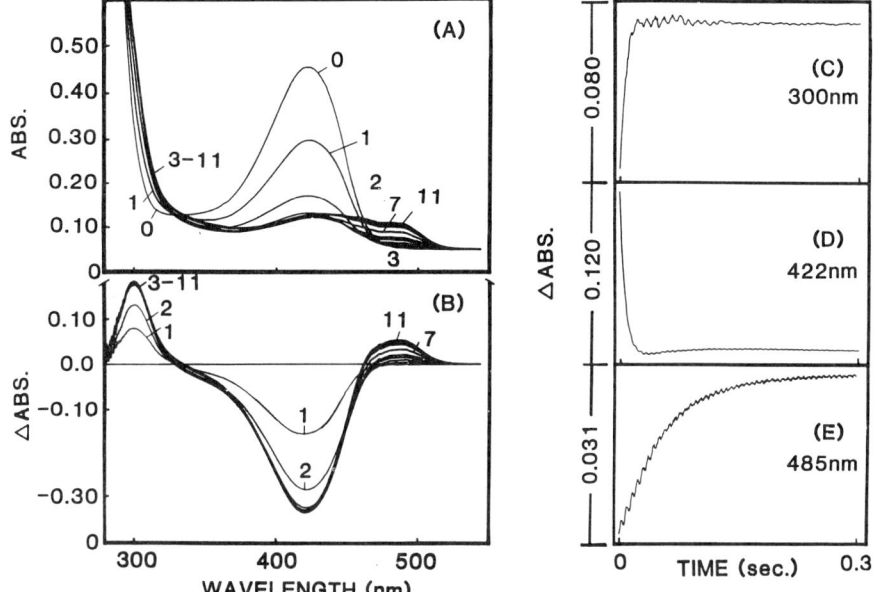

Fig. 20. Rapid-scanning, stopped-flow spectra (A), difference spectra (B), and single-wavelength, stopped-flow time courses at 300, 422, and 485 nm (C, D, E) for the γ-elimination reaction of CGS with OSHS. Data were obtained as described in Figure 19. All concentrations refer to conditions immediately after mixing: [OSHS] = 10 mM, [CGS] = 6.25 μM, 0.1 M potassium phosphate, 1 mM EDTA, pH 7.2 at 25°C. (A) RSSF spectra, Trace 0 is the spectrum of the enzyme in the absence of substrates. The initiation of scanning occurred 1.3 ms after flow stopped. Spectra shown were collected at 1.3, 10.2, 19.1, 28.0, 36.9, 72.5, 117.0, 161.5, 250.5, 392.9, and 695.5 ms after flow stopped. (B) Difference spectra were computed as $(scan)_t - (scan)_o$ from the data presented in (A). Single-wavelength time courses (C–E) were collected in SWSF experiments under conditions identical with those described for (A). [Taken from Brzović et al. (107) with permission.]

2.3. Investigation of Enzyme Structure–Function Relationships by Site-Directed Mutagenesis and RSSF Spectroscopy

Site-directed mutagenesis has become an important and widespread technique for the elucidation of structure–function relationships in proteins. However, the repercussions of mutations on both protein structure and catalysis are often subtle and, particularly in the case of mechanisms that require multiple catalytic steps, not always easily interpretable. Classical comparison of catalytic rate parameters between mutant and native enzymes where an amino acid substitution results in a change in the the rate-limiting step of a reaction are not necessarily valid (109). Thus, direct detection of reaction intermediates is an important means for assessing the effect of mutations on the mechanism and for accurately determining the role of various protein residues in catalysis.

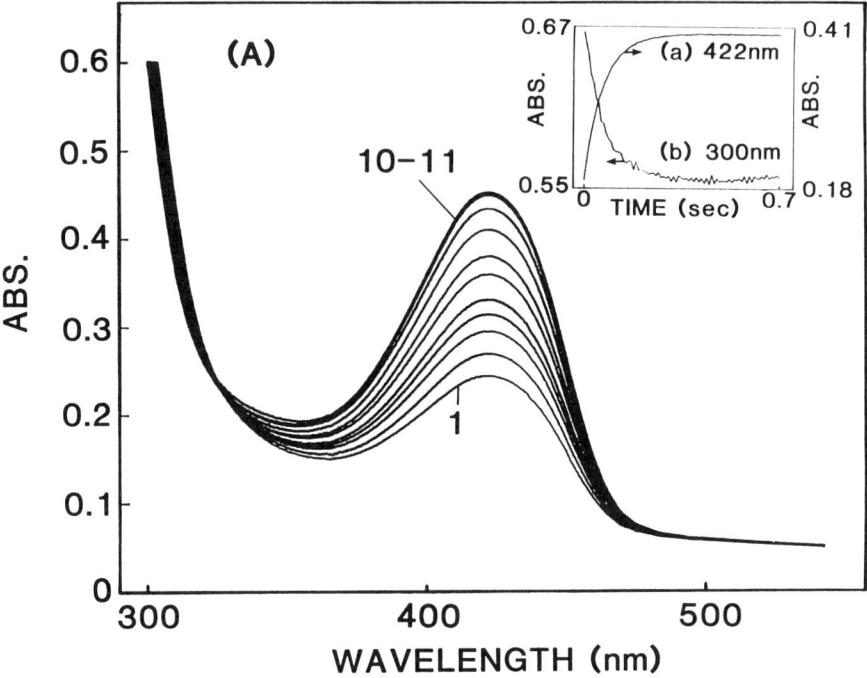

Fig. 21. Rapid-scanning, stopped-flow spectra for the reaction of L-cysteine with CGS pre-incubated with L-allylglycine. Concentrations refer to conditions immediately after mixing: [CGS] = 7.08 μm, [L-allylglycine] = 28mM, [L-cysteine] = 10 mM, 0.1 M potassium phosphate, 5 mM DTE, 1 mM EDTA, pH 7.2 and 25°C. Initiation of scanning occurred 2.3 ms after flow stopped. Data shown are representative scans from a sequential 79-scan data set collected with a repetitive scan rate of 8.9 ms/scan. The spectra shown were acquired at 2.3, 11.2, 20.1, 29.0, 37.9, 55.7, 73.5, 109.1, 162.5, 289.6, and 696.5 ms after flow stopped. The inset to (A) represents single-wavelengths time courses taken from the entire set of 79 scans at (a) 300 nm and (b) 420 nm. The left and right ordinates correspond to the absorbance values at 300 and 420 nm, respectively. [Taken from Brzović et al. (107) with permission.]

2.3.1. MUTATIONS IN THE β-SUBUNIT OF TRYPTOPHAN SYNTHASE

An interesting example of the use of RSSF spectroscopy to study the effects of mutations on catalytic activity involves the substitution of both aspartate and alanine for glutamate at position 109 in the β-subunit of tryptophan synthase. The X-ray structure suggests that βE109 is suitably poised to activate indole as a nucleophile during stage II of the β-reaction (81). Substitution of Asp at position 109 results in an enzyme with decreased steady-state activity in the β-reaction. RSSF analysis of the reaction of βE109D with L-Ser indicates that the mutation has only a moderate affect on stage I of the β-reaction [data not shown (83, 87)]. If the allosteric effector GP is bound at the α-active site, the reaction of the mutant with L-Ser is nearly indistinguishable from the wild-type catalyzed reaction.

Comparison of the spectral bands of the transient $E(A_{ex1})$, which is observed to accumulate in both the native and βE109D reactions with L-Ser, shows no apparent changes in the spectral characteristics of this species. Therefore, substitution of Glu by Asp at position 109 has only a minor effect on stage I of the β-reaction.

In marked contrast to the wild-type reaction (Fig. 22A, B), rapid mixing of the βE109D E(A-A) complex with indole results only in very small changes in the E(A-A) spectrum with no apparent accumulation of subsequent stage II intermediates (Fig. 22C, D). Even in the presence of GP, which when bound at the α-site generally results in the stabilization of quinonoids during stage II of the β-reaction (Fig. 22B), no spectral changes were evident (Fig. 22D). However, other nucleophiles, in particular the indole analog indoline, undergo a facile reaction with the βE109D E(A-A) complex resulting in the accumulation of a

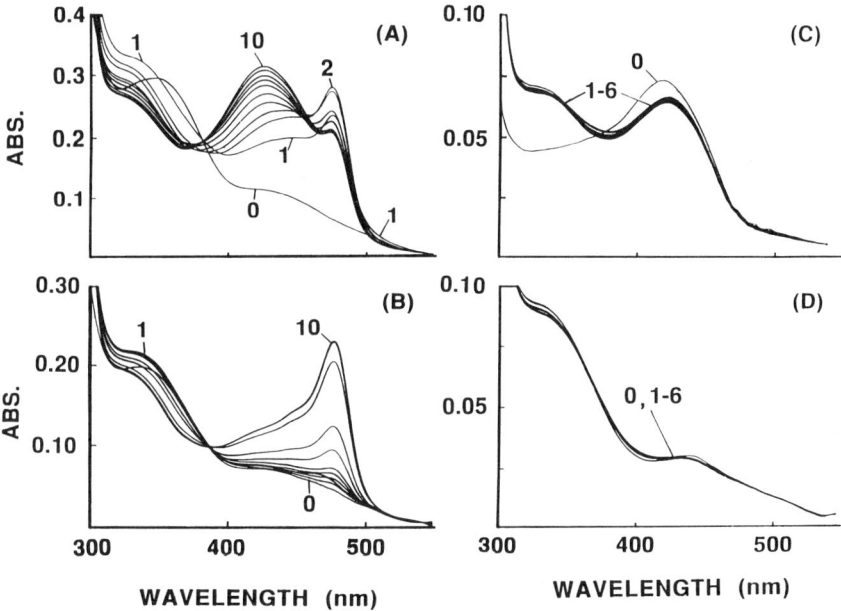

Fig. 22. RSSF spectra for the reaction of the wild-type and βE109D tryptophan synthase E(A-A) complexes with 1 mM indole in both the absence and presence of 100 mM D,L-GP. Enzyme and L-Ser were in one syringe and indole and L-Ser were in the other. GP, when present, was also placed in both syringes to prevent unwanted concentration changes. In each case, spectrum 0 represents the enzyme E(A-A) in the absence or presence of GP before mixing with indole. (A) Reaction of 10 μM wild-type E(A-A) complex with indole. (B) Reaction of 10 μM wild-type E(A-A) complex with indole in the presence of D,L-GP. (C) Reaction of 5 μM βE109D E(A-A) complex with indole. (D) Reaction of 5 μM βE109D E(A-A) complex with indole in the presence of GP. Spectra shown in (A) were collected at 8.53, 17.1, 25.6, 34.1, 42.6, 59.7, 85.3, 170.6, 255.8, and 341.1 ms after flow stopped; (b) 8.53, 17.1, 34.1, 42.6, 76.8, 127.9, 383.8, 852.8, and 1705.6 ms after flow stopped; (C, D) 8.53, 17.1, 34.1, 42.6, 76.8, 127.9, 383.8, 852.8, and 1705.6 ms after flow stopped. [Taken from Brzović et al. (87) with permission.]

quinonoidal intermediate, which is slowly converted to the new amino acid. The maximum of the quinonoid band is shifted by 2 nm to shorter wavelengths, indicating that the mutation has caused some alteration in the environment of the nucleophile binding site (Fig. 23).

A completely unexpected finding was that replacement of E109 by Alanine resulted in an enzyme, which was defective during stage I of the β-reaction (see scheme IB). Reaction with L-Ser terminated at the $E(A_{ex1})$ intermediate. Exchange of deuterium for hydrogen at the α-position of L-Ser was not observed in D_2O, presumably because the binding of L-Ser to the βE109A complex is very tight (~0.4 μM). This roadblock in the reaction mechanism could be circumvented by utilizing β-Cl-L-Ala instead of L-Ser as the amino acid substrate. Whereas the hydroxyl functional group likely requires protonation for facile elimination,

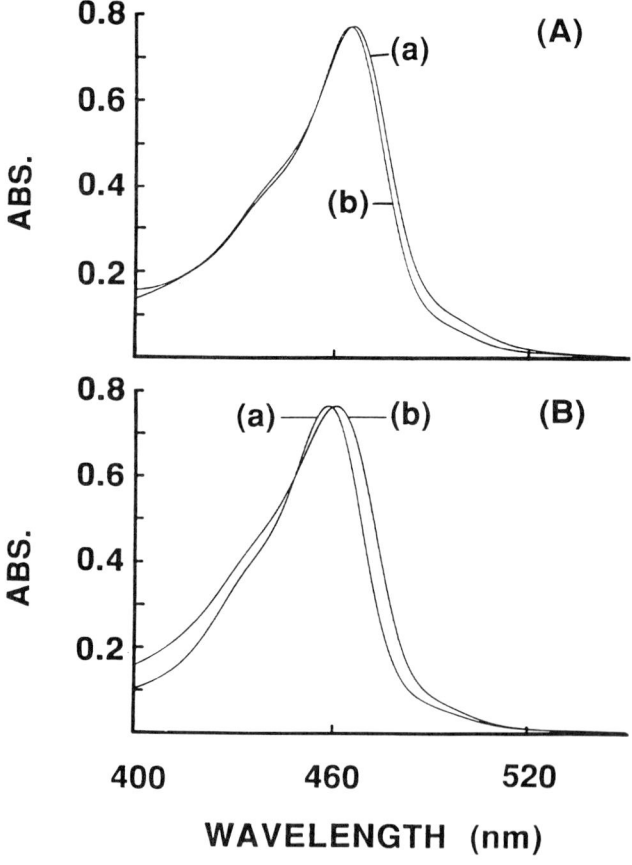

Fig. 23. Comparison of wild-type and βE109D quinonoid spectral bands derived from the reaction of (A) indoline and (B) N-methylhydroxylamine with the E(A-a) complex of each enzyme. For both panels, trace (a) is the wild-type enzyme and trace (b) is the βE109D enzyme. Spectra were normalized to the same maximum absorbance for the purposes of comparison. [Taken from Brzović et al. (87) with permission.]

chloride is a much better leaving group and may be eliminated directly after formation of an α-carbanion quinonoid intermediate without requiring general acid catalysis. If indole, or IGP, is also present in the reaction mixture, then the bienzyme complex is fully functional in catalyzing the synthesis of L-Trp. Spectral bands, which correspond to the L-Trp quinonoid, are observed and are very similar to those seen in the wild-type enzyme (data not shown).

Therefore, mutations at a single locus of the β-subunit in tryptophan synthase may alternatively affect either stage I or stage II of the β-reaction. Because the mutation of βE109 to L-Ala yields a mutant that is perfectly functional in the reaction of indole with the E(A-A) complex, E109 is not essential for the activation of indole as a nucleophile during stage II of the β-reaction. However, it is possible that E109 assists either in protonation of the hydroxyl leaving group of L-Ser or by providing the proper environment for elimination to occur. Particularly in the case of βE109D, mutation of a residue in the active site of the β-subunit perturbs the active site environment. This is evidenced both by the small spectral shifts found for structurally identical quinonoidal intermediates for the mutant and wild-type enzymes, and also by the drastic alterations in the accumulation of species during the β-reaction. Though the effect of the βE109D mutation on the steady-state turnover rates is rather modest (27-fold decrease in the β-reaction and four-fold decrease in the αβ-reaction as compared to wild type), clearly the rate determining steps for the βE109D, βE109A, and wild-type enzymes occur at very different positions in the catalytic pathway. The acquisition of such information is critical if reasonable comparisons and conclusions are to be made about the nature of specific active site residues on catalysis. Thus, in certain cases, RSSF spectroscopy can provide a direct, efficient, and economical means for determining this information.

2.3.2. MUTATIONS IN THE α-SUBUNIT OF TRYPTOPHAN SYNTHASE

The presence of a suitable chromophore, such as the PLP cofactor, makes it possible to investigate protein structure–function relationships that may be far removed from the chromophoric site. Tryptophan synthase is considered to be a prototype multienzyme complex in which metabolites are channeled directly between successive metabolic enzymes. As described above, allosteric interactions serve to coordinate catalytic events between the heterologous active sites in the complex. Such close interactions suggest that mutations in one enzyme may affect the reactivity of the other. We have found it possible to study the consequences of mutations in the α-subunit by looking for changes in the presteady state behavior of reactions catalyzed at the β-site (88, 89). Since amino acid replacements in the α-subunit will not affect the primary amino acid sequence of the β-subunit, alterations in the reactivity of the $\alpha_2\beta_2$ complex will be due primarily to differences in the reactivity of the α-subunit and/or αβ-subunit interactions.

This approach has been used to study the role of α-subunit structural elements in catalysis and allosteric interactions. One example is the replacement of Arg by Leu at position 179 (αR197L) of the α-subunit. This residue resides within a flexible loop that connects strand 6 with helix 6 in the α-subunit of tryptophan synthase. The native α-subunit shares the same folding topology with triose-phosphate isomerase (TIM), and both proteins have an extended loop region between strand 6 and helix 6 in the αβ-barrel structure. In TIM, this loop closes over the bound substrate and serves to sequester a labile intermediate within the confines of the active site (110). This conformational change prevents an irreversible side-reaction and promotes the efficient throughput of substrate to product. Substitution of Arg 179 in the α-subunit for Leu does not affect the steady-state kinetics of the β-reaction, and this mutation causes only a modest change in the turnover rate of the αβ-reaction. The K_m and k_{cat} for the αβ-reaction are decreased by only four-fold and five-fold respectively (99). Because the mutant bienzyme complex retains significant catalytic activity, this happenstance allows rapid-kinetic investigation of the effect of the mutation on the overall α-reaction. Consistent with steady-state kinetic studies, RSSF experiments (data not shown) show that the mutation does not alter the series of spectral changes observed for either stage I or II of the β-reaction. However, the accumulation of intermediates during the overall αβ-reaction are significantly different from the wild-type reaction. Subsequent investigation of the kinetics of the αβ-reaction show that the mutation decreases the affinity of IGP for the α-subunit by at least 15-fold. By analogy with the flexible loop in TIM, it is likely that loop 6 is involved in the conformational transition from an open to a closed structure that occurs upon ligand binding to the α-active site. Modeling studies (88, 89) suggest that movement of loop 6 would close off the active site from the solvent, thus sequestering the substrate and initial products of the α-reaction within the confines of the multienzyme complex.

Similar RSSF comparisons have been used to analyze effects caused by other mutations in the α-subunit (88). The most highly conserved region of the α-subunit is an inserted sequence between β-strand 2 and helix 2 containing an additional loop and helical segment. This region apparently serves several functions. It contains the catalytic residue Asp 60, a residue that also forms part of the tunnel wall within the α-subunit. Other residues in this segment have extensive interactions with the β-subunit. Substitution of Phe for Glu at position 49, Leu for Gly at position 51 and Tyr for Asp at position 60, residues within this highly conserved structural region of the α-subunit, result in enzymes that have either lost or have greatly reduced activity in the α-reaction. However, the kinetic behavior of these mutant enzyme complexes in reactions catalyzed at the β-site is qualitatively very similar to that observed for the αR179L mutant, suggesting that both the loop 2 and the loop 6 regions of the α-subunit are involved in the conformational transition from an open to a closed structure (88). The length of loop 6 makes it possible for these two regions of the α-subunit to directly interact upon the binding of ligands. These studies establish that certain mutations in either loop 2 or loop 6 could perturb such an interaction, altering the equilibrium between open and closed forms of the α-subunit. Furthermore, the extensive β-subunit contacts within loop 2 provide a mechanism for the transmission of allosteric information between the heterologous subunits in response to substrate-induced conformational changes in the α-subunit (89).

2.3.3. TRYPTOPHANASE

RSSF spectroscopy has also been used to study the effect of active site mutations on the β-elimination reaction catalyzed by tryptophanase. In many PLP-dependent enzymes, the Lys residue that forms the E(Ain) with the cofactor is preceded by a basic residue in the primary amino acid sequence. Phillips et al. (106) have examined the effect of changing Lys 269 to Arg on the formation and accumulation of reaction intermediates. The activity of the mutant enzyme is only 10% of the native enzyme. Secondly, the mutant enzyme exhibits an altered pH dependence both in the spectrum of the native enzyme and in the catalytic rate profile. RSSF studies of the reaction of L-alanine, L-Trp, S-methyl-L-cysteine, S-benzyl-L-cysteine (SBC), and oxindolyl-L-alanine show that all these various substrates react with the enzyme to form covalent intermediates. However, the rate and extent of quinonoid accumulation is greatly reduced. Analysis of quinonoid bands formed in the reactions of SBC and oxindolyl-L-alanine with tryptophanase show that mutation effects the equilibrium distribution of intermediates, but does not perturb either the band shape or the λ_{max} of the observed quinonoid intermediates. Therefore, the structure of the quinonoid intermediate and the surrounding active site environment are similar to the wild-type enzyme. SWSF characterization of these reactions show that the K_{eq} for $E(A_{ex})$ formation with each substrate is similar to that found for the wild-type enzyme. Instead, the primary effect of the Lys 269 Arg mutation is at the catalytic step in which the α-proton is removed from $E(A_{ex})$ to form a quinonoid. These studies show that Lys 269 is not a critical catalytic residue; nevertheless it does contribute to the conformational and/or electrostatic environment of the active site that is necessary for the formation and breakdown of quinonoidal species.

2.4. RSSF Studies of Heme Containing Proteins and Enzymes

Heme prosthetic groups are natural chromophores that are incorporated into a functionally diverse group of proteins that are involved in the binding and transport of dioxygen (myoglobin and hemoglobin), electron transport (cytochromes), reduction of peroxides (catalases and peroxidases), and in the catalysis of various hydroxylation reactions (68). Experiments utilizing rapid-scanning stopped-flow techniques have been productively applied to hemoproteins to gain insight into catalytic mechanism, allostery, and structure-function relationships within this broad group of proteins. A few examples are briefly described as follows.

2.4.1. HEMOGLOBIN AND MYOGLOBIN

Belleli et al. (111) have utilized RSSF to investigate the influence of the distal histidine residue on the dissociation of cyanide from ferrous (Fe^{2+}) myoglobin. Rapid mixing of a ferric-cyanide complex with dithionite results in the rapid formation of a spectroscopically distinguishable ferrous-cyanide complex, which slowly decomposes to yield reduced myoglobin and HCN. In this study, the RSSF spectral changes and kinetic time courses of both horse heart and sperm whale muscle myoglobin, which have histidine in the distal E7 position, were compared with mutant proteins in which His-E7 is substituted with either Gly or Val. The results demonstrated that the nature of the distal residue in the heme pocket not only affects the rate of cyanide dissociation but also influences the spectral properties of the ferrous-cyanide intermediate.

A similar system has been used to study the influence of allosteric interactions on cyanide dissociation in hemoglobin (112). The results of these RSSF studies demonstrate that the cyanide complex of the α- and β-chains are spectroscopically distinguishable and that dissociation is a cooperative process that is modulated by allosteric effectors.

2.4.2. ELECTRON TRANSPORT PROTEINS

Several investigators have utilized RSSF to study the oxidation and reduction of cytochromes involved in electron transport (113–116). An interesting example involves the use of RSSF spectroscopy and principal component analysis (see below) to analyze the spectral changes observed in the Soret and α-band regions of cytochrome c oxidase during reduction by 5,10-dihydro-5-methyl phenazine (MPH) (115, 116). Cytochrome c oxidase contains both α and $α_3$-type cytochromes with associated Cu atoms. Although the spectral bands of the cytochromes strongly overlap, it was possible to distinguish between reduction of cytochrome α and cytochrome $α_3$ based upon the time-resolved spectral changes during the reaction of the oxidase with MPH. The authors found that cytochrome a is the initial target reduced by MPH, whereas cytochrome a_3 is reduced in a slower, subsequent step that is likely controlled by protein conformational changes. Furthermore, using principal component analysis (see below), it was possible to determine (1) the minimum number of components necessary to describe the RSSF reaction profile and to resolve strongly overlapping spectral bands, (2) the spectral shapes for each of the individual components, and (3) the concentration versus time profile for each spectral component.

2.4.3. RSSF INVESTIGATION OF HYDROPEROXIDASES

RSSF techniques have been used extensively to study the catalytic mechanisms of both peroxidases and catalases (117–127). This class of enzymes are ferri(Fe III)-hemoproteins that exhibit characteristic α, β, and Soret spectral bands in the 400- to 600-nm region of the visible spectrum. These spectral bands are generally indicative of a high-spin nonplanar Fe heme prosthetic group. Peroxidases and catalases utilize either alkyl peroxides (Equation 5, where AH_2 represents a wide variety of reducing substrates) or hydrogen peroxide (Equation 6) as substrates, respectively.

$$AH_2 + ROOH \rightarrow ROH + H_2O + A \qquad (5)$$

$$2H_2O_2 \rightarrow O_2 + 2H_2O \tag{6}$$

At least two quasistable reaction intermediates, which differ in the nature of the Fe-bound ligands and the formal Fe oxidation state, referred to as compounds I and II, have been identified and characterized by a number of biophysical techniques (Walsh, 1979). Both of these species are spectrally distinct from the resting state of the native enzyme. In favorable cases, suitable reaction conditions or substrates may be chosen to conduct rapid kinetic investigations of the individual catalytic steps in the peroxidases or catalase mechanisms.

One example involves the use of RSSF to investigate the primary events in the reaction of H_2O_2 with horseradish peroxidase (HRP) to form compound I. Although the observed rates for this reaction are rapid, the calculated second order rate constants are below the diffusion limit. This finding suggests that another rapid process precedes the chemical reaction that results in the formation of compound I. Baek and Van Wart (1989) have used stopped-flow cryoenzymological techniques in conjunction with RSSF spectroscopy to investigate the elementary steps involved in the formation of compound I. Exploiting the large reductions in rate observed at lower temperatures, these authors found that the reaction of H_2O_2 with HRP under pseudo first-order conditions displays saturation kinetics. This kinetic behavior is indicative of the formation of an enzyme-substrate (ES) complex, which precedes the chemical redox transformations to form compound I. The RSSF spectral changes in the Soret band region between 300 and 450 nm, which accompany the initial phase of the reaction at $-25°C$ are shown in Fig. 24. The native enzyme in the absence of substrates displays a characteristic absorption band with $\lambda_{max} = 406$ nm (Fig. 24, trace a). Compound I (not depicted) exhibits a distinct spectral band at 405 nm, while compound II (the final species formed under the experimental conditions described) displays an absorption band with $\lambda_{max} = 420$ nm (trace b). The first scan (trace c) initiated at the start of mixing (mixing time is typically between 20 30 ms) shows some residual compound II ($\lambda_{max} = 420$ nm) present in the observation cell. Also evident in trace c are previously undetected spectral bands in the 330-nm and 400-nm regions of the spectrum. The next spectrum, initiated 21 ms after the start of mixing, shows decay of the 330-nm and 400-nm bands (trace d). Decay of these species is followed by the complete formation of compound I (data not shown). Thus, a transient intermediate with distinct spectral properties, designated compound 0, was found to precede the formation of compound I. Based upon the spectral properties of the new intermediate, the authors have postulated that compound 0 may represent the formation of a "hyperporphyrin" complex arising from the formation of a charge-transfer complex between the Fe-porphyrin and some form of the peroxide ligand.

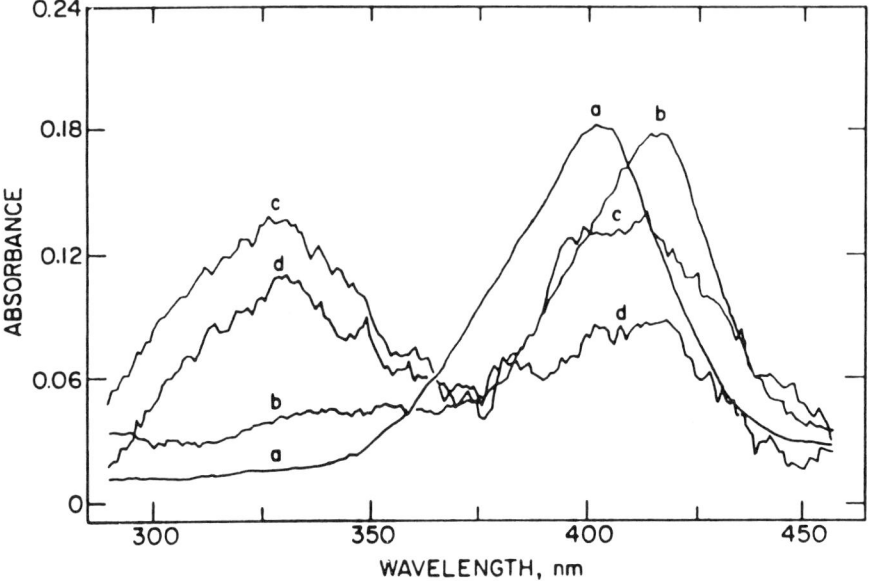

Fig. 24. Rapid-scan optical spectra of the reaction of 1 μM HRP with 1 mM H_2O_2 in 50% v/v methanol/10 mM phosphate, pH 7.3. Traces a and b are spectra of HRP and compound II, respectively. Traces c and d are consecutive 21 ms scans initiated at the start of mixing, which show the formation of compound 0 and subsequent conversion to compound I. [Taken from Baek and Van Wart (120) with permission.]

3. BASIC PRINCIPLES AND INSTRUMENTATION

3.1. Time Domains, Wavelength Domains, and Biological Processes

3.1.1. TIME DOMAINS

In the classical experiments of Hartridge and Roughton on the reaction of dioxygen with hemoglobin (128), rapid-mixing technology was first introduced as a strategy for the investigation of the dynamic properties of biological macromolecules. This technology has been thoroughly explored and extensively developed. There are now more than half a dozen commercially available rapid-mixing, stopped-flow spectrometers. Most are designed for aqueous solution work, but can be used with nonaqueous solutions as well. There appears to be no well-defined theoretical limit on how quickly the efficient mixing of two (or more) solutions can be accomplished. Considerations involving maximum flow velocity, rapid stopping, cavitation, and cuvette geometries which allow immediate optical interrogation of the mixed solution, give a lower practical limit for stopped-flow, rapid-mixing of about 200 μs (129–131). Most commercially available instruments advertise mixing dead times in the vicinity of 1 ms. Consequently, if the system provides a decent signal, then reactions with apparent first-order rate constants as fast as 700 to $1000 s^{-1}$ can be routinely measured. Certain modes of rapid-scanning detection can place additional time constraints on the rate of data acquisition, or data readout to computer memory. With solid-state linear array detectors, the combination of the time interval required to read out all the array elements and the obligatory exposure time can limit the fastest repetitive scanning rate to 1-2 ms.

Although the limitation to an ms experiment deadtime imposed by rapid-mixing technology precludes study of some physical-chemical phenomena of biological interest, the examples presented herein make it evident that the rates of many types of biological processes are amenable to interrogation by RSSF spectroscopies.

3.1.2. WAVELENGTH DOMAINS

The spectroscopy of biological molecules relevant to the investigation of structure-function relationships includes most of the electromagnetic spectrum. However, RSSF methodology has been centered on the use of UV-visible absorbance measurements in the 300- to 800-nm range. Almost no work has been published outside this wavelength range. In view of the extensive use of fluorescence (both in static and rapid kinetic measurements) to study a wide variety of biological molecules and biological phenomena and the extensive use of time-resolved vibrational spectroscopy in photobiology (132), it is somewhat surprising that RSSF emission spectroscopy has not yet emerged as a tool for the investigation of biological systems. This is particularly so since a rapid-scannng Raman spectrometer with scan rates up to 10000 cm-1/sec was described in 1968 (133). The low levels of light emitted appear to be the major impediment to the development of a rapid-scanning, stopped-flow Raman system. Whereas signal averaging methods and Fourier Transform technology have made it possible to extract time-resolved IR, Raman, and Resonance Raman spectra on ns time scales for intermediates in the photocycles of plant and bacterial reaction centers and the rhodopsins, so far as we are aware, this technology has not been extended to RSSF applications. Furthermore, neither RSSF fluorometry nor RSSF CD spectroscopy have become important tools for biological investigations.

3.2. Detection Strategies: Phototubes Versus Solid-State Detectors

Strategies for signal detection in RSSF UV-visible absorbance spectroscopy have developed along two different lines (134). Both rely on the dispersion mode wherein a grating (or, rarely, a prism) is used to separate the spectrum into its component wavelengths (or bands), which are then individually measured. In the classical approach (Fig. 25A), the dispersed spectrum is mechanically scanned across an exit slit before passing through the sample on to a single detector (usually a phototube). In the second approach, incident ("white") light is passed through the sample prior to dispersion (Fig. 25B). The dispersed spectrum is then imaged onto an array detector (usually a photodiode array or a charge coupled device) and spectral information is generated by the signals derived from the individual elements of the array. These two approaches are designated *single element detector* and *array detector* methods throughout.

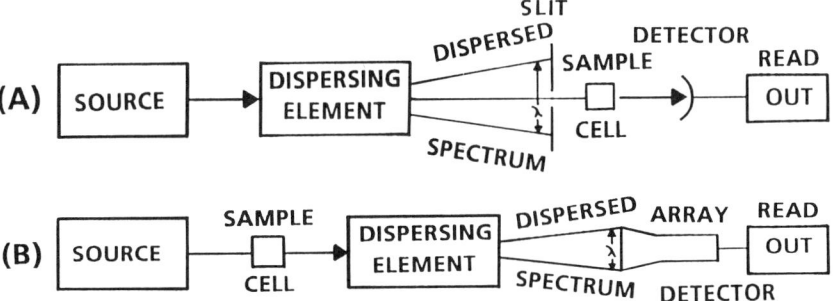

Fig. 25. Schematic diagrams of (A) conventional and (B) diode array scanning spectrophometers operated in the dispersion mode. (A) The spectrum is dispersed by a grating or prism (the dispersing element) and scanned across an exit slit. A single-element detector (usually a photomultiplier tube) is used to measure the intensity of a monochromatic beam after it passes through the sample. (B) The sample is illuminated with white light prior to dispersion. The dispersed spectrum is imaged on a linear array detector, and the signals from individual elements provide the information necessary to generate spectral information. [Redrawn from Santini et al. (134) with permission.]

The multiplex methods employing either Fourier transform or Hadamard transform technology have not yet been applied to RSSF UV-visible spectrometers. These methods are widely used in IR and Raman spectrometers. For the classic multiplex methods (135), frequency domain information is received by a single detector and transforms based on Fourier or Hadamard algorithms are employed to convert the data into spectral information. However, systems employing a moving mirror as part of a two mirror interferometer to create constructive or destructive interference at each wavelength contained in the incident beam are subject to mechanical tolerances, which compromise the use of transform methods in the UV-visible region of the spectrum. Thus, the signal-to-noise (SIN) advantages, which make FT IR and FT Raman viable spectroscopic approaches do not apply to the UV-visible region (136–138). To circumvent these limitations, Okamoto et al. (139) constructed a triangle, common-path interferometer with source doubling (Fig. 26) for use in the 300-nm to 1.2-µm region. This system generates a spatially resolved interference pattern, which can be imaged onto a linear array detector. Fast Fourier transform computer analysis of the digital interferogram then yields a reconstructed spectrum. This system requires no slits, apertures, moving mechanical parts. It is characterized by a large optical throughput and a resolving power limited by the number of array elements. Such an instrument would seem to have considerable potential for RSSF applications. However, the system presented by Okamoto et al. (139) requires a detector with a large dynamic range because most of the radiant energy of the source is concentrated in the central fringes of the interferogram. Okamoto et al. (139) propose that the insertion of a dispersive element into the optical path could be used to better distribute the energy of the interferogram over the linear diode array.

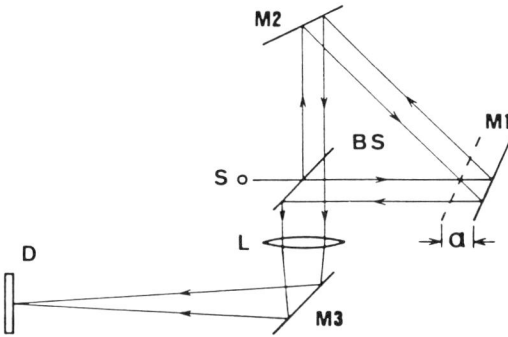

Fig. 26. Block diagram of the optics of the triangle common-path interferometer Fourier transform spectrometer: S, light source: BS, beam splitter; M1,M2,M3, plane mirrors; L, lens; D, self-scanning photodiode array. [Redrawn from Okamoto et al. (139) with permission.]

3.2.1. SINGLE ELEMENT DETECTORS

The commercially available, mechanically scanned instruments employ photomultiplier tube detectors. The utility of photomultiplier tube detectors stems from the wide linear dynamic range, high sensitivity and rapid (essentially instantaneous) response. In the absence of incident light, a residual current flows in the phototube anode lead. This dark current is the result of thermionic emission, emission caused by natural radioactivity (e.g., due to traces of ^{40}K in the glass envelope), field emission, and ohmic leakage. Because this dark current results from ejected electrons and, therefore, produces an amplified current, it sets a lower limit to the light intensity that can be directly detected. Thermal dark current can be minimized by cooling the photomultiplier tube. Dark current is a steady component that may be offset by potentiometer zeroing. Photomultiplier tube detectors are usually considered to be shot noise limited. Using a long electronic time constant in the amplification circuitry tends to average shot noise to zero. However, the need to make measurements on ms time scales places limits on the extent to which shot noise can be reduced by introducing a long electronic time constant. Clearly, this approach to noise reduction only works if the electronic rise time of the amplifier is fast relative to the rate of change of the optical signal under interrogation. Time-averaging or ensemble-averaging techniques provide viable alternatives for minimizing noise. However, rapid-scanning stopped-flow places severe constraints on the use of signal averaging techniques to reduce shot noise.

3.2.2. ARRAY DETECTORS

A wide variety of solid-state detectors consisting of multiple elements (multiple channels) have been developed over the past 20 years (140). Devices that fall into this category include silicon photodiode (SPD) arrays, charge injection devices (CID), charge coupled devices (CCD), microchannel plate (MCP) image intensifier-coupled silicon photodiode (MCP-SPD) arrays and lens-coupled intensified charge-coupled devices (LCI-CCD). The currently available array detector rapid-scanning instrumentation employs SPD arrays, CCD arrays, MCP-SPD arrays, and LCI-CCD arrays (136, 140, 141). The SPD, CID, and CCD devices approach photon-counting capabilities, the intensified array detectors achieve single photon counting abilities. The wide spectral sensitivity, linearity of response, large dynamic range, low noise, and temporal stability of these solid-state devices make them appropriate for RSSF UV-visible absorbance spectroscopy.

3.2.3. SPD AND MCP-SPD ARRAY DETECTORS

SPD detectors are reverse biased and, therefore, are charged capacitors (136, 140-144). At the start of the measuring cycle, each diode element is fully charged. Electron hole pairs are produced when light strikes an SPD. These electron hole pairs discharge the capacitance of the diode. During readout, each SPD element is sequentially accessed via the array shift registers and multiplex switches, and recharged by the video line. The voltage change on the video line measures the residual capacitance of the photodiode. This change is sensed by the preamplifier, digitized, and transmitted as data to computer memory. Signal in the form of electron hole pairs accumulates during the time between the start pulse (when the diode is recharged) and read out. Thus, the signal level is determined by the intensity of light and the dark current, integrated over this exposure time interval. The sensitivity of an SPD array is limited by thermal (1/f) noise and by systematic pattern noise resulting from diode-to-diode variations in current dc leakage (144).

The MCP-SPD array detectors, currently marketed by EG&G Princeton Applied Research, are variable gain, gateable devices that consist of a proximity-focused microchannel plate intensifier that is connected by fiber optics to a SPD array (140). A semitransparent photocathode in front of the MCP emits electrons when struck by light. These electrons are accelerated by a potential across the space between the photocathode and the microchannel plate. The electrons enter the microchannels (bundles of fine glass tubes with partially conducting walls) of the MCP. Collisions with the microchannel walls liberate additional electrons, causing the microchannels to act as electron multipliers. The shower of electrons exiting from the MCP is accelerated toward a phosphor screen by a high voltage potential (as much as 5kV) Photons are emitted when the phosphor is struck by these energetic electrons. The light generated is coupled via fiber optic links to the individual elements of the SPD array.

Because the dark current properties of SPD and MCP-SPD arrays are low and much of the dark current has thermal origins, the dark current can be substantially reduced by detector cooling. Most devices are equipped with Peltier-effect thermoelectric cooling systems coupled to cold liquid coolant systems to achieve temperatures as low as -40 to $-80°C$ in some designs.

3.2.4. CHARGE INJECTION DEVICES

The CID array is an integrating device, which consists of an x–y addressed array of charge storage capacitors capable of storing photon-generated charge (140, 143). Each element is made up of a pair of charge storage capacitors, one modulated by an x-row drive line, the other by a y-column drive line. Each storage capacitor collects thermal charge and photon-induced charge during light exposure. The charge generated during integration is read out by first removing voltage from one capacitor of the pair. This transfers the charge to the second capacitor. Removal of the voltage from the second capacitor injects the charge into the substrate. The capacitive coupling of the drive voltage to the substrate gives a video signal, which measures the charge stored on that element. Charge injection can be speeded up by using an epitaxial junction to form a charge collector under the array.

3.2.5. THE CCD AND LCI-CCD ARRAYS

CCD arrays are made up of metal-insulator–semiconductor capacitors, each composed of a conductive electrode and a thin layer of oxide insulator placed on a semiconductor p-type silicon wafer (143, 145). Application of a positive voltage to the conductive electrode generates a charge depleted region in the silicon layer. These charge-depleted potential wells are storage elements for thermal and photon generated electrons. As for the SPD and CID devices, the amount of charge accumulated is a linear function of the intensity of the light and the integration period. The gaps between electrodes are filled with undoped polysilicon to give highly resistive dielectric barriers that prevent coupling of charge between electrode gaps. There are several strategies for reading out signals from CCD devices (143, 145). These depend upon whether or not the array is linear or two dimensional, and whether or not readout can be accomplished rapidly enough to make "smearing" negligible. In general, the charge stored on a given element of the array during exposure is advanced sequentially from one element to the next via a high-speed shift register and then to the readout preamplifier. This provides a serial raster scan of each row of the individual charges stored in the array. The CCD devices are characterized by a very wide, dynamic range and very low, dark noise.

The LCI-CCD arrays employ a lens-coupled intensifier system composed of an image intensifier tube and a high quality floating element lens (140, 143). This design allows the performances of the CCD array and the intensifier tube to be optimized independently. Dynamic range is limited by the strip current capability of the intensifier at high light levels and by the system noise at low light levels. In the two-dimensional LCI-CCD array system provided by EG&G Princeton Applied Research, the detector system has variable gain and programmable pulse width and delay. The gated intensifier can function as a nanosecond time-scale electro-optic shutter. These features make the system very useful for time-resolved spectroscopy applications and for imaging.

For simple RSSF UV-visible spectroscopy, the SPD linear array is adequate for a wide variety of applications (see Section 2). The MCP-SPD linear array should extend the useful wavelength range and signal detection limits to make possible studies in the UV region between 200 and 300 nm, studies at higher scan rates and studies where the signal intensity is low. Gating on a nanosecond time scale also renders these detectors suitable for use in a variety of time-resolved spectroscopies.

Due to low, dark current and rapid readout characteristics, their large dynamic range ($\geqslant 18$ bit for the CCD) and the two-dimensional array feature, the CCD and LCI-CCD are more versatile detectors. These detectors and the MCP-SPD array detector are particularly useful for picosecond, time-resolved emission, absorption, and Raman spectroscopy, and for imaging applications where signal averaging is required.

3.3. Rapid-Mixing, Rapid-Scanning Stopped-Flow Systems

3.3.1. SINGLE ELEMENT DETECTOR SYSTEMS

Most designs for single element RSSF UV-visible spectrophotometers have relied upon mechanically scanned monochromators. Early instrumentation employed either rotating mirrors or a vibrating mirror to scan the spectrum across the monochromator exit slit. One design incorporates 24 corner mirrors fixed to a rotating drum (146). To achieve acceptable photometric accuracy, this system requires that the mirrors be exactly matched. The advantage achieved by the incorporation of the 24 corner mirrors is a high scan rate with low drum velocity.

In systems employing a vibrating mirror, the angular velocity of the mirror must be closely controlled. In one version (149), this is accomplished using a triangular wave to provide the drive current. This instrument also incorporates a beam splitter to divide the light from the exit slit into sample and reference beams. The appropriate amplifier system then gives a direct absorbance signal output. Scan rates ranging from 0.01 to 4000 scans/s for the wavelength range 370 to 700 nm were achieved. Linearity for the 0.02–2.0-Å absorbance range and a wavelength resolution of 2 nm at 1000 scans/s were reported.

Coolen et al. (148) and Papadakis et al. (149) have described a rapid-scan, stopped-flow spectrometer interfaced to a PDP-8/I computer. In this system, rapid scanning is achieved through a design, which utilizes a scanning monochromator with a rotating mirror system that presents the sample with monochromatic light of variable wavelength. A photomultiplier tube detector was used to monitor light intensity. With such systems, it is critically important to be able to correlate the rotation of the mirror with the beginning of the scan cycle. This was accomplished by attaching a 136-tooth gear to the mirror shaft. During every revolution, a beam of light is reflected from a polished gear tooth to signal the beginning of the scan pulse. The gear tooth signal provides the triggering pulse for sampling. A sampling frequency of 20.4 kHz was achieved with this system.

Although the hardware used in the design of the instrument and the computer system have been superseded by recent advances in RSSF technology and computer hardware (see below), many of the features built into the data acquisition and analysis software of this instrument represent innovations that should be incorporated into more modern instruments. The data acquisition strategy employed by Coolen et al. (148) was based on the realization that in a rapid-scanning mode of data collection, the time dependence of the signal change is not only a function of the reaction rate but also the nature of the spectral changes and the scan rate. At early times, the spectrum changes rapidly with time and this places severe limits on the extent to which signal averaging can be employed without introducing distortions in the individual spectra. However, at longer times, extensive signal averaging with negligible signal distortion is possible. Therefore, Coolen et al. (148) designed protocols into their data acquisition software that provide a variable frequency bandpass to achieve an increase in S/N. This was accomplished by averaging consecutive spectra. The time course was divided into groups such that each group contains the same number of spectra but each of the spectra stored from a particular group is the average of a certain number of consecutive spectra (C) determined by the equation

$$C_n = g^{n-1} \tag{7}$$

where n is the group number and g is the grouping factor. For example, using a group factor of 2, 10 spectra per group, and 5 groups gives 50 stored spectra from a database of 310 measured spectra. Such a data collection strategy should be useful both for single element detector systems or for array detector systems. One advantage of the variable bandwidth strategy is that reactions that involve both rapid and slow relaxations may be monitored during the same kinetic run while achieving the best compromise between scan rate and S/N throughout the time course.

A very recent innovation to the single element detector approach to RSSF UV-visible spectrophotometry (manufactured by On-Line Instrument Systems, Inc.) has just become commercially available. This system, the OLIS RSM, employs a mechanical moving slit mechanism based on a subtractive double monochromator. To achieve scanning, the intermediate slit between the two monochromators is moved. The double monochromator design gives very low stray light characteristics, a spectrally homogeneous output beam, spectral resolution determined by the intermediate slit and no temporal dispersion. Scanning speeds of 1000 scans/s for a 220-nm span are claimed for routine operation, and higher scan rates (3000 to 10,000 scans/s) are possible (specification sheet, OLIS, 1992). With suitable grating and optical components, this instrument is capable of scanning within the 200- to 1000-nm region. An extensive software package is offered, which places the RSSF unit under computer control for operation, data acquisition, display, and data analysis. The analysis package includes algorithms for singular value decomposition to determine the number of components involved in the reaction, robust global fits, and data compression. Because this instrument has only just appeared on the market, it has not yet been given extensive use in the field, and no published work derived from this instrument has yet appeared. The key design innovation is the use of a subtractive double monochromator and a moving slit mechanism at the position of the intermediate slit. By moving the slit across the dispersed spectrum from the first monochromator, the wavelength of light entering the second monochromator is scanned. The slit motion is achieved by placing the slit on a rotating disk. The rate of rotation is precisely controlled by a stepper motor mechanism so that high wavelength accuracy can be achieved.

Single detector and array detector RSSF systems, which acquire a family of time-resolved, UV-visible spectra from a single, stopped-flow, rapid-mixing event have been the two methods of choice. However, one manufacturer of stopped-flow, rapid-mixing instrumentation (Applied Photophysics) has recently introduced a computer controlled system, which collects a set of single wavelength time courses at programmable wavelength intervals. From these time courses, time-resolved spectra are constructed from "time slices" taken at selected times and a predetermined set of 100% T values for each wavelength. Because a microvolume flow system is used, a 1-ml solution is sufficient for the acquisition of up to 400 spectra from 40 traces spread over a 200-nm range with 5-nm resolution and with a minimum time interval limited by the speed of the A/D converter (Specifications sheet, Applied Photophysics, 1992). The entire system is automated so that the operator initiates data collection by setting the start parameters via a workstation after loading the drive syringes. The manufacturer provides a multiparametric software package for analysis of the kinetic time courses and analysis of the number of components.

3.3.2. ARRAY DETECTOR SYSTEMS

Because much of the RSSF work presented in this review is derived from experiments carried out in the author's laboratory, the instrumentation employed are described below in some detail. The author's RSSF system is assembled, with minor modifications, from commercially available components. The schematic diagram presented in Fig. 27 gives an overview of the optical, electronic, and mechanical elements of the system. The rapid-mixing system consists of a Durrum D110 stopped-flow spectrometer with the grating/prism monochromator removed so that the "white light" output of an Xenon or tungsten–iodide lamp is passed through the sample. A system of first surface mirrors is employed to provide a columnated beam of light through the observation cuvette. The intensity of light reaching the sample is modulated at three points: the output of each lamp is controlled by variable voltage/current power supplies, the light exits the lamp housing through an aperture of variable diameter, and then passes through a slit of selectable width located in the optical path just before the columnating mirrors. After passing through the observation cuvette, the beam is refocused on the polychromator entrance slit (a point where light intensity can be further modulated). The polychromator consists of a Model HR-320S spectrograph (ISA Instruments, S.A. Inc.). The spectrograph has a 0.32-m, coma corrected Czerny–Turner configuration with F/5.0 aperture and a 590

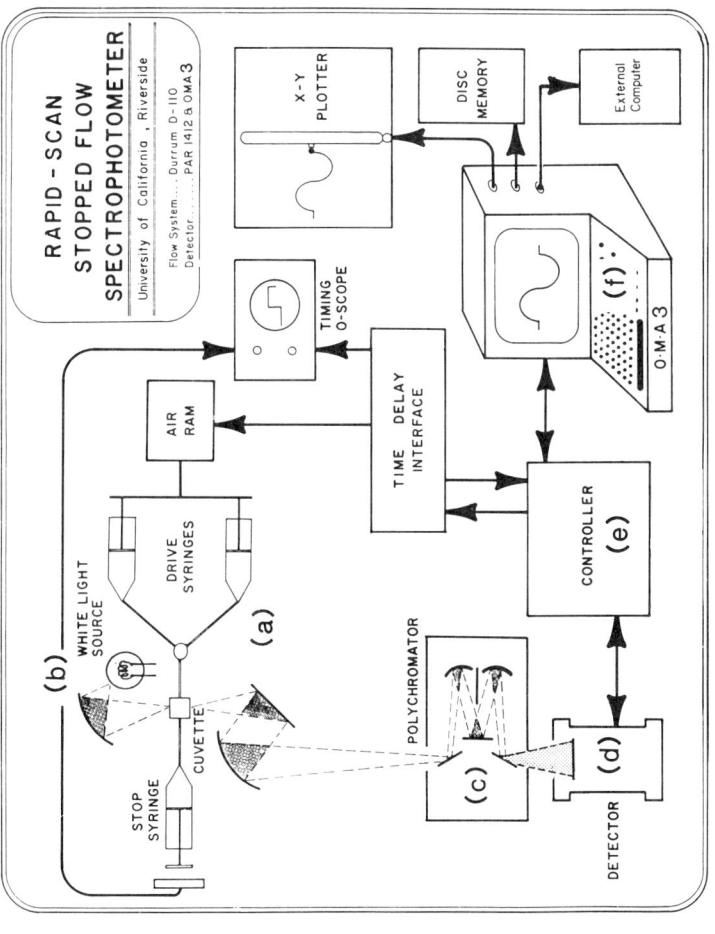

Fig. 27. Diagram of the RSSF system used in the author's laboratory. (a) Durrum D110 stopped-flow, rapid mixing unit modified to illuminate the observation cuvette with white light from tungsten or Xenon lamps. (b) Timing device consisting of a time delay interface and oscilloscope for controlling and measuring the time interval between the initiation of flow, flow stopping and the initiation of scanning. (c) Model HR-320S spectrograph (ISA Instruments, S.A. Inc.). (d) PAR Model 1412 1024-element linear SPD array. (e) and (f) PAR Model 1460 OMA III with 14 bit A/D converter. The A/D converter board resides inside the OMA III. Data transfer modes to an X-Y plotter, to hard disk memory or to an external computer are indicated. From S. C. Koerber and M. F. Dunn, unpublished work.

grooves/mm grating blazed at 300 nm to give efficient throughput of light in the UV region. The exit slit is removed so that the rainbow of light from the first spectral order containing wavelengths from 200 to 800 nm can be imaged on a linear photodiode array detector. Extensive internal masking of the focusing mirrors and central zone of the polychromator with black photographic paper reduces stray light artifacts and unwanted overlap of spectral orders. An adjustable, wavelength-variable mask is situated just in front of the diode array so that the intensity of light at any wavelength can be selectively attenuated. With a suitably tailored mask, partial compensation for the large variation in signal intensity due to the wavelength dependent emission of the lamp and the wavelength-dependent sensitivity of the diode array can be achieved with an acceptable loss of wavelength resolution. In our system, this masking is of critical importance for obtaining spectra of reasonable quality for the wavelength range below 360 nm. To achieve sufficient illumination of the diodes receiving light in this wavelength region but not exceed the dynamic range of the array A/D converter at longer wavelengths, the intensity of light at the longer wavelengths is partially masked. The relationship of such a mask to the diode array is depicted in Fig. 28.

The diode array employed depends upon the wavelength region of interest. For spectra measured between 290 and 800 nm, a simple PAR 1024-element linear SPD array (Model 1412) is used; because the sensitivity of the model 1412 SPD array is insufficient for rapid-scanning applications in the region between 200 and 360 nm, an intensified MCP-SPD (with 512 elements, PAR model 1420B-512-HQ) is used. The model 1420B-512-HQ detector is designed for optimal detection in the blue region of the UV-visible spectrum. This intensified array employs a proximity focused, multichannel plate (MCP) intensifier coupled by fiber optics to an SPD array (see Section 3.23). Both arrays are run at reduced temperatures to reduce thermal noise and are operated under the control of an optical multichannel analyzer, the OMA III model 1460. The model 1460 OMA III is a microprocessor controlled instrument that stores and displays spectral data. The OMA III has a high-speed, 14 bit A/D converter for digitization of the SPD array signals. The digitization rate is 16 μs/diode plus 0.5 μs/diode for read/reset operations and 201 ps overhead time. Consequently, scanning all 1024 diodes of the 1412 array with the minimum exposure time possible gives a maximum scan rate of 16.663 ms/scan. Grouping diodes provides increased sensitivity at the expense of resolution and allows the group to be read as a single data point. Grouping diodes allows faster reading. A group of two diodes requires a minimum read time of 32 μs. Each additional diode in the group requires an additional 0.5 μs. Thus, if the set of 1024 diodes are divided into groups of four, the maximum scan rate is 8.826 ms/scan. The OMA III software and hardware also can operate in a "fast access" mode, which allows diodes to be processed without reading at a rate of 0.5 μs/diode. Diodes designated for fast access are, in effect, skipped over at this rate. Areas of the spectrum, which do not give useful signals, can be skipped over by designating the corresponding diodes for fast access to achieve a faster scan rate. For example, if the fast access mode is used to skip 512 diodes, then the remaining 512 diodes can be scanned at a rate of 8.826 ms/scan. Consequently, by the appropriately selected

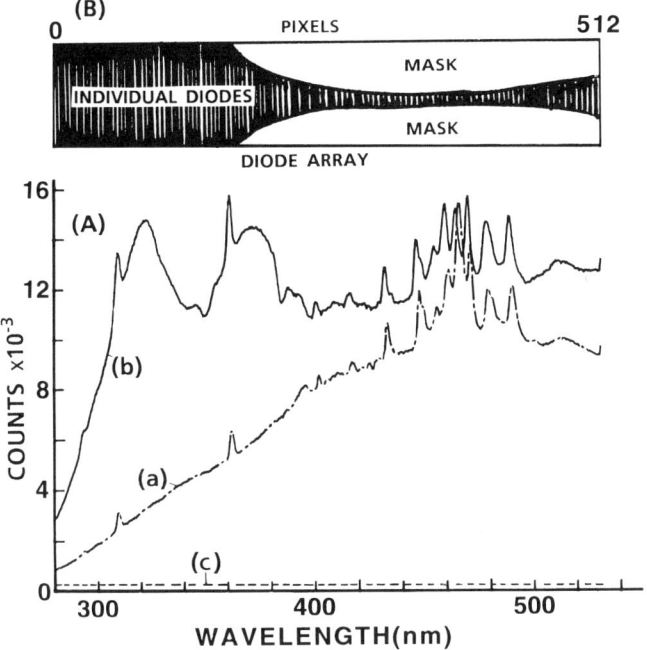

Fig. 28. (A) Comparison of the Xenon lamp emission spectrum 280 to 540 nm) in the absence (a) and presence (b) of the mask (see B) positioned in front of the photodiode array. Trace (c) is the dark current spectrum of the diode array. Only the signals from 512 of the 1024 pixels are shown. To achieve the improved signal to noise in the 280–400 nm region, compare (a) and (b), the mask is adjusted to reduce signal intensity above 380 nm (see B), and the lamp intensity is increased. Since no masking is used below ~380 nm, the increased lamp intensity gives a net increase in signal intensity below 380 nm but, due to the masking at longer wavelengths, the signal levels do not exceed the dynamic range of the A/D converter. (B) Carton depicting a portion of the photodiode array and the mask.

grouping of diodes and designation of fast access for unwanted signals, it is possible to achieve scan rates of 3 to 10 ms/scan with acceptable wavelength resolution and range for rapid-scanning applications.

The OMA III software is an extensive menu-based system with many submenus that either allow the user to initiate specific system functions, or allow the user to create customized routines for data acquisition and analysis. The OMA III computer system also allows the operator to create user-designed software programs using Hemenway BASIC for data acquisition and analysis. The data acquisition options available through the software of the operating system are extensive. The authors have created an interactive, menu-based software program for data acquisition and analysis that includes wavelength calibration, selection of the subset of diodes to be scanned, the grouping of diodes, diode exposure time, the pattern of saved scans, the designation of memory files for data storage, the collection of 100% transmission reference spectra and dark current spectra, the conversion of intensity data to absorbance, the display of spectra, the display of "time-slices" at selected wavelengths, and the collection of single wavelength time courses. A multiparameter algorithm is used to correct a modest nonlinearity in the wavelength scale.

The outstanding features of this instrument are the reliability of operation and ease with which the optics can be adjusted to optimize performance in the wavelength range of interest. The more modern electronics of the commercial instruments that have recently appeared (see below) provide for faster scan rates. However, we have yet to encounter a system where increasing the scan rate from the 3 to 10 ms/scan we routinely achieve to 1 or 2 ms/scan would prove critical to the investigation. The OMA III software and data analysis capabilities for RSSF applications are limited. Software that provides for multicomponent analysis and single value decomposition would make possible a more sophisticated and quantitative analysis of the data.

The OMA III-diode array system provides a highly integrated hardware-software package designed for the acquisition and limited analysis of spectral data on fast time scales. As an alternative to the OMA, EG&G currently offers application software packages for PC-style computers for data acquisition and analysis of array spectrographs. This software includes spectral calibration for wavelength and intensity, data storage, full access to complex triggering modes of array operation, display of spectra as absorbance, transmission or as the raw spectral data, automatic background subtraction, and some data analysis via user defined software. These software packages make it possible to substitute an IBM-compatible computer for the optical multichannel analyzer with a resulting savings in overall price for a RSSF spectrometer system.

Several investigators have published descriptions of diode array detector systems designed for rapid-scanning applications (150–153) Carter et al. (150) describe the construction of a low cost, 512-element diode array detection system interfaced and controlled by an IBM-compatible computer capable of collecting multiple spectra with a spectral window of 300 nm and a scan rate of 5 ms/scan. Schlemmer and Mächler (151) present an optimized design for a SPD array spectrometer (Fig. 29) that employs a concave holographic grating designed to achieve a flat focal plane for spectral measurements for use with a 512-element SPD (the EG&G Reticon G series array) to measure spectra between 380 and 780 nm. This instrument employs a large aperture (f/2.3) and is designed for spectral measurements on areas as small as 0.1-mm dia with a spectral resolution better than 2.5 nm and scan rates in the 10-ms range. A multiprocessor circuit for the fast correction of spectra to eliminate the influence of the wavelength dependent emission of the light source and the dark current also is described. This SPD array spectrometer should be economical to build, the optimized design appears well suited for RSSF applications. With slight modifications, it should be possible to construct an instrument that includes the UV-region of the spectrum and incorporates diode arrays with scan rates of 1–2 ms/scan (see below).

The paper of Diem and Ludl (152) describes a low-cost microcomputer-controlled diode array detector data acquisition device (interface) designed for low-level optical spectroscopies. While these authors are interested in Raman

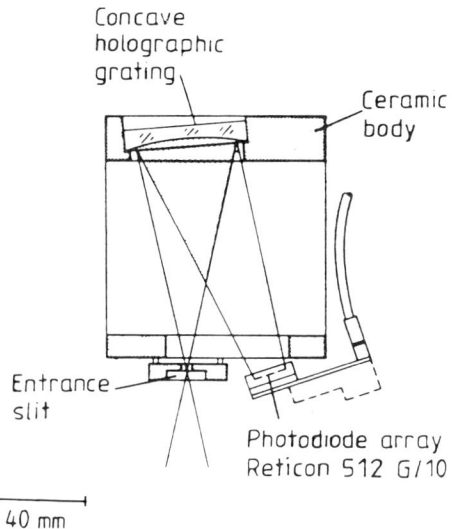

Fig. 29. Compact diode array spectrophotometer consisting of the minimum number of optical elements possible. [Taken from Schlemmer and Mächler (151) with permission.]

spectroscopy applications, their instrumentation, with suitable modifications, may also be appropriate for RSSF applications where low signal levels (e.g., fluorescence or CD) complicate the acquisition of spectral data on ms time scales.

Ryan-Hotchkiss and Ingle (153) evaluate an intensified diode array system for low light level applications for luminescence measurements with the emphasis placed on applications involving the monitoring of reactions involving fluorescent species. The intensified array examined employs a MCP intensifier and phosphor system (see Section 3.23). They found that if a large, dark signal is allowed to accumulate, several scans are required to reach a constant level for successive identical integration periods. This lag phenomena can be circumvented if the array is scanned periodically so that the array is not allowed to become saturated with dark current signal. We have seen similar effects with simple SPD arrays. Lag effects were also detected when the intensifier was switched on and the array is briefly exposed to a saturating level of light (153). This lag appeared to be a component of the phosphor decay.

At least three manufacturers of stopped-flow, rapid-mixing spectrometers offer fully integrated RSSF systems which employ diode array detectors, the Atago Bussan Co., Ltd. (Japan), HIGH-TECH Scientific Ltd. (England), and BIO-LOGIC Co. (France). Atago Bussan manufactures the Union Giken RA-401 Stopped Flow Analysis Unit, which can be used with their RA-415 Rapid Scan Attachment for RSSF applications. When configured for rapid scanning measurements, the sample is illuminated by white light either from a 50-W tungsten lamp or a 25-W deuterium lamp The light passes through the observation cuvette and enters a 25-cm focal length, F/4 Czerny–Turner type monochromator equipped with a 600 grooves/mm grating. A 1024 element photodiode array replaces the exit slit. The spectral range is 200–800 nm with a wavelength resolution of 512 points over 184 nm. A 12-bit A/D converter is used to digitize the diode array signals. The maximum scan rate is 97 nm/ms. The high speed data buffer used to capture scans limits the collection of spectra to 16 spectra per run. Exposure times range from 2 ms to 200 ms and the scanning interval is 0 to 2 s (Specification sheet, Atago Bussan Co., 1993). The high speed data buffer provides rapid, temporary storage of the digitized diode signals, which then are transferred to an IBM-compatible PC/AT style computer. The computer software controls data acquisition and analysis. The data acquisition program allows the operator to select operating conditions, establish wavelength calibration, collect reference 100% transmission and dark current spectra, initiate data collection, and create storage files for the data collected. The analysis of data by the manufacturer's software is limited to the calculation of difference spectra and the smoothing of data. The RA401 flow system is equipped with two flow cells, a 30-µl 2-mm path cell and a 42-µl, 10-mm path cell. A mixing deadtime of 0.5 ms is claimed for the 2-mm flow cell. The outstanding feature of this instrument is the rapid repetitive scan rate. The scanning interval of 0 to 2 s, the limit of 16 spectra per run, and the modest data analysis software package place some limits on the usefulness of the system for RSSF spectroscopy.

The model MG-6000 by HI-TECH Scientific, which employs the SF-60 series HIGH-TECH stopped flow spectrometers, is designed to collect spectral data between 250 and 1000 nm at a maximum scan rate of 1.25 ms/scan with wavelength resolution <2 nm and selectable integration times from 1.25 ms to 20 s. The array contains 512 diodes, the polychromator uses a Czerny–Turner configuration with a 100 lines/mm grating, an aperture of f/4, and a 0.015-mm slit. Up to 200 scans per experiment can be collected. Data acquisition and analysis are under the control of IBM-compatible hardware. Any 100% IBM PC/AT clone running at ⩾33 MHz with at least 8-Mb memory will suffice. Hard disk memory of 100 Mb is recommended (Specification sheet, model MG-6000, HI-TECH Scientific). Details of the data analysis software are not yet available. While there are no published studies at this writing using the MG-6000, the specifications of this instrument indicate it has the potential to be a very useful system for RSSF applications.

The BIO-LOGIC diode array spectrometer, the KINSPEC, employs a 512 element diode array with a minimum scan rate of 1.3 ms/scan. This rate can be reduced to 0.8 ms/scan by reducing the array to 256 pixels. Spectra are captured in a data buffer that can accommodate up to 1000 spectra. A 16-bit A/D converter is used to digitize the array signals. A 486 based PC-compatible 25 or 33 MHz computer with at least 4 Mbyte RAM and 100 Mbyte hard disk and VGA graphics board is recommended for use with this system. The optical system consists of a polychromator specified for operation between 200 and 620 nm, 300 and 720 nm, 360 and 780 nm or 600 and 1010 nm with wavelength resolution of 0.80 nm that is coupled to the stopped-flow rapid mixing unit via fiber optics (Specification sheet KINSPEC array spectrometer, BIO-LOGIC, 1993). The RSSF detector system is designed for use with the BIO-LOGIC SFM-3 and SFM-4 stopped-flow instruments. White light sources are provided with lamps and optics suitable for the wavelength range selected. The software provided controls the RSSF system and provides for data acquisition and storage. Data analysis includes a routine for bi-exponential analysis of the spectra to obtain rate constants and amplitudes using a Marquardt algorithm. At this writing, there are no published examples of RSSF studies with the KINSPEC, and the available literature does not give detailed information on the manufacturer's software. This system also appears to have the potential for useful applications in RSSF spectroscopy. Like the Union Giken RA-415 and the HI-TECH model GM-6000, the relatively rapid repetitive scanning rate achieved by the KINSPEC and the integrated design of the RSSF spectrometer are important characteristics of the system.

4. DATA ANALYSIS STRATEGIES

RSSF spectroscopy provides a three-dimensional data set where, in the absorbance mode of detection, the three dimensions are absorbance, wavelength, and time. Compared to SWSF spectroscopy, RSSF spectroscopy significantly expands the set of information contained in a single experiment. This wealth of information has the potential not only for defining the kinetic pathway of a reaction, but also for allowing spectroscopic characterization and identification of transient intermediates that accumulate and decay along the pathway.

4.1. Qualitative and Semiquantitative Analyses

RSSF spectroscopy has proven to be a valuable tool for the investigation of complex rapid reactions in biological systems. Inspection of the time-resolved spectra or simple manipulation of the spectra to generate difference spectra and first or second-derivative spectra often has proven sufficient to characterize and identify transient species. Such inspections rely on the observation of isosbestic points or the lack thereof, and the detection of multiphasic processes involving the appearance and decay of transient spectral bands to infer the minimum number of kinetic processes involved. The appearance and decay of new spectral bands establishes the involvement of transient intermediates, and the characteristics of the spectrum of the intermediate (when well resolved) can provide enough information to allow an unambiguous assignment of a chemical structure to the intermediate. Further quantitative work on the system usually is undertaken through a systematic set of SWSF kinetic studies at wavelengths chosen after inspection of the RSSF data. Hence, the RSSF data are used for qualitative analysis of the system, while SWSF are used for a more quantitative treatment of the system that may involve analysis of rates and amplitudes within the framework of relaxation kinetic theory (154).

The semiquantitative analysis of RSSF data sets has proceeded along two lines. One approach has been to remove unwanted contributions to the spectra by subtraction of reference spectra from the set. In this fashion, contributions from the reactants or products can be eliminated. Oftentimes, the calculation of such "difference spectra" reveals the band shapes of spectra derived from the spectrum of a transient intermediate. Examples where this approach has been used to good advantage are the investigations of Zn(II)- and Co(II)-substituted liver alcohol dehydrogenase (3, 20, 22, 23, 26, 155, 156), tryptophan synthase (85, 86, 91, 157), cystathionine γ-synthase (107), and the insulin hexamer (55). The study of an intermediate formed in the T- to R-transition of the Co(II)-substituted insulin hexamer by difference RSSF UV-visible spectroscopy provides a particularly clear example of how simple difference spectra calculated from scaled subtractions can be used to decompose the overlapping spectral bands of reactant, intermediate, and product to obtain the spectrum of the intermediate (viz., Fig. 8).

A second approach has been to manipulate the set of RSSF spectra by differentiation. Because the large bandwidths of electronic transitions make resolution of closely spaced bands difficult to resolve, it is sometimes useful to calculate the first and second derivatives of the spectra with respect to wavelength (i.e., $d^n A/d\lambda^n$, where $n = 1$ or 2). This manipulation of the spectra can provide information about the number of component electronic transitions and the locations of bands. The derivative spectra exhibit narrower bandwidths and thus give an enhancement of resolution, and points of constant wavelength-first derivative with respect to time (isotachic points) can aid in determining the relationship between individual absorbers in the set of RSSF spectra (157). According to theory, the locations of the isotachic points provide information about interconverting species. Drewe et al. (157) developed a theory of isotachic points for RSSF spectroscopy. The practical corollaries, which emerged from this treatment, are as follows: (1) by definition, an isotachic point, λ_{iso}, is a wavelength where the observed derivative spectra do not change as the concentrations change; (2) if an isotachic point, λ_{iso}, exists, then the change in the derivative spectra due to the ith species must be balanced exactly by contributions from all other species, (3) when the concentration of a species does not change during a reaction, the positions of any isotachic points are unaffected by the spectrum of that species, and (4) if the concentration of a species changes slowly with respect to other species present, apparent isotachic points will be present in the spectrum while the more rapid process(es) occur. Furthermore, if two absorbing species interconvert, isotachic points will lie outside the region bounded by the λ_{max} of the two species. If only one of the two species does not exhibit a spectral band in the region under study, then the isotachic point will be located at the λ_{max} of the absorbing species. The RSSF data shown in Fig. 30A–D compare the time course of the spectral changes that occur in the reaction of D-Trp with the *E. coli* tryptophan synthase bienzyme complex (A) with the first (B) and second (C) derivative spectra for the same RSSF data set. There are two isotachic points, one located at 404 nm (positive value) and one at 463 nm (negative value). The 463-nm isotachic point lies between the two absorbers (λ_{max} 450 nm and 478 nm) that form in the reaction, the 404-nm isotachic point is blue-shifted relative to the λ_{max} of the initial absorber (the internal aldimine, λ_{max} 410 nm). The location of the 463-nm isotachic point between the product bands implies these bands (λ_{max} 450 and 478 nm) represent species that are in very rapid equilibrium relative to their rate of formation (157). These isotachic points support the conclusion that the 423-, 450-, and 478-nm bands arise from three different species.

4.2. Quantitative Analysis of RSSF Data

As described above, the three-dimensional data sets generated from RSSF experiments potentially contain a wealth of information regarding enzyme reaction mechanism. This information includes data concerning the presence, covalent structure, and the kinetics of the accumulation and decay of transient reaction intermediates. Qualitative inspection of two-dimensional plots of RSSF data is often sufficient both to determine the existence of transient species and to develop strategies for detailed SWSF analysis of a particular enzymatic reaction. However, in certain cases the large amount of spectral data may be more fully exploited by applying quantitative analysis techniques. With the aid of low-cost computers capable of storing and handling large amounts of data and the ready availability of efficient algorithms for processing and analyzing multivariant absorption data, it is possible to extract information regarding the number of components necessary to describe the observed spectral changes, the spectral characteristics of individual intermediates, and to determine rates for the accumulation and decay of species with a high degree of confidence. Two approaches, which will undoubtedly receive greater utilization in the future, are the application of both global analysis and singular value decomposition (SVD) for the analysis of RSSF spectra. Only a brief description of these approaches as they relate to the analysis of multivariant absorption data is presented here. However, complete discussions concerning the general application of these methods to the analysis of biophysical data have been presented elsewhere (159, 160, 161).

Global analysis of data involves the simultaneous analysis of multiple experiments in which certain constraints have been imposed upon fitting parameters. RSSF data is essentially a multiple set of single wavelength "time-slices" collected at regular wavelength intervals. If the experiment has been conducted under pseudo first-order conditions, the rates of the various observed relaxations are obtained by fitting the individual time courses to a sum of exponentials using some sort of nonlinear least squares type of algorithm. Global analysis of RSSF data would involve the *simultaneous* analysis of multiple time courses in which the same rate constants are imposed for each time course but vary with respect to amplitude. The advantage of this approach is that it is possible to accurately determine from a single experiment the rates of poorly resolved or closely spaced relaxations, which may differ by a factor of three-fold or less in rate. Faced with a similar problem in conventional SWSF spectroscopy, experiments may be required, which alter the physical conditions of the reaction temperature, pH, pressure, etc.) in order to demonstrate that a single time

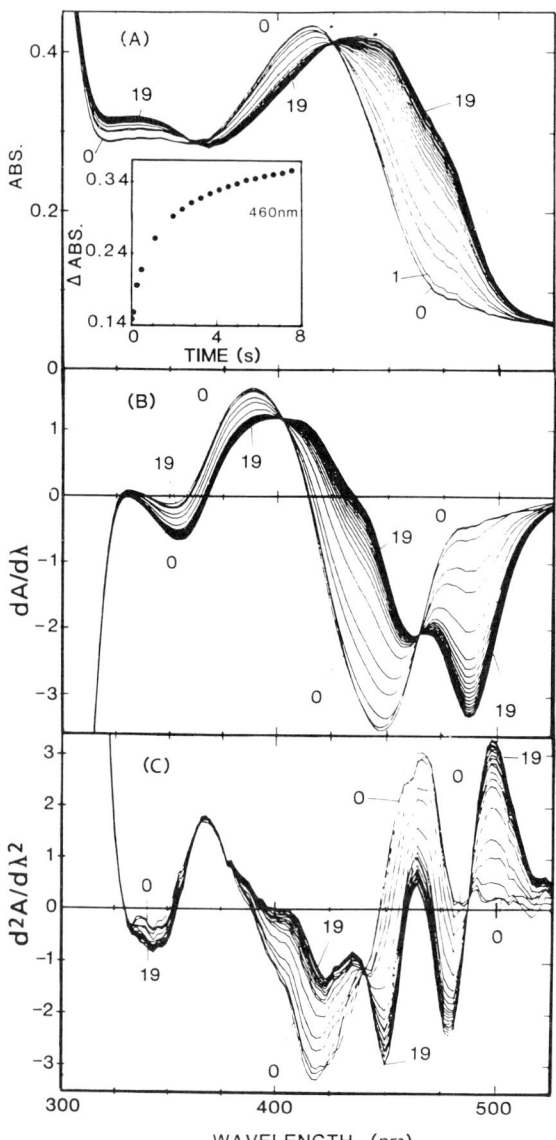

Fig. 30. Rapid-scanning, stopped-flow spectra (A), first derivative (B), and second derivative (C) spectra for the reaction of $\alpha_2\beta_2$ with D-Trp. Scan 0 is the reconstructed spectrum of the reactants. Collection of spectrum 1 was initiated 0 ms after flow stopped. Inset, 460-nm time course reconstructed from the RSSF data. Conditions after mixing: $[\alpha_2\beta_2] = 26.7$ µM; [D-Trp] = 1 mM; 0.1 M potassium phosphate, 1 mM EDTA, pH 7.80; 25°C. First (B) and second (C) derivative spectra calculated from the RSSF spectra in (A) for the reaction of $\alpha_2\beta_2$ with D-Trp. The first and second derivatives of each spectrum in (A) were evaluated pointwise over a 22-nm bandwidth by the method of Savitzsky and Golay (158). The ordinate scale is given in arbitrary units. Inset: Time-course at 476 nm reconstructed from the changes in the second derivative spectra presented in (C). The first derivative spectra (B) are characterized by isotachic points at 404 nm and 463 nm. The second derivative spectra exhibit four well-defined minima located at 340, 423, 450, and 478 nm. [Redrawn from Drewe et al. (157) with permission.]

course is actually composed of multiple relaxations. In these situations, closely spaced relaxations may not be determined with a high degree of confidence.

Strategies for the implementation of global analysis of biophysical data are described in detail elsewhere (159). However, it should be kept in mind that global analysis of an entire RSSF data set consisting of 100 or more spectra collected over a 250-nm wavelength region may become cumbersome due to the large number of data points, which must be processed. In some cases, it may be necessary to choose a subset of the data for simultaneous analysis. Secondly, relaxations that correspond to relatively fast processes may not be adequately determined due to limitations imposed by the time resolution (typically, 1–5 ms/scan) of a RSSF experiment (162).

Digitization of data collected during a RSSF experiment allows the data to be represented by a matrix in which the columns represent the individual time-resolved spectra collected, and the rows represent the wavelengths at which absorbance values are measured. Two related approaches, principal component analysis (PCA) and singular value decomposition (SVD), may be used to decompose the data matrix to obtain model independent information regarding both the minimum number of components present in the system under investigation and the contribution of each component with respect to both the wavelength and the time domains. With regard to RSSF spectroscopy, this information can be used to determine the number of different chromophores necessary to describe the observed spectral changes, their spectral shapes, and the concentration versus time profiles for each species. The primary requirement for the application of PCA and SVD is that the absorbance at each wavelength is a linear function of the concentration of each absorbing species (i.e., that Beer's Law is obeyed). Halaka et al. (115) have described in detail the use of principal component analysis as applied to RSSF spectroscopy. More recently, SVD-based analysis has become more widespread since the SVD algorithm is more robust and a number of efficient programs are readily available (160, 161, 163). Utilization of either PCA or SVD has the added advantage of greatly reducing the volume of data, which must be both stored and processed further, increasing the efficiency of data analysis.

The quantitative data analysis protocols described herein may be generally applied to multivariant data sets and promise to be of increasing utility for the analysis of RSSF spectral data in the near future. It is notable that many of the commercially available RSSF units incorporate data analysis software based upon both global analysis and/or SVD into their data processing software packages.

5. CONCLUDING REMARKS

Historically, enzymologists have had to rely on a variety of techniques in order to infer the existence of intermediates along a plausible reaction pathway. Rapid reaction techniques present the opportunity to directly observe certain catalytic events, but generally have been limited to monitoring a single species at a single wavelength. Therefore, the presence of other intermediates in a reaction pathway must be determined indirectly both by the number of observed relaxations required to describe a particular time course and the dependence of each relaxation on the concentration of one or more reactants. The principal advantage of RSSF spectroscopy is that this technique provides a means to directly detect the presence of chromophoric reaction intermediates and to characterize (a) the sequence of catalytic events, (b) the spectra of transient species, and (c) the covalent structure of multiple intermediates, which accumulate at an enzyme catalytic site. This information is invaluable for the investigation of enzyme reaction mechanism. RSSF spectral data aids and simplifies both the collection and interpretation of SWSF kinetic data. Secondly, characterization of both the spectral changes and the kinetics of an enzyme catalyzed reaction by RSSF spectroscopy provides a solid foundation to study both allosteric interactions and protein structure function relationships. We have emphasized RSSF spectroscopy is not necessarily limited to enzyme systems containing intrinsic chromophores, but may easily be adapted to other dynamic systems by the judicious choice of chromophoric substrate or cofactor analogs. With the introduction of both a number of commercially available RSSF spectrometers complete with data analysis software and a growing literature describing low-cost, home-built RSSF spectrophotometers we anticipate much greater utilization of RSSF technology by enzymologists in the immediate future.

6. REFERENCES

1. G. C. Pimentel, *Applied Optics* **7**, 2155–2160 (1968).
2. B. Chance, *J. Biol. Chem.* **179**, 1331–1339 (1949).
3. S. C. Koerber, A. K. H. MacGibbon, H. Dietrich, M. Zeppezauer, and M. F. Dunn, *Biochemistry* **22**, 3424–3431 (1983).
4. S. A. Bernhard, M. F. Dunn, P. L. Luisi, and P. Schack, *Biochemistry* **9**, 185–192 (1970).
5. M. F. Dunn and J. S. Hutchison, *Biochemistry* **12**, 4882–4892 (1973).
6. M. F. Dunn, *Biochemistry* **13**, 1146–1151 (1974).
7. J. D. Shore, H. Gutfreund, R. L. Brooks, D. Santiago, and P. Santiago, *Biochemistry* **13**, 4185–4190 (1974).
8. S. C. Koerber, P. Schack, A. M.-J. Au, and M. F. Dunn, *Biochemistry* **19**, 731–738 (1980).
9. R. N. Cochran, F. H. Horne, J. L. Dye, J. Ceraso, and C. H. Suelter, *J. Phys. Chem.* **84**, 2567–2575 (1980).
10. A. C. Stover, D. J. Phelps, and P. R. Carey, *Biochemistry* **20**, 3454–3461 (1981).
11. P. V. Argade, G. K. Gerke, J. P. Weber, and W. L. Peticolas, *Biochemistry* **23**, 299–306 (1984).
12. P. J. Tonge and P. R. Carey, *Biochemistry* **28**, 6701–6709 (1989).
13. S. A. Bernhard and S.-J. Lau, *Cold Spring Harbor Symp. Quant. Biol.* **36**, 75–83. (1971).
14. S. Kunugi, H. Hirohara, and N. Ise, *J. Am. Chem. Soc.* **101**, 3640–3646 (1979).
15. S. Kunugi, H. Hirohara, E. Nishimura, and N. Ise, *Arch. Biochem. Biopphys.* **189**, 298–308 (1978).
16. A. Himoe, K. G. Brandt, R. J. DeSa, and G. P. Hess, *J. Biol. Chem.* **244**, 3483–3492 (1969).

17. T. E. Barman and H. Gutfreund, *Biochem. J.* **101**, 411–416 (1966).
18. S. A. Bernhard, B. F. Lee, and Z. H. Tasjian, *J. Mol. Biol.* **18**, 405–420 (1966).
19. I. V. Berezin, N. F. Kazanskoya, and A. A. Klyosov, *FEBS Lett.* **15**, 121–125 (1971).
20. A. K. H. MacGibbon, K. Peace, S. C. Koerber, and M. F. Dunn, *Biochemistry* **26**, 3058–3067 (1987).
21. M. F. Dunn, A. K. H. MacGibbon, and K. Pease, in *Comparative Analysis of Catalytic Mechanisms of Zinc Enzymes: Progress in Inorganic Biochemistry and Biophysics* (eds. I. Bertini and M. Zeppezauer), Birkhauer, Basel, Switzerland, 1987, pp. 485–505.
22. K. H. Dahl and M. F. Dunn, *Biochemistry* **23**, 4094–4100 (1984).
23. K. H. Dahl, and M. F. Dunn, *Biochemistry* **23**, 6829–6839 (1984).
24. M. Zeppezauer, I. Andersson, H. Dietrich, M. Gerber, W. Maret, G. Schneider, H. Schneider-Berhlöhr, *J. Mol. Catalysis* **23**, 377 (1984).
25. M. F. Dunn, H. Dietrich, A. K. H. MacGibbon, S. C. Koerber, and M. Zeppezauer, *Biochemistry* **21**, 354–363 (1982).
26. C. Sartorius, M. Gerber, M. Zeppezauer, and M. F. Dunn, *Biochemistry* **26**, 871–882 (1987).
27. M. F. Dunn, J.-F. Biellmann, and G. Branlant, *Biochemistry* **14**, 3176–3188 (1975).
28. C. T. Angelis, M. F. Dunn, D. C. Muchmore, and R. W. Wing, *Biochemistry* **16**, 2922–2931 (1977).
29. H. Dietrich, W. Maret, L. Wallén, and M. Zeppezauer, *Eur. J. Biochem.* **100**, 267–270 (1979).
30. R. G. Morris, G. Saliman, and M. F. Dunn, *Biochemistry* **19**, 725–731 (1980).
31. E. Cedergren-Zeppezauer, J.-P. Samama, and H. Eklund, *Biochemistry* **21**, 4895–4908 (1982).
32. P. Andersson, J. Kvassmann, B. Oldén, and G. Pettersson, *Eur. J. Biochem.* **133**, 651–655 (1983).
33. M. F. Dunn, in *Enzymology of Carbonyl Metabolism: Aldehyde Dehydrogenase, Aldo/Keto Reductase and Alcohol Dehydrogenase, Vol. II* (eds., T. G. Flyn and H. Weiner), 1985, pp. 151–168, Alan R. Liss, Inc., New York.
34. H. Dietrich and M. Zeppezauer, *J. Inorg. Biochem.* **17**, 227–235 (1982).
35. C. Sartorius, M. Zeppezauer, and M. F. Dunn, *Eur. J. Biochem.* **117**, 493–499 (1988).
36. H. Eklund and C.-I. Brändén, *J. Biol. Chem.* **254**, 3458 (1979).
37. H. Dietrich, W. Maret, H. Kozlowski, and M. Zeppezauer, *J. Inorg. Biochem.* **14**, 297–311 (1981).
38. M. Gerber, M. Zeppezauer, and M. F. Dunn, unpublished results.
39. J. Kvassman, A. Larsson, and G. Pettersson, *Eur. J. Biochem.* **114**, 555–563 (1981).
40. G. Pettersson, *CRC Crit. Rev. Biochem.* **21**, 349–389 (1987).
41. G. D. Smith, D. C. Swenson, E. J. Dodson, G. G. Dodson, and C. D. Reynolds, *Proc. Natl. Acad. Sci. USA* **81**, 7093–7097 (1984).
42. E. N. Baker, T. L. Blundell, J. F. Cutfield, S. M. Cutfield, E. J. Dodson, G. G. Dodson, D. C. Hodgkin, R. E. Hubbard, N. W. Isaacs, C. D. Reynolds, K. Sakabe, N. Sakabe, and N. M. Vijayan, *Phil Trans. Roy. Soc. (London) B*, **319**, 369–456 (1988).
43. U. Derewenda, Z. Derewenda, E. J. Dodson, G. G. Dodson, C. D. Reynolds, G. D. Smith, C. Sparks, and D. Swensen, *Nature* **338**, 594–596 (1989).
44. G. D. Smith and G. G. Dodson, *Biopolymers* **32**, 441–445 (1992).
45. G. D. Smith and G. G. Dodson, *Proteins-Structure Function and Genetics* **14**, 401–408 (1992).
46. A. Wollmer, B. Rannefeld, B. R. Johansen, K. R. Hejnaes, P. Balschmidt, and F. B. Hansen, *Biol. Chem. Hoppe-Seyler* **368**, 903–912 (1987).
47. N. C. Kaarsholm, H.-C. Ko, and M. F. Dunn, *Biochemistry* **28**, 4427–4435 (1989).
48. M. Roy, M. L. Brader, R. W.-K. Lee, N. C. Kaarsholm, J. Hansen, and M. F. Dunn, *J. Biol. Chem.* **264**, 19081–19085 (1989).

49. M. L. Brader, N. C. Kaarsholm, and M. F. Dunn, *J. Biol. Chem.* **265**, 15666-15670 (1990).
50. M. L. Brader, N. C. Kaarsholm, and M. F. Dunn, *Biochemistry* **30**, 6636-6645 (1991).
51. M. L. Brader, D. Borchardt, and M. F. Dunn, *J. Am. Chem. Soc.* **114**, 4480-4486 (1992).
52. M. L. Brader, D. Borchardt, and M. F. Dunn, *Biochemistry* **31**, 4691-4696 (1992).
53. M. L. Brader and M. F. Dunn, *J. Am. Chem. Soc.* **112**, 4585-4587 (1990).
54. W. E. Choi, M. L. Brader, V. Aguilar, N. C. Kaarsholm, and M. F. Dunn, *Biochemistry* **32** (1993).
55. L. Gross and M. F. Dunn, *Biochemistry* **31**, 1295-1301 (1992).
56. M. F. Dunn, S. E. Pattison, M. C. Storm, and E. Quiel, *Biochemistry* **19**, 718-725 (1980).
57. M. C. Storm and M. F. Dunn, *Biochemistry* **24**, 1749-1756 (1985).
58. N. C. Kaarsholm and M. F. Dunn, *Biochemistry* **26**, 883-890 (1987).
59. G. Gould and M. F. Dunn, unpublished results.
60. F. D. Coffman and M. F. Dunn, *Biochemistry* **27**, 6179-6187 (1988).
61. C. P. Hill, Z. Dauter, E. J. Dodson, G. G. Dodson, and M. F. Dunn, *Biochemistry* **30**, 917-924 (1991).
62. O. P. Malhotra and S. A. Bernhard, *J. Biol. Chem.* **243**, 1243-1252 (1968).
63. P. D. Buckley and M. F. Dunn in Enzymology of Carbonyl Metabolism: Aldehyde Dehydrogenase and Aldo/Keto Reductase (eds. H. Weiner), Alan R. Liss, New York 1982, pp. 23-25.
64. M. F. Dunn and P. D. Buckley, in *Enzymology of Carbonyl Metabolism: Aldehyde Dehydrogenase, Aldo/keto Reductase and Alcohol Dehydrogenase*, Vol. II (ed. H. Weiner), 1985, pp. 15-27, Alan R. Liss, New York.
65. S. A. Bernhard, S.-J. Lau, and H. F. Noller, *Biochemistry* **4**, 1108-1118 (1965).
66. O. P. Malhotra and S. A. Bernhard, *Proc. Natl. Acad. Sci. USA* **70**, 2077-2081 (1973).
67. E. E. Snell and S. J. Dimari in *The Enzymes* (ed. P. D. Boyer), 1970, pp. 335-370, Academic Press, New York.
68. C. Walsh, *Enzymatic Reaction Mechanisms*, pp. 777-827, Freman, San Francisco (1979).
69. H. C. Dunathan, *Adv. Enzymol.* **35**, 79-134 (1971).
70. C. Yanofsky and I. P. Crawford in *Enzymes*, Vol. VIII, 34th ed., 1-31, 1972.
71. D. E. Metzler, *Adv. Enzymology* **50**, 1-40, 1979.
72. E. W. Miles, *Pyridoxal phosphate enzymes catalyzing β-elimination and β-replacement reactions*, in *Pyridoxal Phosphate and Derivatives*, Vol. I in the series "Coenzymes and Cofactors," (eds. D. Dolphin, R. Poulson, and O. Avramovic), John Wiley and Sons, New York, 1986, pp. 253-310.
73. E. W. Miles, in *Biochemmical and Medical Aspects of Tryptophan Metabolism* (eds. O. Hayaishi, Y. Ishimura, and R. Kido), Elsevier/North-Holland Biomedical Press, Amsterdam, 1980, pp. 137-147.
74. P. S. Brzović and M. F. Dunn in *Biosynthesis and Molecular Regulation of Amino Acids in Plants* (eds. H. Flores, J. Shannon, and B. Singh), American Society of Plant Physiologists, 1992, pp. 37-51.
75. D. E. Metzler, C. M. Harris, R. J. Johnson, D. B. Siano, and J. A. Thompson, *Biochemistry* **12**, 5377-5392 (1973).
76. C. M. Metzler, A. E. Cahill, and D. E. Metzler, *J. Am. Chem. Soc.* **89**, 1322-1330 (1980).
77. C. M. Metzler, A. E. Cahill, S. Petty, D. E. Metzler, and L. Lang, *Applied Spectroscopy* **39**, 333-339 (1985).
78. R.G. Kallen, T. Korpella, A. E. Martell, Y. Matsushima, D. E. Metzler, T. V. Morozov, I. M. Ralston, F. A. Savin, Y. M. Torchinshky, and H. Ueno, in *Transaminases* (eds. P. Christen, and D. W. Metzler), 1985, pp. 37-108, John Wiley, New York.
79. Y. Karube and Y. Matsushima, *Chem. Parm. Bull.* **25**, 2568-2575 (1977).

80. R. J. Ulevitch and R. G. Kallen, *Biochemistry* **16**, 5350–5354 (1977).
81. C.C. Hyde, A. Ahmed, E. A. Padlan, E.W. Miles, and D. R. Davies, *J. Biol. Chem.* **263**, 17857–17871 (1988).
82. M. F. Dunn, V. Aguilar, P. S. Brzović, W. F. Drewe, Jr., K. F. Houben, C. A. Leja, and M. Roy, *Biochemistry* **29**, 8598–8607 (1990).
83. K. S. Anderson, E. W. Miles, and K. A. Johnson, *J. Biol. Chem.* **266**, 8020–8033 (1991).
84. A. N. Lane and K.Kirschner, *Biochemistry* **30**, 479–484 (1991).
85. W. F. Drewe and M. F. Dunn, *Biochemistry* **24**, 3977–3987 (1985).
86. W. F. Drew and M. F. Dunn, *Biochemistry* **25**, 2494–2501 (1986).
87. P. S. Brzović, A. M. Kayastha, E. W. Miles, and M. F. Dunn, *Biochemistry* **31**, 1180–1190 (1992).
88. P. S. Brzović, Y. Sawa, C. C. Hyde, E. W. Miles, and M. F. Dunn, *J. Biol. CHem.* **267**, 13028–13038 (1992).
89. P. S. Brzović, C. C. Hyde, E. W. Miles, and M. F. Dunn, *Biochemistry* (in press) (1993).
90. E. W. Miles and P. McPhie, *J. Biol. Chem.* **249**, 2852–2857 (1974).
91. M. Roy, E. W. Miles, R. S. Phillips, and M. F. Dunn, *Biochemistry* **27**, 8661–8669 (1988).
92. K. D. Schnackerz, J. H. Ehrlich, W. Giesemann, and T. A. Reed, *Biochemistry* **18**, 3557–3563 (1979).
93. A. N. Lane and K. Kirschner, *Eur. J. Biochem.* **120**, 379–387 (1981).
94. A. N. Lane and K. Kirschner, *Eur. J. Biochem.* **129**, 571–582.
95. W. O. Weischet and K. Kirschner, *Eur. J. Biochem.* **65**, 365–373 (1976).
96. A. N. Lane and K. Kirschner, *Eur. J. Biochem.* **129**, 561–570 (1983).
97. K. F. Houben and M. F. Dunn, *Biochemistry* **29**, 2421–2429 (1990).
98. P. S. Brzović, K. Ngo, and M. F. Dunn, *Biochemistry* **31**, 3831–3839 (1992).
99. H. Kawasaki, R. Bauerle, G. Zon, S. Ahmed, and E. W. Miles, *J. Biol. Chem.* **262**, 10678–10683 (1987).
100. K. Kirschner, A. N. Lane, and A. W. M. Strasser, *Biochemistry* **30**, 472–478 (1991).
101. D. S. June, C. H. Suelter, and J. L. Dye, *Biochemistry* **10**, 2707–2713 (1981).
102. D. S. June, C. H. Suelter, and J. L. Dye, *Biochemistry* **10**, 2714–2719 (1981).
103. M. N. Kazarinoff and E. E. Snell, *J. Biol. Chem.* **255**, 6228–6233 (1980).
104. R. S. Phillips, *Biochemistry* **30**, 5927–5934 (1991).
105. R. S. Phillips, I. Richter, P. Gollnick, P. S. Brzović, and M. F. Dunn, *J. Biol. Chem.* **266**, 18642–18648 (1991).
106. R. S. Phillips, S. L. Bender, P. S. Brzovic', and M. F. Dunn, *Biochemistry* **29**, 8608–8614 (1990).
107. P. S. Brzović, E. L. Holbrook, R. C. Greene, and M. F. Dunn, *Biochemistry* **29**, 442–451 (1990).
108. M. Johnston, P. Marcotte, J. Donovan, and C. Walsh, *Biochemistry* **18**, 1729–1738 (1979).
109. J. R. Knowles, *Science* **236**, 1252–1258 (1987).
110. D. L. Pompliano, P. Anusch, and J. R. Knowles, *Biochemistry* **29**, 3186–3194 (1990).
111. A. Bellelli, G. Antonini, M. Brunori, B. A. Springer, and S. G. Silgar, *J. Biol. Chem.* **265**, 18898–18901 (1990).
112. M. Brunori, G. Antonini, M. Castagnola, and A. Bellelli, *J. Biol. Chem.* **267**, 2258–2263 (1992).
113. Y. Orii, I. Yumoto, Y. Fukumori, and T. Yamanaka, *J. Biol. Chem.* **266**, 14310–14316 (1991).
114. G. Antonini, F. Malatesta, P. Sarti, and M. Brunori, *J. Biol. Chem.* **266**, 13193–13202 (1991).
115. F. G. Halaka, G. T. Babcock, and J. L. Dye, *Biophys. J.* **48**, 209–219 (1985).
116. F. G. Halaka, Z. K. Barnes, G. T. Babcock, and J. L. Dye, *Biochemistry* **23**, 2005–2010 (1984).

117. M. M. Palcic and H. B. Dunford, *J. Biol. Chem.* **255**, 6128–6132 (1983).
118. A. M. Lambier and H. B. Dunford, *Eur. J. Biochem.* **147**, 93–96 (1985).
119. L. Marquez, H. Wariishi, H. B. Dunford, and M. H. Gold, *J. Biol. Chem.* **263**, 10549–10552 (1988).
120. H. K. Baek and H. E. Van Wart, *Biochemistry* **28**, 5714–5719 (1989).
121. J. E. Erman, L. B. Vitella, J. M. Mauro, and J. Kraut, *Biochemistry* **28**, 7992–7995 (1989).
122. Y. Orii and H. Anni, *FEBS Lett.* **267**, 117–120 (1990).
123. W. Sun and H. B. Dunford, *J. Biol. Chem.* **267**, 17649–17657 (1992).
124. Y. Hsuanyu and H. B. Dunford, *J. Biol. Chem.* **267**, 17649–17657 (1992).
125. Y. Hsuanyu and H. B. Dunford, *Arch. Biochem. Biophys.* **292**, 213–220 (1992).
126. Y. Hsuanyu and H. B. Dunford, *Arch. Biochem. Biophys.* **281**, 282–286 (1990).
127. K. Kikuchi, Y. Kawamura-Konishi, and H. Suzuki, *Arch. Biochem. Biophys.* **296**, 88–94 (1992).
128. H. Hartridge and F. J. W. Roughton, *Proc. Roy. Soc. (London) A104*, 376–385 (1923).
129. B. Chance, in *Techniques of Chemistry*, Vol. VI, 3rd ed. (ed. G. G. Hammes) Wiley Interscience, New York, 1974, pp. 5–61.
130. R. L. Berger, in *Rapid Mixing and Sampling Techniques in Biochemistry* (eds. B. Chance, R. Eisenhardt, Q. H. Gibson, and K. K. Longberg-Holm), Academic Press, New York, 1964, pp. 57–65.
131. R. L. Berger, B. Balko, and H. F. Chapman, *Rev. Sci. Instr.* **39**, 493–502 (1968).
132. R. E. Hester and R. B. Girling, in *Spectroscopy of Biological Molecules* (eds. R. E. Hester and R. B. Girling), The Royal Society of Chemistry, Thomas Graham House, Cambridge, p. 464.
133. M. Delhaye, *Applied Optics* **7**, 2195–2219 (1968).
134. R. E.Santini, M. J. Milano, and H. L. Pardue, *Anal. Chem.* **45**, 915A–927A (1973).
135. H. J. Caulfield, *Spectroscopy*, in *Handbook of Optical Holography* (ed. H. J. Caulfield), Academic Press, New York, 1979.
136. D. G. Jones, *Anal. Chem.* **54**, 1057A–1073A (1985).
137. T. J. Hirschfield, *J. Appl. Spectrosc.* **30**, 68–69 (1976).
138. G. D. Boutitier, B. D. Pollard, and J. D. Winefordner, *Spectrochim. Acta* **33B**, 401–415 (1978).
139. T. Okamoto, S. Kawata, and S. Minami, *Applied Optics* **23**, 269–273 (1984).
140. Y. Talmi, *Multichannel Image Detectors, Vol. 2* (ed. Y. Talmi) *Amer. Chem. Soc.*, Washington, D.C., pp. 332 (1983).
141. D. G. Jones, *Anal. Chem.* **57**, 1207A–1214A (1985).
142. Y. Talmi, *Anal. Chem.* **47**, 658A–670A (1975).
143. Y. Talmi, *Anal. Chem.* **47**, 669A–709A (1975).
144. Y. Talmi and R. W. Simpson, *Applied Optics* **19**, 1401–1414 (1980).
145. M. B. Denton, H. A. Lewis, and G.R. Sims, in *Multichannel Image Detectors, Vol. 2* (ed. Y. Talmi), *Amer. Chem. Soc.,* Washington, D.C., pp. 135–154 (1983).
146. H. J. Babrov and R. H. Tourin, *Applied Optics* **7**, 2171–2177 (1968).
147. J. W. Strojek, G. A. Gruver, and T. Kuwana, *Anal. Chem.* **41**, 481–484 (1969).
148. R.B. Coolen, N. Papadakis, J. Avery, C. G. Enke, and J. L. Dye, *Anal. Chem.* **47**, 1649–1655 (1975).
149. N. Papadakis, R. B. Coolen, and J. L. Dye, *Anal. Chem.* **47**, 1644–1649 (1975).
150. T. P. Carter, H. K. Baek, L. Bonninghausen, R. J. Morris, and H. E. Van Wart, *Anal. Biochem.* **190**, 134–140 (1990).
151. H. H. Schlemmer and M. Mächler, *J. Phys. E.: Sci. Instrum.* **18**, 914–919 (1985).
152. M. Diem and H. Ludl, *Computer Enhanced Spectro.* **3**, 23–28 (1986).
153. M. Ryan-Hotchkiss and J. D. Ingle, *Talanta* **34**, 619–627 (1987).

154. C. F. Bernasconi, *Relaxation Kinetics*, Academic Press, New York, 1976.
155. S.C. Koerber and M. F. Dunn, *Biochimie* **63**, 97–102 (1981).
156. M. Gerber, M. Zeppezauer, and M. F. Dunn, *Inorg. Chim. Acta* **79**, 161–164 (1983).
157. W. F. Drewe, Jr., S. C. Koerber, and M. F. Dunn, *Biochimie* **71**, 509–519 (1989).
158. A. Savitzsky and M. J. E. Golay, *Anal. Chem.* **36**, 1627–1638 (1964).
159. J. M. Beecham, *Meth. In Enzy.* **210**, 37–54 (1992).
160. E. R. Henry and J. Hofrichter, *Meth. in Enzym.* **210**, 179–192 (1992).
161. M. Maeder and A. D. Zuberbuhler, *Anal. Chem.* **62**, 2220–2224 (1990).
162. K. Hiromi, *Kinetics of Fast Enzyme Reactions*, Halsted Press, New York, 1979.
163. W. H. Press, B. P. Flannery, S. A. Tenkolsky, and W. T. Vetterhig, *Numerical Recipes*, Cambridge University Press, Cambridge (1986).

AUTHOR INDEX

Numbers in parentheses are reference numbers and indicate that the author's work is referred to although his name is not mentioned in the text. Numbers in *italics* indicate pages on which the complete reference appears.

Abbé, E., 2(2), *69*
Acharya, K. R., 68(97), *72*
Adreesen, R., 165(55), *185*
Adrian, M., 109(163), *115*
Aebi, U., 79(15), 91(15), 96(102), 98(112–113), 100(119), 106(151), 107(151), *110*, *113*, *114*
Agar, A. W., 74(4), 76(4), 77(4), *110*
Agrawal, S., 133(97), *141*
Aguilar, V., 208(54), 217(82), *270*, *271*
Ahmed, A., 217(81), 240(81), *271*
Ahmed, S., 227(99), 244(99), *271*
Akerman, K. E. O., 128(68), 130(68), *140*
Alber, T., 59(82), *72*
Alberti, K. G. M. M., 167(87), *186*
Alderson, R. H., 74(4), 76(4), 77(4), *110*
Aldred, P., 93(95), *113*
Allain, P., 152(30), *185*
Allewell, N. N., 68(98), *72*
Alpert, N. L., 144(1), *184*
Alvarez, R., 181(166), *189*
Ames Division, Miles Laboratories Ltd., 167(80), *186*
Ammitzboll, T., 86(53), *111*
Amrein, M., 108(153), *114*
Anders, J. J., *141*
Anderson, C. A., 81(22), *110*
Anderson, K. S., 217(83), 227(83), 240(83), *271*
Anderson, N. G., 144(5), *184*
Andersson, I., 198(24), 200(24), *269*
Andersson, P., 198(32), *269*
Andrews, S. B., 80(16), 82(16), *110*
Angelis, C. T., 198(28), *269*
Anni, H., 246(122), *272*
Antonini, G., 245(111), 246(112, 114), *271*
Anusch, P., 244(110), *271*
Apell, H. J., 128(66), *140*
Appleton, T. C., 91(85), *112*

Arakawa, H., 107(152), *114*
Argade, P. V., 195(11), *268*
Armarego, W. L. F., 63(93), *72*
Arndt, U. W., 27(33), *70*
Arnsdorf, M. F., 106(146), 107(146), *114*
Assaf, Y., 119(1), 122(1), *138*
Aszalos, A., 130(86), *141*
Atmadja, J., 98(114), *113*
Au, A. M.-J., 194(8), 198(8), 204(8), *268*
Avery, J., 254(148), 255(148), *272*
Avramovic, O., 215(72), *270*

Baadenhuijsen, H., 144(15, 18), *184*, *185*
Baarslag, M. W., 137(102), *141*
Babcock, G. T., 246(115–116), 267(115), *271*
Babrov, H. J., 254(146), *272*
Bachmann, I., 104(132), *114*
Bachmann, L., 91(86), *112*
Bachovchin, W. W., 50(71), *71*
Backer, E. T., 144(16), *184*
Baek, H. K., 246(120), 247(120), 248(120), 260(150), *272*
Baeyens, W. R. G., 165(56), *186*
Bagaringo, T., 178(115), *187*
Bailey, R. H., 124(47), *139*
Bailly, M., 180(154), *188*
Bais, R., 181(171), *189*
Baker, E. N., 208(42), *269*
Balhorn, R., 108(156), 109(156), *114*
Balint, E., 130(86), *141*
Balko, B., 248(131), *272*
Balschmidt, P., 208(46), *269*
Banner, D. W., 59(82), *72*
Barak, L. S., 122(23), 128(23), 130(23), *139*
Barford, D., 68(97), *72*
Barlow, I. M., 153(37), *185*
Barman, T. E., 195(17), *269*
Barnes, Z. K., 246(116), *271*

Barnoumi, R., 124(47), *139*
Baron-Epel, O., 129(71), 130(71), 133(71), *140*
Barry, P. L., 178(126), *187*
Barth, M., 104(132), *114*
Bartsch, H. H., 29(35), *70*
Bartunik, H. D., 29(35), *70*
Bas, B. M., 166(67), *186*
Bassnet, S., 127(61), *140*
Bauerle, R., 227(99), 244(99), *271*
Baumeister, W., 80(19), 95(101), 104(132), 108(157), *110*, *113*, *114*
Baumgärtner, K. H., 3(9), *70*
Baxter Diagnostics Inc., 167(84), *186*
Bayer, E. A., 85(39), 89(39), *111*
Bayer, P. M., 144(15, 18), *184*, *185*
Bayse, D. D., 181(161), *189*
Beals, T. F., 91(78-79), *112*
Bear, D. G., 108(154), *114*
Beatty, B. R., 104(133), *114*
Beaudet, A., 91(83), *112*
Becker, R., 104(132), *114*
Beckman Instruments, 166(65), *186*
Beddell, C. R., 63(90), *72*
Bednarek, S. Y., 93(92), *113*
Beebe, D. C., 127(61), *140*
Beebe, T. P., Jr., 106(149), 108(156), 109(156), *114*
Beecham, J. L., 180(155), *188*
Beecham, J. M., 265(159), 267(159), *273*
Beer, M., 79(11), *110*
Behnke, O., 86(53), *111*
Behring Diagnostics, 166(66), *186*
Bellelli, A., 245(111), 246(112), *271*
Bendayan, M., 86(44-50), 87(45), 88(44), *111*
Bender, S. L., 244(106), *271*
Benson, D. M., 122(17), *139*
Berezin, I. V., 195(19), *269*
Berger, G. M. B., 178(122), *187*
Berger, R. L., 248(130-131), *272*
Bergmeyer, H.-U., 162(47), 163(48), 179(138-139), *185*, *188*
Bergmeyer, H. U., 179(129), *187*
Bergmeyer, J., 170(106), *187*
Bernal, J. D., 3(13), *70*
Bernasconi, C. F., 263(154), *273*
Bernhard, S. A., 194(4), 195(13, 18), 197(13), 198(4), 213(62), 214(62, 65-66), *268*, *269*, *270*
Bernstein, F. C., 54(76), 55(76), *71*
Berreth, R. C., 168(94), *187*
Bersch, B., 128(66), *140*
Berthold Analytical Instruments, 165(64), *186*
Beutler, H. O., 178(113-114), *187*
Beyer, A. L., 104(134), *114*

Bhave, A. S., 135(100), *141*
Bhave, S. V., 135(100), *141*
Biellmann, J.-F., 198(27), 208(27), *269*
Bijvoet, J. M., 21(30), 35(44), 41(55-56), *70*, *71*
Bilderback, D., 68(96), *72*
Billings-Gagliardi, S., 89(66), *112*
Binder, M., 91(73-74), *112*
Binnig, G., 2(4), *69*, *74*
Bissell, M. G., 144(23), *185*
Blijenberg, B. G., 144(10), *184*
Bloomer, A. C., 59(82), *72*
Blow, D. M., 40(50), 41(52), 65(94), 67(94), *71*, *72*
Blundell, T. L., 3(5), *69*, 208(42), *269*
Boeckx, R. L., 165(53), *185*
Boehringer Mannheim, Corp., 167(81), *186*
Boezeman, B. M., 127(58), *140*
Bohner, J., 181(176), *189*
Bokhoven, C., 35(44), 41(56), *71*
Boland, J., 144(24), *185*
Bolin, J. T., 63(88-90), *72*
Bombick, D. W., *141*
Bonig, I., 90(72), 93(72), *112*
Bonnard, C., 85(40), *111*
Bonninghausen, L., 260(150), *272*
Boon, P. I., 178(123), *187*
Borchardt, D., 208(51-52), *270*
Boutitier, G. D., 250(138), *272*
Boutwell, J. H., 180(153), *188*
Bowen, I. D., 105(139), 106(140), *114*
Bowers, G. N., Jr., 179(135), 180(135, 144, 146-149), 181(166, 178), *188*, *189*
Bowers, W. C., 180(148), *188*
Bracewell, R. N., 22(32), *70*
Bracker, C. E., 106(142), *114*
Braconnier, F., 144(10), *184*
Brader, M. L., 208(48-54), 209(48-50), *269*, *270*
Bragg, W. L., 14(27), *70*
Brakenhoff, G. J., 121(14), 137(101-102), *138*, *141*
Branden, C., 51(72), *71*
Brändén, C.-I., 200(36), *269*
Brandt, K. G., 195(16), *268*
Branlant, G., 198(27), 208(27), *269*
Branson, H. R., 56(78), *72*
Branton, D., 97(103), 100(117), 101(117), *113*
Brauner, T., 128(67), 130(67), *140*
Bresser, J., 133(96), *141*
Bretaudiere, J.-P., 180(154, 158), *188*, *189*
Brice, M. D., 54(76), 55(76), *71*
Bricogne, G., 43(63, 66), *71*
Bright, G. R., 127(59, 64), *140*

AUTHOR INDEX

Brimacombe, R., 98(114), *113*
Brodie, M. J., 178(116), *187*
Brooks, R. L., 194(7), *268*
Broom, M. B., 4(20), *70*
Broughton, P. M. G., 168(97), *187*
Brown, N. M. D., 106(145), *114*
Brown, S. S., 178(120), *187*
Brünger, A. T., 48(70), *71*
Brunori, M., 245(111), 246(112, 114), *271*
Bruschi, B., 168(101), *187*
Brzovic, P. S., 215(74), 217(82, 87–89), 219(88), 226(87, 98), 227(87–89, 98), 228(98), 231(105), 234(105), 236(107), 237(107), 238(107), 239(107), 240(87, 107), 241(87), 242(87), 243(88–89), 244(88–89, 106), 264(107), *270, 271*
Buchholz, C., 137(103), *141*
Buchsbaum, R. N., 127(63), *140*
Buckley, P. D., 214(63–64), 216(63), *270*
Buckmire, F. L. A., 95(100), *113*
Budd, G. C., 91(81, 88), *112*
Bugg, C. E., 4(20), 63(92), *70, 72*
Buhle, E. L., Jr., 79(15), 91(15), *110*
Buitenhuis, S., 181(164), *189*
Bullock, D. G., 179(134), 180(134), *188*
Bullock, G. R., 89(68), 93(91), *112, 113*
Bump, E. A., 124(44–45), *139*
Burdick, B. A., 168(101), *187*
Burgess, M. S., 144(8), *184*
Burghardt, R. C., 124(47), *139, 141*
Burr, J. G., 165(59), *186*
Burridge, J. M., 63(90), *72*
Burtis, C. A., 144(5), 153(36), 157(39), *184, 185*
Busa, W. B., 127(52), *140*
Bustamante, C., 106(144), 108(154), *114*
Butt, H.-J., 109(162), *115*

Cahill, A. E., 215(76–77), *270*
Cahn, R. S., 63(91), *72*
Cali, J. P., 181(166), *189*
Campbell, A. K., 129(74), *140*
Campbell, B. C., 178(116), *187*
Campbell, J. W., 47(69), *71*
Campbell, R. S., 167(90), 175(90), *186*
Campbell, T., 130(87), *141*
Cantino, M., 81(25), 82(25), *110*
Cardullo, R. A., 133(97), *141*
Carey, G., 144(24), *185*
Carey, P. R., 195(10, 12), *268*
Carragher, B. O., 98(111), *113*
Carrico, R. J., 175(108), *187*
Carter, D., 4(20), *70*
Carter, T. P., 260(150), *272*

Castagnola, M., 246(112), *271*
Casteels, R., 135(99), *141*
Caulfield, H. J., 250(135), *272*
Cedergren-Zeppezauer, E., 198(31), *269*
Ceraso, J., 194(9), 204(9), *268*
Chambard, J. C., 127(54), *140*
Champness, J. N., 63(90), *72*
Chan, D. W., 167(85), *186*
Chance, B., 193(2), 248(129–130), *268, 272*
Chang, C. C., 126(48), 133(93), *139, 141*
Chang, S. L., 34(41), *71*
Chapman, H. F., 248(131), *272*
Chen, W.-T., 85(37), *111*
Chescoe, D., 74(4), 76(4), 77(4), *110*
Childs, G. V., 86(43), *111*
Chipman, D. M., 119(1), 122(1), *138*
Chiu, K.-P., 93(93), *113*
Choi, W. E., 208(54), *270*
Chothia, C., 58(81), *72*
Chow, A. T., 144(12), *184*
Christen, P., 215(78), *270*
Chu, G. L., 127(62), *140*
Chu, T. M., 181(172), *189*
Clapper, F. R., 173(107), 175(107), *187*
Clark, P. M. S., 168(97), *187*
Clarke, A. E., 90(72), 93(72), *112*
Clarke, J., 106(141), 107(141), 108(141), *114*
Clement, B. A., 124(47), *139*
Clemmer, C. R., 106(149), *114*
Clinical Products Division, Eastman Kodak Company, 168(91), *186*
Cobbold, P. H., 129(76), *140*
Cochran, R. N., 194(9), 204(9), *268*
Coffman, F. D., 211(60), *270*
Coghlan, J. P., 93(95), *113*
Cohen, S. H., 93(93), *113*
Colinet, E., 179(143), 180(159), 181(143), *188, 189*
College of American Pathologists, 152(28), *185*
Colliex, C., 79(14), *110*
Collins, D. M., 43(65), *71*
Collins, J. M., 123(31–32), *139*
Columbus, R. L., 153(33), 168(102), 175(102), *185, 187*
Cook, J. A., 124(46), *139*
Coolen, R. B., 254(148–149), 255(148), *272*
Copeland, W. H., 157(38), 180(151), *185, 188*
Corey, R. B., 56(78), 57(79), *72*
Cork, J. M., 35(43), *71*
Cornish, E. C., 90(72), 93(72), *112*
Cosslett, V. E., 77(9), *110*
Coster, D., 41(54), *71*
Costin, K. A., 178(119), *187*

Coulombe, P. A., 86(50), *111*
Cox, J. M., 4(15), 39(15), *70*
Craik, C. S., 67(95), *72*
Crang, R. F. E., 84(35), *110*
Crawford, I. P., 215(70), *270*
Crewe, A. V., 79(12), *110*
Crowfoot, D., 3(13), *70*
Crowther, R. A., 41(52), *71*
Cuatrecasas, P., 130(82), *140*
Curme, H. G., 168(102), 175(102), *187*
Cutfield, J. F., 208(42), *269*
Cutfield, S. M., 208(42), *269*

Dahl, K. H., 198(22-23), 199(22), 200(22-23), 264(22-23), *269*
Dahl, R., 84(33), *110*
Dappen, G. M., 168(101-102), 175(102), *187*
Das, M. K., 129(75), *140*
Dauter, Z., 211(61), *270*
Davies, D., 4(14), 39(14), *70*
Davies, D. R., 217(81), 240(81), *271*
Davies, S. N., 181(170), *189*
Dawes, C., 83(28), *110*
Deanin, G. G., 127(57), *140*
deFeiiter, A. W., 126(51), *140*
deFeiner, A. W., 133(94), *141*
deFeiter, A. W., 126(48), *139*
Deisenhofer, J., 4(17), *70*
deJongh, G. J., 127(58), *140*
Delhaye, M., 249(133), *272*
DeLucas, L. J., 4(20), *70*
Denton, M. B., 253(145), *272*
De Pree, L., 144(21), *185*
Derewenda, U., 208(43), *269*
Derewenda, Z., 208(43), *269*
Derrick, K. S., 98(109), *113*
DeSa, R. J., 195(16), *268*
Descouts, P., 109(163), *115*
DeSmedt, H., 135(99), *141*
Dewey, T. G., 119(8), 122(8), 123(26, 28), *138*, *139*
Dewey, W. C., 127(62), *140*
Dianoux, A.-C., 119(12), 122(12), *138*
Dicke-Evinger, M. J., 133(95), *141*
Dickerson, R. E., 4(14), 39(14), *70*
Diego, J., 178(112), *187*
Diem, M., 260(152), *272*
Dietrich, H., 194(3), 198(3, 24-25), 198(29), 200(3, 24-25, 29, 34), 201(3), 203(3), 204(25, 37), 264(3), *268*, *269*
Dietzler, D. N., 181(175), *189*
Dimari, S. J., 215(67), *270*
Dinchuk, J. E., 103(128), *114*
DiPietro, M., 126(49), *139*
Dixon Northern, B., 109(162), *115*

Dodson, E. J., 208(41-43), 209(41), 211(61), *269*, *270*
Dodson, G. G., 208(41-45), 209(41), 211(61), *269*, *270*
Dolphin, D., 215(72), *270*
Dombrowicz, D., 91(77), *112*
Dominey, R. N., 123(31), *139*
Donovan, J., 238(108), *271*
Doscher, M., 68(98), *72*
Doumas, B. T., 144(22), *185*
Drewe, W. F., Jr., 217(82, 85-86), 221(85), 222(85), 223(85-86), 224(86), 225(86), 226(85-86), 264(85-86, 157), 265(157), 266(157), *271, 273*
Droogmans, G., 135(99), *141*
Duch, D. S., 103(128), *114*
Ducruix, A., 4(18), 5(18), 6(18), *70*
Dumont, G., 180(154), *188*
Dunathan, H. C., 215(69), *270*
Duncan, P. H., 181(160-161, 163), *189*
Dunford, H. B., 246(117-119, 123-126), *272*
Dunn, M. F., 194(3-6, 8), 198(3-5, 8, 20-23, 25-28, 30), 199(22), 200(3, 5, 20-23, 25-26, 33, 35), 201(3), 203(3, 26), 204(8, 20, 25-26, 38), 205(20-21, 33), 207(20), 208(5, 21, 26-27, 47-55), 209(47-48, 50), 210(47, 55-61), 212(55), 214(63-64), 215(74), 216(63), 217(82, 85-89), 219(88), 221(85), 222(85, 91), 223(85-86), 224(86), 225(86), 226(85-87, 97-98), 227(87-89, 97-98), 228(98), 231(105), 234(105), 236(107), 237(107), 238(107), 239(107), 240(87, 107), 241(87), 242(87), 243(88-89), 244(88-89, 106), 264(3, 20, 22-23, 26, 85-86, 91, 107, 155-157), 265(157), 266(157), *268, 269, 270, 271, 273*
Durr, R., 108(153), *114*
Durrenberger, M., 89(61), *111*
Dutton, A. H., 85(37), *111*
Dworetzky, S. I., 89(64), *111*
Dye, J. L., 194(9), 204(9), 227(101-102), 231(102), 233(102), 246(115-116), 254(148-149), 255(148), 267(115), *268, 271, 272*
Dyer, C., 135

Eberhardt, K. R., 181(166), *189*
Echlin, P., 74(3), 77(3), 81(3), *109*
Eckert, S. E., 150(27), *185*
Edelmann, L., 85(36), *110*
Eder, T. W., 168(102), 175(102), *187*
Edstrom, R. D., 108(158-159), 109(159), *115*

Edwards, G. C., 179(135), 180(135, 144, 149), 181(178), *188*, *189*
Edwards, J. B., 181(171), *189*
Eftink, M. R., 119(6), 122(6), 123(6, 28), *138*, *139*
Egger, M., 108(161), *115*
Ehrlich, J. H., 222(92), *271*
Eisenberg, B. R., 89(70), *112*
Eisenhardt, R., 248(130), *272*
Eklund, H., 198(31), 200(36), *269*
Elder, M., 47(69), *71*
Elings, V. B., 106(142, 144), 108(155, 159), 109(159), *114*, *115*
Ellis, P. B., 175(108), *187*
Elser, R. C., 179(142), *188*
Emch, R., 109(163), *115*
Eng, A. E., 168(98), *187*
Engel, A., 79(13), 80(19), 106(150-151), *107(150-151)*, *110*, *114*
Enke, C. G., 254(148), 255(148), *272*
Enkvist, M. O. K., 128(68), 130(68), *140*
Epp, O., 4(17), *70*
Erickson, M., 79(11), *110*
Erman, J. E., 246(121), *272*
Ernst, M., 165(55), *185*
Escaig-Haye, F., 89(67), *112*
Evans, D. F., 108(158), *115*
Evinger-Hodges, M. J., 133(96), *141*

Fakan, J., 91(89), *112*
Fakan, S., 91(89), *112*
Farrant, J. L., 94(96), *113*
Farr-Jones, S., 50(71), *71*
Fasce, C. F., 180(151), *188*
Fast, D. M., 181(160-161), *189*
Faulkner, W. R., 153(35), *185*
Favilla, J. T., 91(80), *112*
Feder, R., 2(3), *69*
Feldherr, C., 89(64), *111*
Fell, G. S., 178(117), *187*
Feurstein, B. G., 129(75), *140*
Filman, D. J., 63(88-90), *72*
Finer-Moore, J., 67(95), *72*
Finzel, B. C., 62(86), *72*
Fiori, C., 74(3), 77(3), 81(3), *109*
Fischer, S., 178(113), *187*
Fisher, G. W., 127(64), *140*
Fisher, H. W., 98(106), *113*
Fisher, K. A., 106(141), 107(141), 108(141), *114*
Fitzgerald, P. M. D., 41(53), *71*
Flannery, B. P., 267(163), *273*
Fletterick, R. J., 67(95), *72*
Flood, J., 144(17), *184*
Flores, C., 133(97), *141*

Flores, H., 215(74), *270*
Fontecilla-Camps, J. C., 63(92), *72*
Förster, T., 123(29), *139*
Fourier, J. B. J., 19(28), *70*
Fournier, J.-G., 89(67), *112*
Fowler, J. S., 178(110), *187*
Fowler, W. E., 79(15), 91(15), 98(113), 100(119), *110*, *113*
Franzini-Armstrong, C., 100(120), *113*
Freeman, R., 80(18), *110*
Friedel, G., 33(38), *70*
Friedrich, W., 2(1), *69*
Fritz, M., 106(143), *114*
Fuchs, R., 127(56), *140*
Fuji Photo Film Co., Ltd., 167(82), *186*
Fukumori, Y., 246(113), *271*
Fuller, B. H., 178(122), *187*
Fulwyler, M. J., 127(62), *140*
Fung, B. K., 123(27), *139*

Gabel, D., 119(4), *138*
Gahan, P. B., 91(81, 85), *112*
Galteau, M. M., 180(155), *188*
Garcia, R., 108(154), *114*
Garcia-Webb, P., 168(98), 178(119), *187*
Garland, T. B., 130(87), *141*
Garzon, S., 86(49), *111*
Gassmann, J., 42(60), *71*
Gaub, H. E., 106(143), 108(161), *114*, *115*
Gawehn, K., 162(47), *185*
Gehring, W. J., 91(73), *112*
Gerber, C., 2(4), *69*
Gerber, M., 198(24, 26), 200(24, 26), 203(26), 204(26, 38), 208(26), 264(26, 156), *269*, *273*
Gerhardt, W., 179(142), *188*
Gerke, G. K., 195(11), *268*
Geuskens, M., 93(90), *113*
Geuze, H. J., 88(58), *111*
Gharyal, P. K., 128(69), 130(69), 133(69), *140*
Gibson, Q. H., 248(130), *272*
Giegé, R., 4(18), 5(18), 6(18), *70*
Giesemann, W., 222(92), *271*
Gilkey, J. C., 84(32), 103(32), *110*
Gillett, J., 99
Gilroy, S., 123(35), *139*
Girling, R. B., 249(132), *272*
Gitler, C., 119(12), 122(12), *138*
Gladen, B. C., 178(118), *187*
Glaeser, R. M., 98(107), *113*
Glauert, A. M., 74(4), 76(4), 77(4), 83(30-31), 84(30-31, 34), 91(82), *110*, *112*
Glick, D., 81(22), 85(39), 89(39), *110*, *111*
Glick, M. L., 160(43), *185*
Glusker, J. P., 3(7), *70*

Goaman, L. C. G., 4(15), 39(15), *70*
Godwin, A. K., 126(49-50), *139*, *140*
Golay, M. J. E., 266(158), *273*
Gold, M. H., 246(119), *272*
Goldberg, D. M., 179(131), *188*
Goldman, Y. E., 123(39), *139*
Goldspink, P. H., 89(71), *112*
Goldstein, J. I., 74(3), 77(3), 81(3), *109*
Gollnick, P., 231(105), 234(105), *271*
Goodman, J. I., 126(48), *139*
Gorus, F. K., 144(21), *185*
Gould, G., 210(59), *270*
Gould, S. A. C., 108(160-161), 109(162), *115*
Grace, D. E. P., 56(77), *71*
Grassi, M., 170(106), *187*
Green, D. W., 35(45), *71*
Greene, R. C., 236(107), 237(107), 238(107), 239(107), 240(107), 264(107), *271*
Greyson, J., 168(92), *187*
Griffin, R. G., 50(71), *71*
Griffith, J. D., 85(37), 101(122), *111*, *113*
Griffiths, G., 88(57), *111*
Griffiths, J. C., 181(170), *189*
Grigoriev, V., 89(67), *112*
Grimley, P. M., 130(86), *141*
Griswald, A., 119(7), 122(7), *138*
Grogan, W. M., 123(31-32), *139*
Gross, H., 102(126-127), 108(153), *114*
Gross, L., 208(55), 210(55), 212(55), *270*
Gruber, W., 165(62), *186*
Grynkiewicz, G., 129(80), *140*
Guckenberger, R., 104(132), 108(157), *114*
Guern, J., 127(53), *140*
Gurley, J., 108(155), *114*
Gustafsson, M. G. L., 106(141), 107(141), 108(141), *114*
Gutfreund, H., 194(7), 195(17), *268*, *269*

Haase, A. T., 93(94), *113*
Hadju, J., 68(97), *72*
Haeckel, R., 144(13), *184*
Hagen, S. J., *87*
Hahn, M., 95(101), *113*
Hahn, T., 6(24), 8(24), 9(24), 31(24), *70*
Hainfeld, J. F., 80(17, 21), *110*
Halaka, F. G., 246(115-116), 267(115), *271*
Hall, C. E., 94(97), 95(97), *113*
Hallgren, R. C., 137(103), *141*
Ham, W., 181(165), *189*
Hamlin, R. C., 63(88), 67(95), *72*
Hammes, G. G., 248(129), *272*
Han, F., 34(42), *71*
Hanseler, E., 144(13), *184*
Hansen, B. F., 208(46), *269*
Hansen, J., 208(48), 209(48), *269*

Hansma, H. G., 106(144), 108(161), 109(164), *114*, *115*
Hansma, P. K., 106(142, 144), 108(155, 160-161), 109(162, 164), *114*, *115*
Haralambidis, J., 93(95), *113*
Hardman, K. D., 68(98), *72*
Harff, G. A., 144(16), *184*
Harker, D., 38(49), *71*
Harootunian, A. T., 129(79), *140*
Harris, C. M., 215(75), *270*
Harrison, S. P., 153(37), *185*
Hart, R. G., 4(14), 39(14), *70*
Hartmann, T., 108(157), *114*
Hartridge, H., 248(128), *272*
Haschemeyer, R. H., 80(21), *110*
Hasse, A. T., 89(66), *112*
Haugland, R. P., 127(59), *140*
Hauptman, H., 34(42), *71*
Hayaishi, O., 215(73), *270*
Hayat, M. A., 74(6), 83(29), 86(42), 87(56), 89(61), 91(86, 88), 93(90), 95(99, 101), 96(29), 104(131), 105(136, 139), *110*, *111*, *112*, *113*, *114*
Haydon, P. G., 109(165), *115*
Hayes, W. J., 178(121), *187*
Hazlett, T. L., 119(8), 122(8), *138*
Hecht, E., 11(25), *70*
Heckman, J. W., 100(118), 101(118), *113*
Hedley, D., 127(60), *140*
Hefti, A., 106(151), 107(151), *114*
Hejnaes, K. R., 208(46), *269*
Helenius, A., 127(56), *140*
Helliwell, J. R., 29(34), *70*
Helm, C. A., 108(160), *115*
Henderson, E., 109(165), *115*
Hendrickson, W. A., 41(58), 42(59), *71*
Henny, J., 180(155), *188*
Henry, E. R., 265(160), 267(160), *273*
Henry, N. F. M., 6(24), 8(24), 9(24), 31(24), *70*
Henry, S., 131
Herman, E. H., 87(55), 89(63), *111*
Hermans, I. T. H., 166(67), 186
Hernaez, L., 130(82), *140*
Hernandez, D., 129(71), 130(71), 133(71), *140*
Herren, B., 4(20), *70*
Herriott, J. R., 41(57), *71*
Hess, G. P., 195(16), *268*
Hester, R. E., 249(132), *272*
Heuser, J., 99(116), 103(116), *113*
Heyn, S.-P., 108(161), *115*
Higgins, S., 166(68), *186*
Hildebrandt, J., 144(25), *185*
Hill, C. P., 211(61), *270*
Hillebrand, A., 108(157), *114*
Himoe, A., 195(16), *268*
Himpens, B., 129(77), 135(99), *140*, *141*

Hinshaw, J. E., 98(111), *113*
Hirohara, H., 195(14), 195(15), *268*
Hiromi, K., 267(162), *273*
Hirs, C. H. W., 4(19), *70*
Hirschfield, T. J., 250(137), *272*
Hockenborn, M., 166(76), *186*
Hodgkin, D. C., 208(42), *269*
Hofrichter, J., 265(160), 267(160), *273*
Hoh, J. H., 106(146), 107(146), *114*
Holbrook, E. L., 236(107), 237(107), 238(107), 239(107), 240(107), 264(107), *271*
Holen, J. T., 166(79), *186*
Holopainen, I., 128(68), 130(68), *140*
Homburger, H. A., 179(137), *188*
Hoppe, W., 42(60), *71*
Hoppeler, H., 88(57), *111*
Horan, P. K., 124(41), *139*
Horder, M., 179(137-139, 142), 180(156), *188*
Horisberger, M., 85(38), *111*
Horn, M. J., 144(26), *185*
Hornby, W. E., 167(89), *186*
Horne, F. H., 194(9), 204(9), *268*
Houben, K. F., 217(82), 226(97), 227(97), *271*
Howanitz, J. H., 178(127), *187*
Howanitz, P. J., 178(127), *187*
Howard, A. J., 62(86-87), *72*
Howard, K. S., 74(2), *109*
Hsuanyu, Y., 246(124-126), *272*
Hubbard, R. E., 208(42), *269*
Huber, R., 4(17), *70*
Hui, S. W., 103(128), *114*
Hulser, D. F., 128(67), 130(67), *140*
Hutchison, J. S., 194(5), 198(5), 200(5), 208(5), *268*
Hutchison, N. J., 89(65), 93(65), *112*
Hyde, C. C., 217(81, 88-89), 219(88), 227(88-89), 240(81), 243(88-89), 244(88-89), *271*
Hyun, W. C., 127(62), 129(75), *140*

Ikai, A., 107(152), *114*
Imai, K., 165(56), *186*
Inagami, T., 68(98), *72*
Ingle, J. D., 260(153), 261(153), *273*
Ingold, C. K., 63(91), *72*
Ingram, V. M., 35(45), *71*
Innes, D. F., 178(122), *187*
Isaacs, N. W., 208(42), *269*
Ise, N., 195(14-15), *268*
Isenhour, T. L., 150(27), *185*
Ishimura, Y., 215(73), *270*
Iype, S. N., 124(46), *139*
Izutsu, K., 81(25), 82(25), *110*

Jack, A., 47(68), *71*
Jacob, J., 91(88), *112*

Jacobs, B. L., 106(147), 109(147), *114*
Jacobson, K., 130(84), *140*
Jaderlund, B., 165(63), *186*
James, M. N. G., 60(84), *72*
Jameson, D. M., 119(8), 122(8), *138*
Jansen, A. P., 179(133), *188*
Jansen, M. J. J. M., 127(58), *140*
Jansen, R. T. P., 179(133), *188*
Janson, P. C. W., 166(67), *186*
Jarnik, M., 98(112), *113*
Jatlow, P., 166(70, 74), *186*
Jaynes, E. T., 43(64), *71*
Jeanguillaume, C., 79(14), *110*
Jenny, R. W., 180(158), *189*
Jensen, L. H., 3(6), 41(57), *69*, *71*
Jessen, H., 86(53), *111*
Jett, S. D., 104(135), *114*
Jiang, L.-W., 119(13), 122(21), 128(13, 21, 69), 129(71), 130(13, 21, 69, 71), 133(69, 71), *138*, *139*, *140*
Jobin, M., 109(163), *115*
Johansen, B. R., 208(46), *269*
John, S. A., 106(146), 107(146), *114*
Johnson, A. H., 74(2), *109*
Johnson, D., 81(25), 82(25), *110*
Johnson, K. A., 80(20), *110*, 217(83), 227(83), 240(83), *271*
Johnson, L. N., 3(5), 68(97-98), *69*, *72*
Johnson, P., 130(87), *141*
Johnson, R. J., 215(75), *270*
Johnson, T. J. A., 103(128), *114*
Johnston, M., 238(108), *271*
Jones, D. G., 250(136), 252(136, 141), *272*
Jones, R. W., 152(31), *185*
Jones, T. A., 44(67), 45(67), *71*
Jordan, G. W., 93(93), *113*
Joy, D. C., 74(3), 77(3), 81(3), *109*
Judson, H. F., 36(48), *71*
June, D. S., 227(101-102), 231(102), 233(102), *271*
Jung, K., 178(112), *187*

Kaarsholm, N. C., 208(47-50, 54), 209(47-50), 210(47, 58), *269*, *270*
Kallen, R. G., 215(78, 80), 222(80), *270*, *271*
Kalow, W., 178(120), *187*
Kan, F. W. K., 86(48, 50), *111*
Kao, J. P. Y., 123(37), 129(78-79), *139*, *140*
Kapp, A., 165(55), *185*
Karmen, A., 168(96), *187*
Karplus, M., 48(70), *71*
Karube, Y., 215(79), 235(79), *270*
Kaspar, P., 166(77), *186*
Katchalski, E., 119(4), *138*
Katz, J. E., 108(156), 109(156), *114*
Kaufman, B. T., 63(90), *72*

Kawamura-Konishi, Y., 246(127), *272*
Kawasaki, H., 227(99), 244(99), *271*
Kawata, S., 250(139), 251(139), *272*
Kay, D., 98(108), 101(108), *113*
Kayastha, A. M., 217(87), 226(87), 227(87), 240(87), 241(87), 242(87), *271*
Kazanskoya, N. F., 195(19), *269*
Kazarinoff, M. N., 228(103), *271*
Kazmierczak, S. C., 162(46), *185*
Kegelman, J. E., 144(12), *184*
Kelderman, G., 106(144), *114*
Kellenberger, E., 86(42), 89(61), *111*
Keller, D., 108(154), *114*
Keller, H., 144(15, 18), *184*, *185*
Kelly, D. M., 68(98), *72*
Kendrew, J. C., 4(14), 39(14), *70*
Kennard, O., 54(76), 55(76), *71*
Kido, R., 215(73), *270*
Kikuchi, K., 246(127), *272*
Kim, B., 2(3), *69*
Kirschner, K., 217(84), 223(93-95), 226(96), 227(100), *271*
Kirz, J., 2(3), *69*
Kisliuk, R. L., 63(92), *72*
Klauke, R., 181(164), *189*
Klaunig, J. E., 126(48), *139*
Kleinschmidt, A. K., 101(121), *113*
Klokker, M., 86(53), *111*
Klomparens, K. L., 84(35), *110*
Klyosov, A. A., 195(19), *269*
Knight, D., 144(14), *184*
Knipping, P., 2(1), *69*
Knitsch, K. W., 166(75), *186*
Knol, K. K., 41(54), *71*
Knowles, J. R., 239(109), 244(110), *271*
Ko, H.-C., 208(47), 209(47), 210(47), *269*
Kobayashi, Y., 178(124), *187*
Koch, D. D., 144(20), *185*
Koedam, J. C., 181(164-165), *189*
Koerber, S. C., 194(3, 8), 198(3, 8, 20, 25), 200(3, 20, 25), 201(3), 203(3), 204(8, 20, 25), 205(20), 207(20), 264(3, 20, 155, 157), 265(157), 266(157), *268*, *269*, *273*
Koester, C. J., 121(15), *139*
Koetzle, T. F., 54(76), 55(76), *71*
Kohli, C., 144(12), *184*
Kolkmayer, N. H., 21(30), *70*
Koller, P. U., 168(99), *187*
Koppel, D. E., 122(18), 128(18), 130(18, 88-89), *139*, *141*
Korpella, T., 215(78), *270*
Korzun, W. J., 158(42), *185*
Kozlowski, H., 204(37), *269*
Kraehenbuhl, J.-P., 85(40), *111*
Kraut, J., 63(88-90), *72*, 246(121), *272*

Kretsinger, R. H., 60(83), *72*
Kricka, L. J., 165(54, 58), *185*, *186*
Krizman, D. B., 126(51), *140*
Krodel, E., 144(24), *185*
Kropf, J., 144(25), *185*
Kruth, H. S., 124(40), *139*
Kubasik, N. P., 168(105), *187*
Kuhn, E., 79(11), *110*
Kumar, K., 144(9), *184*
Kunugi, S., 195(14-15), *268*
Kurebayashi, N., 130(81), *140*
Kuriyan, J., 48(70), *71*
Kurkdjian, A., 127(53), *140*
Kusel, J. R., 130(87), *141*
Kusnetz, J., 166(78), *186*
Kussisto, A., 123(33), *139*
Kuszinski, C., 133(95), *141*
Kvassmann, J., 198(32), 208(39), *269*

Ladenson, J. H., 181(175), *189*
Lainé-Cessac, P., 152(30), *185*
Lakowicz, J. R., 119(5), 122(5), *138*
Lal, R., 106(146), 107(146), *114*
Lambier, A. M., 246 (118), *272*
Lamperth, L., 91 (80), *112*
Lamport, D. T. A., 100(118), 101(118), *113*
Lamvik, M. K., 77(8), 82(8), *110*
Lane, A. N., 217(84), 223(93-94), 226(96), 227(100), *271*
Lang, H., 181(173-174), *189*
Lang, L., 215(77), *270*
Langevin, G. L., 89(66, 69), *112*
Lanni, F., 122(22), 128(70), 130(22, 70), *139*, *140*
Larison, K. D., 122(24), 123(24), 129(24), *139*
Larsson, A., 208(39), *269*
Larsson, L.-I., 86(51), *111*
Lau, K.-H., 152(29), *185*
Lau, S.-J., 195(13), 197(13), 214(65), *268*, *270*
Laue, D., 179(131), *188*
Laue, M., 2(1), *69*
Lawrence, J. B., 89(66, 68-69), *112*
Leapman, R. D., 80(16), 82(16), *110*
Lebovitz, R. M., 126(49-51), *139*, *140*
Lee, B. F., 195(18), *269*
Lee, C.-L., 181(172), *189*
Lee, R. J., 127(57), *140*
Lee, R. W.-K., 208(48), 209(48), *269*
Leja, C. A., 217(82), *271*
Lemke, G., 91(80), *112*
Lent, R., 168(96), *187*
Lentjes, E. G., 144(16), *184*
Leonard, K. R., 80(18), *110*
Lerner, D. R., 93(92), *113*
Leslie, A. G., 42(62), *71*

Lestienne, P., 89(67), *112*
Leunissen, J. L. M., 85(36), 89(64), *110*, *111*
Leupold, G., 104(132), *114*
Levinson, S. R., 103(128), *114*
Levitt, M., 47(68), 58(81), *71*, *72*
Lewis, F. H., 124(47), *139*
Lewis, H. A., 253(145), *272*
Li, M.-Q., 109(164), *115*
Li, Y., 126(50), *140*
Libeer, J. C., 167(88), *186*
Lichtscheidl, I. K., 122(16), *139*
Lieberman, M. W., 126(49-51), *139*, *140*
Liedtke, R., 144(17), *184*
Lifshin, E., 74(3), 77(3), 81(3), *109*
Lin, M. F., 181(172), *189*
Lindsay, S. M., 106(147-148), 109(147, 166), *114*, *115*
Ling, B. L., 165(56), *186*
Lloyd, R. V., 91(78), *112*
Loeser, K. E., 100(120), *113*
London, J. L., 179(140), *188*
Longberg-Holm, K. K., 248(130), *272*
Longo, M. L., 108(160), *115*
Lonsdale, K., 6(24), 8(24), 9(24), 31(24), *70*
Lott, J. A., 144(11), 152(32), 153(35), 158(41), 162(46), 178(127), 179(128, 130), 180(153), *184*, *185*, *187*, *188*
Lovenberg, W., 41(57), *71*
Lovgren, T., 165(63), *186*
Low, B. W., 33(39), *71*
Luby-Phelps, K., 128(70), 130(70), *140*
Ludl, H., 260(152), *272*
Luisi, P. L., 194(4), 198(4), *268*
Luisoni, E., 98(110), *113*
Lundin, A., 165(63), *186*
Luque de Castro, M. D., 144(6-7), 150(6), *184*
Lyubchenko, U. L., 106(147), 109(147), *114*

Maassab, H. F., 91(78-79), *112*
MacGibbon, A. K. H., 194(3), 198(3, 20-21, 25), 200(3, 20-21, 25), 201(3), 203(3), 204(20, 25), 205(20-21), 207(20), 208(21), 264(3, 20), *268*, *269*
MacGillavry, C. H., 12(26), *70*
Machin, P. A., 47(69), *71*
Mächler, M., 260(151), *272*
Madhukar, B. V., 133(93), *141*
Maeder, M., 265(161), 267(161), *273*
Malatesta, F., 246(114), *271*
Malhotra, O. P., 213(62), 214(62, 66), *270*
Mansberg, H. P., 166(78), *186*
Manuelidis, L., 91(80), *112*
Maramorosch, K., 98(110), *113*
Marcotte, P., 238(108), *271*

Maret, W., 198(24, 29), 200(24, 29), 204(37), *269*
Marquardt, H. F., 166(71), *186*
Marquez, L., 246(119), *272*
Marsden, A. M., 178(110), *187*
Marshall, J. C., 150(27), *185*
Martell, A. E., 215(78), *270*
Marti, O., 106(142), *114*
Marton, L. J., 129(75), *140*
Marx, A. M., 144(25), *185*
Massion, C. G., 179(128, 130), *187*, *188*
Mathews, F. S., 4(15), 39(15), *70*
Mathieu, M., 179(142), *188*
Matsushima, Y., 215(78-79), 235(79), *270*
Mattenheimer, H., 144(17), *184*
Matthews, B. W., 33(40), *71*
Matthews, D. A., 63(88-90), *72*
Matthews, F. S., 30(36), *70*
Mauro, J. M., 246(121), *272*
Maxfield, F. R., 127(55), *140*
Mayer, T. K., 168(105), *187*
Maynard, Y., 181(175), *189*
McBride, J. H., 144(19), 166(68), *185*, *186*
McClellan, G. A., 144(26), *185*
McComb, R. B., 180(144, 146-149), *188*
McCray, J. A., 123(36), *139*
McFadden, G. I., 90(72), 93(72), *112*
McGandy, E. L., 4(15), 39(15), *70*
McGillavry, C. H., 21(30), *70*
McKinnon, C. A., 130(83), *140*
McKneally, S. S., 181(161-162), *189*
McMichael, K. L., 74(2), *109*
McNeil, M. L., 181(161, 163), *189*
McNeil, P. L., 122(22), 130(22), *139*
McPherson, A., 4(20), 6(21), *70*
McPhie, P., 220(90), *271*
Meehan, E. J., 4(20), *70*
Meek, G. A., 74(5), 76(5), 77(5), *110*
Meiners, S., 129(71), 130(71), 133(71), *140*
Meinke, M. H., 108(158-159), 109(159), *115*
Meister, A. J., 124(43), *139*
Meites, S., 153(35), *185*
Melamed, M. R., 124(42), *139*
Mellman, I., 127(56), *140*
Melroy, D., 87(55), *111*
Mendelson, M. L., 124(42), *139*
Merrit, E. A., 42(59), *71*
Metcalf, T. N., III, 122(19-20), 128(20), 130(19-20), *139*
Metzler, C. M., 215(76-77), *270*
Metzler, D. E., 215(71, 75-78), *270*
Metzler, D. W., 215(78), *270*
Meyer, E. F., Jr., 54(76), 55(76), *71*
Michel, H., 4(17), *70*
Miki, K., 4(17), *70*

Milano, M. J., 249(134), 250(134), 272
Miles, E. W., 215(72-73), 217(81, 83, 87-89), 219(88), 220(90), 222(91), 226(87), 227(83, 87-89, 99), 240(81, 83, 87), 241(87), 242(87), 243(88-89), 244(88-89, 99), 264(91), 270, 271
Miller, D. M., 166(79), 186
Miller, O. L., Jr., 104(133), 114
Miller, S. E., 95(99), 113
Miller, W. G., 158(42), 185
Milligan, R. A., 98(111), 113
Milne, R. G., 98(110), 113
Minami, S., 250(139), 251(139), 272
Minta, A., 123(37), 129(78), 139, 140
Mitchell, A. C., 123(34), 139
Mitchell, A. R., 165(51), 185
Mitchell, J. B., 124(46), 139
Mitchison, T. J., 123(38), 139
Moffat, K., 68(96), 72
Montgomery, J. A., 63(92), 72
Moor, H., 102(126), 114
Moore, C. L., 101(122), 113
Morgan, A. J., 105(136), 114
Morgan, C. G., 123(34), 139
Morin, L. G., 181(162), 189
Morozov, T. V., 215(78), 270
Morrell, J. I., 93(91), 113
Morrett, H., 106(144), 114
Morris, D. L., 175(108), 187
Morris, D. W., 93(93), 113
Morris, R. G., 198(30), 269
Morris, R. J., 260(150), 272
Mory, C., 79(14), 110
Mosesson, M. W., 80(21), 110
Moss, D. W., 179(132, 134, 143), 180(134, 155), 181(143), 188
Mrenus, K., 79(11), 110
Muchmore, D. C., 198(28), 269
Muirhead, H., 4(15), 39(15), 70
Muirhead, K., 124(41), 139
Mullaney, P. F., 124(42), 139
Muller, J., 180(159), 189
Muller, T., 102(126-127), 114
Mulvaney, R., 81(27), 82(27), 110
Murbach, N. L., 178(125), 187
Murphy, J. A., 102(124), 114
Murray, J. G., 123(34), 139
Murray, R. G. E., 95(100), 113
Musgrove, E., 127(60), 140

Nagahara, L. A., 109(166), 115
Nakaajima, A., 178(124), 187
Nakashima, K., 165(56), 186
Namba, K., 60(85), 72
Nanci, A., 86(48), 111

Náray-Szabó, S., 6(23), 70
Narju, R., 123(33), 139
National Committee for Clinical Standards, 180(157), 189
National Institutes of Science and Technology, 157(40), 185
National Reference System for the Clinical Laboratory, 180(145), 188
Naugland, R., 122(24), 123(24), 129(24), 139
Naumann, R., 4(20), 70
Nealon, D. A., 157(38), 185
Nelson, B., 4(20), 70
Nermut, M. Y., 104(131), 114
Neumann, K., 85(36), 110
Neumann, U., 166(75, 77), 186
Neumeier, D., 181(174), 189
Newbury, D. E., 74(3), 77(3), 81(3), 109
Newman, M. S., 51(74), 71
Newmark, P., 89(62), 111
Ngo, K., 226(98), 227(98), 228(98), 271
Niall, H. D., 93(95), 113
Nilausen, K., 86(53), 111
Nipper, H. C., 144(26), 185
Nishimura, E., 195(15), 268
Nockolds, C. E., 60(83), 72
Noller, H. F., 214(65), 270
Norris, J. R., 98(108), 101(108), 113
North, A. C. T., 30(36), 70
Northrop, J. H., 3(11), 70
Nuccitelli, R., 127(52), 140
Nuovo, G. J., 85(41), 93(41), 111

Oates, K., 81(27), 82(27), 110
O'Connell, B. T., 178(125), 187
Oden, P. I., 109(166), 115
Ogletree, D. F., 108(156), 109(156), 114
Oh, S. Y., 133(93), 141
O'Hanolon, J. F., 76(7), 110
Ohnesorge, F., 108(161), 115
Okamoto, T., 250(139), 251(139), 272
Oldén, B., 198(32), 269
Oldstone, M. B. A., 89(66), 112
Oliver, J. M., 127(57), 140
Olsson, L., 86(53), 111
Onishi, T., 152(29), 185
Orii, Y., 246(113, 122), 271, 272
Oryall, J. J., 144(20), 185
Osborn, M. J., 119(2), 122(2), 130(88-89), 138, 141
Osheim, Y. N., 104(134), 114

Padlan, E. A., 217(81), 240(81), 271
Pähler, A., 42(59), 71
Palcic, M. M., 246(117), 272
Palmer, H. J., 153(33), 185

Panitz, H., 108(154), *114*
Papadakis, N., 254(148-149), 255(148), *272*
Papermaster, D. S., 85(40), *111*
Pardue, H. L., 249(134), 250(134), *272*
Patel, S. T., 178(127), *187*
Pattanayek, R., 60(85), *72*
Patterson, A. L., 35(46-47), *71*
Pattison, S. E., 210(56), *270*
Pauling, L., 51(73), 56(78), 57(79), *71*, *72*
Pawley, J. B., 74(1), *109*, 121(15), *139*
Pazzagli, M., 165(52), *185*
PB Diagnostic Systems, Inc., 167(83), *186*
Peace, K., 198(20), 200(20), 204(20), 205(20), 207(20), 264(20), *269*
Pearson, K. W., 165(51), *185*
Pease, K., 198(21), 200(21), 205(21), 208(21), *269*
Peerdeman, A. F., 41(55), *71*
Penschow, J. D., 93(95), *113*
Peranzi, G., 89(67), *112*
Perez-Bendito, first name, 144(3), *184*
Perry, G. A., 133(95), *141*
Pershadsingh, H. A., 127(62), *140*
Perutz, M. F., 4(15), 35(45), 39(15), *70*, *71*
Peters, K.-R., 79(10), 102(124), *110*, *114*
Peterson, C. M., 109(162), *115*
Peticolas, W. L., 195(11), *268*
Petrusz, P., 89(68), 93(91), *112*, *113*
Petsko, G. A., 59(82), *72*
Pettersson, G., 198(32), 208(39-40), *269*
Petty, S., 215(77), *270*
Phelps, D. J., 195(10), *268*
Philipp, M., 106(148), *114*
Phillips, D. C., 4(16), 30(36), 56(77), 59(82), 63(16), *70*, *71*, *72*
Phillips, R. S., 222(91), 229(104), 230(104), 231(105), 232(104), 234(105), 244(106), 264(91), *271*
Phillips, T. D., 124(47), *139*
Phizackerley, R. P., 42(59), *71*
Pickel, V. M., 91(83), *112*
Pierce, R. J., 178(110), *187*
Pilz, W., 178(120), *187*
Pimentel, G. C., 192(1), *268*
Pisa, M., 166(68), *186*
Plant, A. L., 122(17), *139*
Plischke, W., 168(95), *187*
Poenie, M., 129(80), *140*
Polak, J. M., 85(38), 88(59), 89(65), 91(83), 93(65), *111*, *112*
Pollard, B. D., 250(138), *272*
Pollard, T. D., 96(102), *113*
Pomerov, M., 89(66), *112*
Pompliano, D. L., 244(110), *271*
Popelka, S. R., 166(79), *186*

Porter, A. B., 21(29), *70*
Poste, G., 124(41), *139*
Postek, M. T., 74(2), *109*
Posthuma, G., 88(58), *111*
Potts, W. T. W., 81(27), 82(27), *110*
Poulos, T. L., 62(86-87), *72*
Poulson, R., 215(72), *270*
Pouyssegur, J., 127(54), *140*
Pravatiner, E. S., 144(23), *185*
Prelog, V., 63(91), *72*
Press, W. H., 267(163), *273*
Price, C. P., 166(69), 167(87, 90), 175(90), *186*
Price, R. G., 178(110), *187*
Prins, J. A., 41(54), *71*
Przywara, D. A., 135(100), *141*
Pusey, M., 4(20), *70*
Pyle, K. A., 124(47), *139*

Qichen, H., 29(35), *70*
Quam, E. F., 144(20), *185*
Quiel, E., 210(56), *270*

Racker, E., 127(63), *140*
Radmacher, M., 106(143), *114*
Raikhel, N. V., 93(92), *113*
Ralston, I. M., 215(78), *270*
Ramachandran, G. N., 53(75), *71*
Rand, R. N., 179(135), 180(135, 149), 181(178), *188*, *189*
Rannefeld, B., 208(46), *269*
Rash, J. E., 103(128), 104(130), *114*
Ray, J. S., 126(48), *139*
Read, N. D., 123(35), *139*
Reddy, S. S., 144(23), *185*
Reed, T. A., 222(92), *271*
Reichelt, R., 79(13), *110*
Reid, A. P., 81(27), 82(27), *110*
Reid, N., 83(30), 84(30), *110*
Reigart, J. R., 178(118), *187*
Reinisch, L., 127(61), *140*
Rej, R., 157(38), 179(138-139), 180(151-152, 154, 156, 158), 181(168-169), *185*, *188*, *189*
Renaud, M., 91(73), *112*
Retzel, E. F., 93(94), *113*
Revel, J.-P., 106(146), 107(146), *114*
Reynolds, C. D., 208(41-43), 209(41), *269*
Reynolds, K. M., 168(104), *187*
Rhoades, R., 136
Rice, G. C., 124(44), 124(45), *139*
Richards, F. M., 33(39), 68(98), *71*, *72*
Richardson, J. S., 57(80), *72*
Richter, I., 231(105), 234(105), *271*
Rick, W., 166(76), *186*
Rieck, G. D., 12(26), *70*

Rill, R. L., 109(166), *115*
Rink, T. J., 129(76), *140*
Rinker, A. D., 157(39), 179(141), *185*, *188*
Ritchie, R. F., 166(78), *186*
Rittersdorf, W., 170(106), *187*
Rivers, P. C., 59(82), *72*
Robards, A. W., 84(34), 103(129), *110*, *114*
Roberts, K., 168(100), *187*
Roche Diagnostic Systems, 153(34), *185*
Rodgers, J. R., 54(76), 55(76), *71*
Rodgerson, D. O., 144(19), 166(68), *185*, *186*
Rogan, W. J., 178(118), *187*
Rogers, A. W., 91(84), *112*
Rogowska, J., 127(64), *140*
Rohrer, H., 2(4), *69*, *74*
Roodyn, D. B., 144(4), *184*
Roomans, G. M., 102(124), 105(137), *114*
Rose, J. D., 63(92), *72*
Ross, A. H., 130(83), *140*
Rossi, E., 178(119), *187*
Rossmann, M. G., 40(50–51), *71*
Roth, J., 86(52), 87(54), 91(73), *111*, *112*
Roth, L. J., 91(87), *112*
Roughton, F. J. W., 248(128), *272*
Rowe, A. J., 101(123), 102(123), *113*
Roy, M., 208(48), 209(48), 217(82), 222(91), 264(91), *269*, *271*
Ruben, G. C., 102(125), *114*
Rubenthaler, G. L., 178(125), *187*
Ruch, R. J., 126(48), *139*
Rugg, C., 127(60), *140*
Ruska, E., *74*
Russell, B., 89(71), 90(70), *112*
Rutter, W. J., 67(95), *72*
Ryan-Hotchkiss, M., 260(153), 261(153), *273*
Ryder, K. W., 160(43), *185*
Ryder, T. A., 105(139), 106(140), *114*

Sakabe, K., 208(42), *269*
Sakaguchi, P. S., 109(165), *115*
Salden, H. J. M., 166(67), *186*
Saliman, G., 198(30), *269*
Salmeron, M. B., 108(156), 109(156), *114*
Salmon, E. D., 130(85), *140*
Salpeter, M. M., 91(86–87), *112*
Samama, J.-P., 198(31), *269*
Sampson, E. J., 179(142), 181(160, 162), *188*, *189*
Sandberg, M., 123(33), *139*
Santiago, D., 194(7), *268*
Santiago, P., 194(7), *268*
Santini, R. E., 249(134), 250(134), *272*
Sarti, P., 246(114), *271*
Sartorius, C., 198(26), 200(26, 35), 203(26), 204(26), 208(26), 264(26), *269*

Sasisekharan, V., 53(75), *71*
Satow, Y., 42(59), *71*
Savin, F. A., 215(78), *270*
Savitzsky, A., 266(158), *273*
Savory, J., 166(73), *186*
Sawa, Y., 217(88), 219(88), 227(88), 243(88), 244(88), *271*
Saxton, W. O., 80(19), *110*
Sayre, D., 2(3), *69*
Scandinavian Committee on Enzymes of the Scandinavian Society for Clinical Chemistry, 179(136), 180(150), *188*
Schack, P., 194(4, 8), 198(4, 8), 204(8), *268*
Schalwijk, J., 127(58), *140*
Schechter, Y., 130(82), *140*
Scheer, U., 91(89), *112*
Schiele, F., 180(159), *189*
Schindler, M., 119(1–2, 13), 122(1–2, 18–21), 128(13, 18, 20–21, 69), 129(71–73), 130(13, 18–21, 69, 71–73, 88–89), 133(69, 71–73), *138*, *139*, *140*, *141*
Schlemmer, H. H., 260(151), *272*
Schlessinger, J., 130(82), *140*
Schlichta, P. J., 6(21), *70*
Schmidt, E., 181(164), *189*
Schnackerz, K. D., 222(92), *271*
Schneider, G., 198(24), 200(24), *269*
Schneider-Berhlöhr, H., 198(24), 200(24), *269*
Schneir, J., 108(155), *114*
Scholmerich, J., 165(55), *185*
Scholz, D., 178(112), *187*
Schoone, J. C., 35(44), 41(56), *71*
Schotters, S. B., 144(19), 166(68), *185*, *186*
Schreiber, G., 178(112), *187*
Schubert, K. R., 122(19), 130(19), *139*
Schuette, I., 178(114), *187*
Schuler, D., 98(114), *113*
Schultz, R. A., *134*
Schwartz, M. K., 144(2), *184*
Schwarz, H., 89(61), *111*
Scott, M. G., 181(177), *189*
Seegan, G., 98(104), *113*
Seligson, D., 166(70, 74), *186*
Sergio, M., 165(52), *185*
Sertic, J., 144(9), *184*
Seveus, L., 123(33), *139*
Seyama, T., 126(49–50), *139*, *140*
Shannon, J., 215(74), *270*
Shapiro, B. M., 94(98), 95(98), 98(98), *113*
Sharon, N., 119(1), 122 (1), *138*
Shaw, L. M., 179(140–141), 181(163), *188*, *189*
Sheetz, M., 122(18), 128(18), 130(18), *139*
Shelburne, J. D., 105(137), *114*

Sherman, J., 51(73), *71*
Shimanouchi, T., 54(76), 55(76), *71*
Shinitzky, M., 119(11-12), 122(11-12), 123(30), *138*, *139*
Shipe, J. R., 166(73), *186*
Shirey, T. L., 168(103), *187*
Shore, J. D., 194(7), *268*
Shotton, D. M., 134(98), *141*
Shrieve, D. C., 124(45), *139*
Shuman, H., 81(23), *110*
Siano, D. B., 215(75), *270*
Siegerist, C., 106(144), *114*
Sieker, L. C., 41(57), *71*
Siekhaus, W. J., 108(156), 109(156), *114*
Siest, G., 179(143), 180(155, 159), 181(143), *188*, *189*
Silgar, S. G., 245(111), *271*
Silva, F. G., 89(66, 68), *112*
Silva, M., 144(3), *184*
Simpson, R. W., 252(144), *272*
Sims, G. R., 253(145), *272*
Singer, J. S., 85(37), *111*
Singer, R., 144(14), *184*
Singer, R. H., 89(66, 68-69), *112*
Singh, B., 215(74), *270*
Sinsheimer, R. L., 106(144), 108(161), 109(164), *114*, *115*
Sitte, K., 85(36), *110*
Sleyter, U. B., 84(34), 103(129), *110*, *114*
Slot, J. W., 88(58), *111*
Smeaton, J. R., 166(71), *186*
Smith, C. A., 98(105), *113*
Smith, G. D., 208(41, 43-45), 209(41), *269*
Smith, J. L., 42(59), *71*
Smith, L. C., 122(17), *139*
Smith, P. R., 79(15), 91(15), 98(113), *110*, *113*
Smith, R. E., 165(51), *185*
Smith, S. O., 50(71), *71*
Snell, E. E., 215(67), 228(103), *270*, *271*
Snyder, R., 4(20), *70*
Somlyo, A. P., 81(23-24), 82(24), *110*, 123(39), 129(77), *139*, *140*
Somlyo, A. V., 123(39), *139*
Sommerville, J., 91(89), *112*
Sparks, C., 208(43), *269*
Spayd, R. W., 168(101), *187*
Spencer, K., 166(69), *186*
Spiller, E., 2(3), *69*
Sprang, S., 67(95), *72*
Springer, B. A., 245(111), *271*
Stadtman, E. R., 94(98), 95(98), 98(98), *113*
Staehlin, L. A., 84(32-33), 103(32), *110*
Stahler, F., 165(62), 168(98), *186*, *187*
Stammers, D. K., 63(90), *72*
Standing, T., 67(95), *72*

Stanley, P. E., 165(54, 61), *185*, *186*
Stanley, W. M., 3(12), *70*
Stasiak, A., 108(153), *114*
Staskus, K. A., 93(94), *113*
Stavljenic, A., 144(9), *184*
Steentjes, G. M., 181(164-165), *189*
Stein, L. S., 141
Stein, W., 181(176), *189*
Steinberg, I. E., 119(7), 122(7), *138*
Steinberg, I. Z., 119(4), *138*
Steinhausen, R. L., 167(87), *186*
Steinrauf, L. K., 6(22), *70*
Stemmer, A., 106(150-151), 107(150-151), *114*
Sternberg, M. J. E., 56(77), *71*
Sternberger, L. A., 87(56), *111*
Stiege, W., 98(114), *113*
Stierhof, Y.-D., 89(61), *111*
Stoeckert, C. J., 79(11), *110*
Stoica, G., 141
Storm, M. C., 210(56-57), *270*
Storz, G., 168(95), *187*
Stout, G. H., 3(6), *69*
Stover, A. C., 195(10), *268*
Strandberg, B. E., 4(14), 39(14), *70*
Strasser, A. W. M., 227(100), *271*
Strasser, R. J., 128(67), 130(67), *140*
Strobelt, V., 178(112), *187*
Stromme, J. H., 179(140), *188*
Stroud, R. M., 67(95), *72*
Stryer, L., 119(3), 122(3), 123(25, 27), *138*, *139*
Strynadka, N. C. J., 60(84), *72*
Stuart, D. I., 68(97), *72*
Stubbs, G., 60(85), *72*
Stump, R. F., 127(57), *140*
Stumpf, W. E., 91(87), *112*
Suddath, F. L., 4(20), *70*
Suelter, C. H., 194(9), 204(9), 227(101-102), 231(102), 233(102), *268*, *271*
Sumner, J. B., 3(10), *70*
Sun, J., 133(95), *141*
Sun, W., 246(123), *272*
Surufka, N., 179(128), *187*
Sussman, J. L., 47(69), *71*
Suzuki, H., 246(127), *272*
Suzuki, S., 178(124), *187*
Swaisgood, M., 119(13), 128(13), 130(13), *138*
Swenson, D. C., 208(41, 43), 209(41), *269*
Syed, D., 180(144), *188*
Syrjanen, S., 123(33), *139*
Szasz, G., 163(49), *185*
Szebenyi, D., 68(96), *72*
Szollosi, J., 129(75), *140*

Tabata, M., 178(124), *187*

Taborelli, M., 109(163), *115*
Talmi, Y., 251(140), 252(140, 142-144), 253(140, 143), *272*
Tasjian, Z. H., 195(18), *269*
Tasumi, M., 54(76), 55(76), *71*
Taylor, D. L., 122(22), 127(59, 64), 128(70), 130(22, 70), *139*, *140*
Taylor, K. A., 98(107), *113*
Taylor, W. H., 6(23), *70*
Teeter, M., 41(58), *71*
Teichberg, V. I., 119(11), 122(11), *138*
Temple, C., Jr., 63(92), *72*
Ten Heggeler-Bourdier, B., 109(163), *115*
Tenkolsky, S. A., 267(163), *273*
Terhune, B. T., 100(118), 101(118), *113*
Terlingen, J. B. A., 181(165), *189*
Tesin, D., 91(80), *112*
Theodorsen, L., 179(140), *188*
Theroit, J. A., 123(38), *139*
Thiry, M., 91(75), 91(76-77), *112*
Tholen, D. W., 179(130), *188*
Thomas, J. A., 127(63), *140*
Thomas, L., 168(95), *187*
Thompson, G. G., 178(116), *187*
Thompson, J. A., 215(75), *270*
Thomson, R. E., 106(141), 107(141), 108(141), *114*
Thorpe, G. H. G., 165(54), *185*
Thundat, T., 109(166), *115*
Tietz, N. W., 157(39), 179(141), 180(152), *185*, *188*
Tiffany, T. O., 144(5), *184*
Tilley, R., 141
Tillmann, R. W., 106(143), *114*
Tilzer, L. L., 152(31), *185*
Timasheff, S. N., 4(19), *70*
Tokuyasu, K. T., 85(37), 89(60), *111*
Tonge, P. J., 195(12), *268*
Tooze, J., 51(72), *71*
Torchinshky, Y. M., 215(78), *270*
Toth, P. P., 85
Tourin, R. H., 254(146), *272*
Tourmente, S., 91(73), *112*
Tranum-Jensen, J., 86(53), *111*
Travaglini, G., 108(153), *114*
Tregear, G. W., 93(95), *113*
Trentham, D. R., 123(36, 39), *139*
Trewavas, A. J., 123(35), *139*
Trosko, J. E., 126(48, 51), 129(72-73), 130(72-73), 133(72-73, 93-94), *139*, *140*, *141*
Troup, C. D., 127(57), *140*
Trueblood, K. N., 3(7), *70*
Tsernoglou, D., 68(98), *72*
Tsien, R. Y., 123(37), 129(78-80), *139*, *140*
Tsukamoto, Y., 165(56), *186*

Turcant, A., 152(30), *185*
Tyler, J. M., 97(103), 100(117), 101(117), *113*

Udenfriend, S., 164(50), *185*
Ueno, H., 215(78), *270*
Ulevitch, R. J., 215(80), 222(80), *271*
Umemura, K., 107(152), *114*
Url, W. G., 122(16), *139*
Usategui-Gomez, M., 181(167), *189*

Vaidya, H. C., 181(175), *189*
Vaisala, M., 123(33), *139*
Valcárcel, M., 144(6-7), *184*
Valentine, R. C., 94(98), 95(98), 98(98), *113*
Vallez, J. M., 144(10), *184*
van Bommel, A. J., 41(55), *71*
Vanderlinde, R. E., 178(111), 180(151-152), 181(178), *187*, *188*, *189*
VanderVoort, H. T. M., 137(102), *141*
van Dreumel, H. J., 181(165), *189*
Van Dyke, K., 165(57, 60), *186*
van Erp, P. E. J., 127(58), *140*
Van Wart, H. E., 246(120), 247(120), 248(120), 260(150), *272*
Varndell, I. M., 85(38), 88(59), 89(65), 91(83), 93(65), *111*, *112*
Ven Katakrishman, G., 130(83), *140*
Verkleij, A. J., 85(36), 89(64), *110*, *111*
Vesenka, J., 106(144), *114*
Vetter, R. D., 178(115), *187*
Vetterhig, W. T., 267(163), *273*
Vijayan, N. M., 208(42), *269*
Villeneuve, D. C., 178(109), *187*
Villiger, W., 89(61), *111*
Visscher, K., 121(14), 137(101), *138*, *141*
Vitella, L. B., 246(121), *272*
Vogel, W. C., 166(72), *186*
Volz, K. W., 63(90), *72*
Vonderschmitt, D., 144(13), *184*
Von Etten, R. L., 181(163), *189*

Wade, M. H., 119(13), 128(13), 129(72-73), 130(13, 72-73), 133(72-73, 93-94), *138*, *140*, *141*
Wadsworth, P., 130(85), *140*
Waggoner, A. S., 119(3), 122(3), 128(65), 130(65), *138*, *140*
Wakade, A. R., 135(100), *141*
Wakade, T. D., 135(100), *141*
Walker, J. W., 123(39), *139*
Wall, J. S., 80(17, 20-21), *110*
Wallén, L., 198(29), 200(29), *269*
Walsh, C., 215(68), 235(68), 238(108), 245(68), 247(68), *270*, *271*
Walter, B., 167(86), 168(94), *186*, *187*
Wang, B. C., 42(61), *71*

Wang, J. L., 122(19), 130(19), *139*
Wang, L., 122(20), 128(20), 130(20), *139*
Wang, Y-Li, 122(22), 130(22), *139*
Wang, Z., 108(157), *114*, 127(62), 133(95), *140*, *141*
Wardlaw, S., 166(70, 74), *186*
Ware, B. R., 122(22), 130(22), *139*
Wariishi, H., 246(119), *272*
Waring, P., 63(93), *72*
Warley, A., 81(26), 82(26), *110*
Warshaw, M., 181(167), *189*
Waser, J., 21(31), *70*
Webb, L. E., 4(15), 39(15), *70*
Webb, W. W., 122(23), 128(23), 130(23), *139*
Weber, G., 119(10, 12), 122(10, 12), *138*
Weber, J. P., 195(11), *268*
Webster, D. deF., 91(80), *112*
Weghorst, C. M., 126(48), *139*
Weibel, E., 2(4), *69*
Weiner, H., 214(63–64), 216(63), *270*
Weinstock, A., 180(152), *188*
Weischet, W. O., 223 (95), *271*
Weisenhorn, A. L., 108(160–161), *115*
Welborn, J. R., 144(8), *184*
Wenderoth, M. P., 89(70–71), 90(70), *112*
Wergedal, J. E., 152(29), *185*
Werner, M., 179(131), *188*
Werner, W., 170(106), *187*
Westgard, J. O., 178(126), *187*
Whiccher, J., 166(69), *186*
Whitaker, J. E., 127(59), *140*
Whitaker, K. B., 180(155), *188*
White, J. M., 144(14), *184*
Whitehead, T. P., 165(54), 179(134), 180(134), *185*, *188*
Whitfield, S. L., 106(141), 107(141), 108(141), *114*
Whitner, V. S., 181(162), *189*
Wicks, R. W., 181 (167), *189*
Wiegrabe, W., 108(157), *114*
Wiggins, J. W., 79(11), *110*
Wilchek, M., 85(39), 89(39), *111*
Wilcox, M., 168(94), *187*
Wildhaber, I., 102(127), *114*
Williams, F. C., 173(107), 175(107), *187*
Williams, G. J. B., 54(76), 55(76), *71*
Williams, J. W., 63(93), *72*
Williams, M. A., 91(82), *112*
Williams, R. C., 98(106), 99(115), *113*
Willison, J. H. M., 101(123), 102(123), *113*
Willon, I. A., 59(82), *72*
Wilson, A. J. C., 32(37), *70*
Wilson, T. E., 108(156), 109(156), *114*
Winefordner, J. D., 250(138), *272*
Wing, R. W., 198(28), *269*
Winkler, H., 102(126–127), *114*

Winokur, T. S., 126(49–50), *139*, *140*
Wojcieszyn, J., 130(84), *140*
Wolber, R. A., 91(78–79), *112*
Wolf, D. E., 130(83), 133(97), *140*, *141*
Wolff, E. K., 109(162), *115*
Wolff, E. W., 81(27), 82(27), *110*
Wollmer, A., 208(46), *269*
Wonacott, A. J., 27(33), *70*
Wong, J., 81(25), 82(25), *110*
Woods, W. G., 165(55), *185*
Wulff, K., 165(62), *186*
Wurst, B., 178(113), *187*
Wurzburg, U., 181(173), *189*
Wyckoff, H. W., 4(19), 68(98), *70*, *72*
Wycoff, R., 99(115), *113*

Xuong, N.-H., 67(95), *72*

Yagminas, Y. P., 178(109), *187*
Yamanaka, T., 246(113), *271*
Yamashiro, D. J., 127(55), *140*
Yanagimoto, K. C., 106(141), 107(141), 108(141), *114*
Yang, R., 108(158–159), 109(159), *115*
Yang, X., 108(158–159), 109(159), *115*
Yanofsky, C., 215(70), *270*
Yasumura, T., 104(130), *114*
Yeager, F. M., 175(108), *187*
Yguerabide, J., 119(9), 122(9), *138*
Yost, V., 4(20), *70*
You, H. X., 106(145), *114*
Young, D. S., 160(45), *185*
Yumoto, I., 246(113), *271*

Zahn, R. K., 101(121), *113*
Zamecnik, P. C., 133(97), *141*
Zanotti, G., 3(8), *70*
Zasadzinski, J. A. N., 108(155, 160), *114*, *115*
Zenhausern, F., 109(163), *115*
Zeppezauer, M., 194(3), 198(3, 24–26, 29), 200(3, 24–26, 29, 34–35), 201(3), 203(3, 26), 204(25–26, 37–38), 208(26), 264(3, 26, 156), *268*, *269*, *273*
Ziegenhorn, J., 166(75, 77), *186*
Zierold, K., 105(138), *114*
Zieve, L., 166(72), *186*
Zimiak, A., 127(63), *140*
Zinterhofer, L., 166(70, 74), *186*
Zipp, A., 167(89), 168(93), *186*, *187*
Zoeten, G. de, 88
Zollinger, M., 86(47), *111*
Zollner, H., 160(44), *185*
Zon, G., 227(99), 244(99), *271*
Zuberbuhler, A. D., 265(161), 267(161), *273*
Zwez, W., 166(75), *186*
Zylstra, U., 82

SUBJECT INDEX

Absolute scale, 30
Absorbance accuracy, enzyme analyzers, 158
Absorption factor, 25, 30
Acousto-optic-modulator, 119
   diagram of, 120
Alcohol dehydrogenase, RSSF analysis of mechanism, 197-208
Aldehyde dehydrogenase:
   acyl enzyme intermediate, 213
   RSSF analysis of intermediates, 211-214
Allosteric transitions, detection by rapid scanning spectroscopy, 226
α Chymotrypsin catalysis, RSSF analysis of mechanism, 195
α helix, 47, 51, 56, 58
Amphipathic helix, 57
Angstrom unit, 6
Anomalous dispersion, 39, 41
Apophyllite crystals, 6
Area detector, 28
Array detectors, use in RSSF spectroscopy, 256
Asymmetric unit, 34, 43
Atomic force microscope:
   advantages and disadvantages, 107
   image formation, 106
   nucleic acids, 109
   proteins, 109
   specimen preparation, 108
Atomic number, 12, 27
Atomic scattering factor, 12, 23
Automated enzyme analyzers, see Enzyme analyzers
   comparison of, 154
   historical development, 144-150
Automation, what is it, 150
Autoradiography, 91
   flat substrate method, 92
   in-situ hybridization, 93
   polymerase chain reaction, 93
   resolution, 92
Avogadro's number, 33

Beta barrel, 59
Beta sheet, 51, 56-58
Bond distances and angles, 54, 55
Bragg's law, 14, 25, 30, 45
Bragg reflections, 3, 13-18, 21-34, 41, 43, 46-48
   multiple, 34
Bravais lattice, 8, 10

Cahn-Ingold notation, 63
Calcium:
   intracellular, measurement of, 129
   oscillations of, 129
Calcium binding motif, 57, 60
Calmodulin, 60
Catalytic triad, 64-67
Cellular activities:
   kinetic analysis of, 128
   kinetics of, 128
Charge injection devices, use in RSSF spectroscopy, 253
Chemical fixation, 83
Chymotrypsin, 64-67
Cis and trans proline, 53
Confocal microscope:
   bilateral laser scanning, 121, 130
   diagram, 121
   pinhole, 121
   real time, 121
   real time imaging, 130
   slit aperture, 121
Confocal microscopy, 119
   advantages of, 135, 138
   three dimensional reconstruction, 134-138
Conformational analysis, 52-56
Contouring electron density maps, 44
Convolution, 6, 7
Crambin crystal structure, 41, 51
Critical point drying, 100
Cryosectioning, 89
Crystal, 3, 25, 28
Crystallation, methods for proteins, 4-6

291

Crystallization method, hanging drop, 5
Crystals:
  apophyllite, 6
  cubic, 8, 10
  density of, 33, 34
  growth of, 4
  hemoglobin, 3, 4, 39
  hexagonal, 8, 10
  lattice, 6-8, 14-17
  lysozyme, 4-6, 56, 63-66
  monoclinic, 8, 10, 31
  mother liquor of, 3, 28, 30
  myoglobin, 4, 39
  nucleation of, 4
  orthorhombic, 8, 10, 31
  protein, 3
  reactions in, 67
  seed, 6
  structure, 3, 7, 22, 50
  symmetry, 8, 32
  systems, 6, 8
  tetragonal, 8, 10
  tobacco mosiac virus, 3
  triclinic, 8, 10
  trigonal, 8, 10
  urease, 3
Cystathionine γ synthase:
  intermediates, detection by RSSF, 235
  key partitioning intermediate, 235
  RSSF analysis of catalytic mechanism, 235
Cytochrome P-450 cam, 61-63
Cytoplasmic components, diffusion of, 130

Data analysis:
  analysis of crossover points in RSSF spectroscopy, 263
  strategies in RSSF spectroscopy, 263
Data reduction in x-ray crystallography, 30
Diffraction:
  neutrons, 3, 12
  pattern, 3, 11-15, 40, 41
  vector, 14, 17
  X-ray(s), 3, 12, 13, 17, 25
Dihydrofolate reductase, 63-65
Disorder, atoms in unit cells, 23
Displacement parameters, 24, 43
DNA binding motif, 57
D-xylose isomerase, 15, 37, 49, 53, 59, 61

EF hand motif, 60
Ektachem:
  data analysis and, 175
  description of, 168-177
  enzymes assayed by, 172

Electron density, 19-22, 42, 43
  map 19-21, 42-47, 52, 61
  modification, 42
Energy dispersive spectroscopy, 81
Energy dispersive X-ray microanalysis:
  elemental distribution, 105
  frozen hydrated samples, 105
  specimen preparation, 104
Energy dispersive X-ray microanalyzer, 79, 81
Energy transfer, fluorescence, 123
Entropy maximization, 43
Enzyme analyzers, see Ektachem, 168
  in animal testing, 177
  absorbance accuracy, 158
  absorbance accuracy, reagents for checking, 159
  analysis of data, 162
  automated, historical development, 144-150
  bioluminescence, 165
  chemiluminescence, 165
  comparison of, 154
  dry reagent systems, 166
    advantages and disadvantages, 167
    the Ektachem system, 168
    evaluation of, 167
    the Reflotron system, 168
    the Seralyzer, 168
  in environmental testing, 178
  enzyme reactions, 156
  film systems, 166
  fluorescence methods, 164
  the ideal system, 181
  identification of enzyme, 152
  interferences, 160
  nephelometry, 166
  pipeting specimen, 153
  quality control, 178
  sample preparation, 152
  specimen integrity, 153
  spectrophotometric methods, 152
  standardization of, 179
  temperature control, 157
  time control, 159
  turbidimetry, 166
  wavelength accuracy, 158
Enzyme reference materials:
  application of, 180
  availability of, 180
  preparation of, 180
Enzymes, standardization as proteins, 181
Ewald sphere, 15, 17, 34

FISH (Fluorescence in situ Hybridization), 133

SUBJECT INDEX

chromosome painting, 133
  metaphase sequences, 133
Flow cell, 28, 68
Fluorescence, 117
  depolarization, 123
  fluorescent probes, 122
  intensity, 119
  lifetime, 119
  polarization, 119, 123
  quantitation, 124
  quenching, 123
  scanning stage, 119
  sensitivity to environment, 122
Fluorescence imaging, applications, 124
Fluorescence in situ hybridization, see FISH
Fluorescence redistribution after
    photobleaching, see FRAP
Fluorescent probes:
  active, 123
  characteristics of, 122
  energy transfer, 123
  passive, 121-123
  photoactivatable, 123
  quenching, 122
Focusing mirrors for monochromatic
    radiation, 27
Fourier:
  series, 18, 19, 50
  synthesis, 19, 20, 48
  transforms, 21, 22, 42, 43
Fourier analysis, 19
FRAP, 130
  diffusion coefficient, 131
  fluorescent, recoverable percentage, 131
  intercellular communication, 133
  oncogene expression, 133
  photobleaching, 130
  use of, 131
Freeze drying, 100
Freeze-etch techniques, 102
  macromolecular assemblies, 103
  macromolecules, 103
  replica interpretation, 103
Freeze substitution, 84
Friedel's Law, 31, 41

Gap junctional communication, 133
Gap junction function, 126
Geiger counter, 27
Globin fold, 59
Glycogen phosphorylase, 68
Goniometer head, 28
Greek key motifs, 57-59

Hanging drop, crystallization method, 5

Harker sections, 37, 38
Heavy atom derivatives, 34, 35, 38-40
Helical wheel, 57
Helix-loop-helix motif, 57, 60
Hemerythrin, 58
Hemoglobin:
  crystals, 3, 4, 39
  RSSF analysis of ligand binding, 245-246
Horseradish peroxidase, RSSF analysis of
    catalysis, 247
Hybridization histochemistry, 89
Hydrogen bonding, 51, 56, 57
Hydroperoxidase, RSSF analysis of catalysis,
    246-247

Icosahedral symmetry, 60
Image formation:
  atomic force microscope, 107
  electron dense, 77
  electron transparent, 77
  scanning tunneling microscope, 106
  transmission electron microscope, 76
Imaging plate, 28
Immunolabeling, 84-86
  colloidal gold, 85
  controls, 86
  cryosectioning, 89
  cryo-ultrathin sectioning, 89
  double-labeling, 88
  post-embedding, 87
  pre-embedding, 86-87
  protein A, 85
  quantitation, 88
Immunosorbent electron microscopy, 98
Inhibitor binding, 44, 48
In situ hybridization, 89
  biotinylated cDNA, 89
  biotinylated nucleotide analogues, 91
  cryosections, 89
  Fab fragments, 91
  high resolution autoradiography, 91
  LR White, 89
Insulin hexamer:
  allosteric transitions in, 208-211
  conformational transitions, 208-211
  reaction with 1-(2-pyridylazo)-2-naphthol,
    210
  RSSF analysis of allosteric transitions,
    208-211
Intensified array detector, use in RSSF
    spectroscopy, 252
Intensified diode array systems, lag
    phenomena in, 261
Intercellular communication, 129, 133
Interference, 13

Intermediates:
  detection in horseradish peroxidase, 247
  detection by rapid scanning spectroscopy, 193, 215
ISEM, see Immunosorbent electron microscopy, 98
Isomorphous material, 3
Isomorphous replacement, 34, 35, 38-40, 61, 67
  multiple, 38
Isotopes, 12

Jelly-roll motif, 58, 59

Labeling:
  immuno-, 85
  target specific, 85
Laser scanning electron microscope, resolving power, 75
Laue photograph, 29, 68
Laue symmetry, 8, 31
Least squares method, 47
Leucine zipper, 58
Ligand binding, 61
Light, visible, 2
Light microscope, resolving power, 75
Lorentz factor, 25, 30
Low temperature measurements, 29
Lysozyme crystals, 4-6, 56, 63-66

Mass loss, 77
Mass mapping, in scanning transmission electron microscope, 80
Mass mapping of molecular weights, 80
Membrane-bound proteins, 5
Membrane components, diffusion of, 130
Membrane potential, measurement of, 127
Metal shadowcasting, 99
Methotrexate, 63, 64
Methylpentanediol (MPD), 5
Microscope, 2, 3
  polarizing, 28
  scanning tunneling, 8
Microscopy, fluorescence, 118
Miller indices, 9, 11, 14, 31
Molecular replacement, 40
Molecular vibration, 23, 24
Monochromators, 27
Monoclinic crystal, 8, 10, 31
Motif:
  DNA binding, 57
  EF hand, 60
  Greek key, 57-59
  helix-loop-helix, 57, 60
  jelly-roll, 58, 59

Mutant enzymes, 66, 67
Myoglobin, crystals, 4, 39

Negative staining, 93-98
  carbon films, 96
  methods, 97
  Miller chromatin spreading, 104
  molybdenum, 94
  support films, 95
  tungsten, 95
  uranium, 95
Neutron sources, 25
Newman projection, 51
Noncrystallographic symmetry, 43
Nucleation of crystals, 4
Nucleophilic attack, 65

Oncogene expression, 133
Open-face sandwich, 59
Orientation matrix, 27

Parvalbumin, 60
Patterson map, 35-37, 40-42
Peptide group, 51
pH, measurement of, 127
Phase, relative, of Bragg reflection, 3, 16-19, 23, 48
Phase determination, direct methods for, 34
Photoactivatable, 123
Photoactivation, 123
Photobleaching, gap junctional communication, 133
Photodiode array detector, use in RSSF spectroscopy, 251
Photographic film, 26, 27
Point groups, 8
Polarization factor, 25, 30
Polarizing microscope, 28
Polyethylene glycol, 5
Polymorphs, 5, 40, 43
Position-sensitive detector, 28
Precision, and protein structure, 50
Primary structure, 51
Principle component analysis, RSSF spectroscopy, 267
Proline, 53, 56
Proportional counter, 27
Protein:
  crystals, 3
  folding, 50-54
  quaternary structure, 51
Protein Data Bank, 54
Protein structure:
  beta barrel, 59
  beta sheet, 51, 56-58

Quantitation:
  fluorescence, 117, 124
    calcium, 118
    free radicals, 124
    glutathione, 124
    in situ, 118
    in situ pH, 127
    membrane potential, 118, 127
    oncogene expression, 126
    pH, 118
Quaternary structure, 51

Ramachandran plot, 51, 53
Rapid scanning spectroscopy:
  alcohol dehydrogenase, 3-hydroxy-4-nitrobenzaldehyde, 197-208, 205
  aldehyde dehydrogenase, 211-214
  aldehyde dehydrogenase, acyl enzyme intermediates, 213
  alpha chymotrypsin, 195
  analysis of crossover points, 263
  analysis by differentiation, 264
  analysis by isotachic points, 264
  applications of, 193
  array detectors, 251
  array detector systems, 256
  artificial substrates, 194
  basic principles, 248-254
  charge injection devices, 253
  Co(II) alcohol dehydrogenase, 200
  Co(II)insulin hexamer, 208
  commercial diode array systems, 261
  cystathionine gamma synthase, 235
  data analysis strategies, 263
  detection of allosteric interactions, 208
  detection of compound O in horseradish peroxidase, 247
  detection of intermediates by, 195, 197-208, 263
  deuterium isotopes on tryptophan synthase, 220
  difference spectra analysis, 263
  electron transport proteins, 246
  global analysis, 265
  heme enzymes, 245-247
  hemoglobin and myoglobin, 245-246
  horseradish peroxidase, 247
  hydroperoxidase, 246
  instrumentation, 251-254
  insulin hexamer, 208
  insulin reaction with 1-(2-pyridylazo)-2-naphthol, 210
  intrinsic chromophores, 215
  lag phenomena in intensified diode arrays, 261
  loop function in alpha-beta barrel enzyme, 243-244
  low cost microcomputer controlled diode array detector, 260
  MCP-SPD array detectors, 252
  modulation of light intensity for array detectors, 258
  Ni(II) alcohol dehydrogenase, 204
  principle component analysis, 267
  pyridoxal phosphates enzymes, 215-245
  quantitative analysis, 265
  signal averaging, 255
  silicon photodiode detectors, 252
  single element detectors, 251
  single wavelength selection, 263
  singular value decomposition, 265
  site directed mutagenesis, study of by, 239
  stopped flow, 191
  stopped flow systems with single element detectors, 254
  strategies for detecting intermediates, 249
  substractive double monochromator, 255
  time domains, 248
  tryptophanase, substrate specificity, 231
  tryptophanase catalysis, 227
  tryptophanase mutation, 245
  tryptophan synthase, 215
  tryptophan synthase, alpha amino acrylate intermediate, 223
  tryptophan synthase mutation, 240-244
  wavelength domains, 249
Rapid-scanning stopped-flow system, *see* Rapid-scanning spectroscopy
Ras oncogene, expression of, 126
Reactions in crystals, 67
Reciprocal lattice, 15-17, 22, 24, 29, 47
Refinement of protein crystal structure, 47
Relative phase angle, 34, 38-43, 50
Replica techniques, 99
Resolution, 24, 39, 43-47, 50, 51
Rotating anode generator of X-rays, 26, 30
Rotation axis, 8, 9, 43
Rotation function, 40-43
RSSF, *see* Rapid-scanning stopped flow

Scanning electron microscope, resolving power, 75
Scanning tunneling/atomic force microscope, resolving power, 75
Scanning tunneling microscope, 106
  advantages and disadvantages, 107
  image interpretation, 106
  specimen preparation, 108
Scintillation counter, 27
Screw axis, 8, 9

Secondary structure, 51
Seed crystals, 6
Selenomethionine, 42
Serine proteases, 64–67
SFP algorithm, 137
Shadowcasting:
  evaporation techniques, 101
  Miller chromatin spreading, 104
  molecular spreading, 101
  snapshot blotting, 104
Sickle cell hemoglobin, 56
Signal averaging, use in RSSF spectroscopy, 255
Silicon photodiode array detectors, use in RSSF spectroscopy, 252
Simulated annealing, 48
Single element detectors, use in RSSF spectroscopy, 254
Site directed mutagenesis:
  RSSF analysis of, 239
  RSSF analysis of tryptophan synthase mutation, 240
Site directed mutatagenesis, RSSF analysis of site directed mutation of tryptophanase, 245
Solvent flattening, 42
Space group, 8, 31–36
Space group ambiguities, 32
Specimen preparation, 83
  chemical fixation, 83–84
  freeze substitution, 84
  ultrarapid freezing, 83–84
  ultrathin sectioning, 83–34
Spectrophotometry, rapid-scanning stopped-flow, 191
STEM, see Scanning transmission electron microscope, 79
STM, see Scanning tunneling microscope
Structure:
  secondary, 51
  tertiary, 51
Structure factor, 22, 23
Structure factor amplitude, 13, 18–22
Substractive double monochromator, use in RSSF spectroscopy, 255
Subtilisin, 59
Synchrotron radiation, 26, 27, 30, 68
Systematic absences in the diffraction pattern, 31, 32

TEM, see Transmission electron microscope, 75
Temperature control, enzyme analyzers, 157

Tertiary structure, 51
Tetragonal crystal, 8, 10
Three dimensional reconstruction, 134–138
Time control, enzyme analyzers, 159
Tobacco mosaic virus crystals, 3
Torsion angle, 51, 52
Tracing the chain, 45
Translation, 9
Translation function, 41, 43
Transmission electron microscope:
  advantages of, 78
  components of, 76
  dedicated, 79
  disadvantages of, 78
  frozen macromolecules, 80
  image formation, 76
  map periodic structures, 79
  mass mapping, 80
  nondedicated, 79
  resolving power, 75
  use in mapping, 79
Triclinic crystal, 8, 10
Trigonal crystals, 8, 10
Triose phosphate isomerase, 59
Troponin C, 60
Tryptophanase:
  RSSF analysis of catalytic mechanism, 227
  RSSF analysis of site directed mutation, 245
  RSSF analysis of substrate specificity, 231
  substrate specificity, 231
Tryptophan synthase:
  RSSF analysis of allosteric behavior, 226
  RSSF analysis of catalytic mechanism, 215
  RSSF analysis of site directed mutation, 240–244

Ultrarapid freezing, 83
Ultrathin sectioning, 83
  epoxy resin, 84
  immunolabeling, 84–85
  modified acrylic resin, 84
Unit cell, 6, 11, 29
  body centered, 10
  dimensions, 3, 6, 14, 30, 35
  face centered, 10
  origin, 9, 18, 22, 39
Urease crystals, 3

Vibration parameters, 55, 56
Viruses, 60

Wavelength accuracy, enzyme analyzers, 158
White radiation, 29

SUBJECT INDEX

Wilson plot, 30

X-ray analysis:
  compositional maps, 82
  mass loss, 82
  quantitation, 82

X-ray crystallography:
  International Tables for, 9, 12, 32
  low temperature measurements, 29
X-ray microanalyzer, 79, 81
X-rays, 2
X-ray sources, 25

CUMULATIVE AUTHOR INDEX, VOLUMES 1–37 AND SUPPLEMENTAL VOLUME

	VOL.	PAGE
Abramson, Fred P., Mass Spectrometry in Pharmacology	34	289
Ackerman, C. J., see Engle, R. W.		
Albertsson, Per-Åke, Interaction Between Biomolecules Studied by Phase Partition	29	1
Albertsson, Per-Åke, Partition Methods for Fractionation of Cell Particles and Macromolecules	10	229
Albertsson, P., Andersson, B., Larsson, C., and Akerlund, H., Phase Partition—A Method of Purification and Analysis of Cell Organelles and Membrane Vesicles	28	115
Alcock, Nancy W., and MacIntyre, Iain, Methods for Estimating Magnesium in Biological Materials	14	1
Amador, Elias, and Wacker, Warren E. C., Enzymatic Methods Used for Diagnosis	13	265
Ames, Stanley, R., see Embree, Norris D.		
Anderegg, Robert J., Mass Spectrometry: An Introduction	34	1
Andersen, C. A., An Introduction to the Electron Probe Microanalyzer and Its Application to Biochemistry	15	147
Anderson, N. G., Preparation Zonal Centrifugation	15	271
Andrews, P., Estimation of Molecular Size and Molecular Weights of Biological Compounds by Gel Filtration	18	1
Arakawa, Tsutomu, Kita, Yoshiko A., Narhi, Linda O., Protein-Ligand Interaction as a Method to Study Surface Properties	35	87
Ariel, Mira, see Grossowicz, Nathan		
Asboe-Hansen, Gustav, see Blumenkrantz, Nelly		
Aspen, Anita J., and Meister, Alton, Determination of Transaminase	6	131
Augustinsson, Klas-Bertil, Assay Methods for Cholinesterases	5	1
Determination of Cholinesterases	Supp.	217
Austin, Robert H., see Chan, Shirley S.		
Awdeh, Z. L., see McLaren, D. S.		
Baker, S. A., Bourne, E. J., and Whiffen, D. H., Use of Infrared Analysis in the Determination of Carbohydrate Structure	3	213
Balis, M. Earl, Determination of Glutamic and Aspartic Acids and Their Amides	20	103
Bannister, Joe V., and Calabrese, Lilia, Assays for Superoxide Dismutase	32	279
Barchas, Jack D., see Faull, Kym F.		
Barenholz, Y., see Lichtenberg, D.		
Barksdale, A. D., and Rosenberg, A., Acquisition and Interpretation of Hydrogen Exchange Data from Peptides, Polymers, and Proteins	28	1
Barman, Thomas F., and Travers, Franck, The Rapid-Flow-Quench Method in the Study of Fast Reactions in Biochemistry: Extension to Subzero Conditions	31	1

	VOL.	PAGE
Bârzu, Octavian, Measurement of Oxygen Consumption by the Spectrophotometric Oxyhemoglobin Method	30	227
Bauld, W. S., and Greenway, R. M., Chemical Determination of Estrogens in Human Urine	5	337
Bayer, Edward A., and Wilchek, Meir, The Use of the Avidin-Biotin Complex as a Tool in Molecular Biology	26	1
Beauregard, Guy, Maret, Arlette, Salvayre, Robert, and Potier, Michel, The Radiation Inactivation Method as a Tool to Study Structure-Function Relationships in Proteins	32	313
Bell, Helen H., see Jaques, Louis B.		
Benesch, Reinhold, and Benesch, Ruth E., Determination of—SH Groups in Proteins	10	43
Benesch, Ruth E., see Benesch, Reinhold		
Benson, E. M., see Storvick, C. A.		
Bentley, J. A., Analysis of Plant Hormones	9	75
Benzinger, T. H., see Kitzinger, Charlotte		
Berg, Marie H., see Schwartz, Samuel		
Berger, Robert L., Clem, Thomas R., Sr., Harden, Victoria A., and Mangum, B. W., Historical Development and Newer Means of Temperature Measurement in Biochemistry	30	269
Bergmann, Felix, and Dikstein, Shabtay, New Methods for Purification and Separation of Purines	6	79
Berson, Solomon A., see Yalow, Rosalyn S.		
Bhatti, Tarig, see Clamp, John R.		
Bickoff, E. M., Determination of Carotene	4	1
Binnerts, W. T., Determination of Iodine in Biological Material	22	251
Bishop, C. T., Separation of Carbohydrate Derivatives by Gas-Liquid Partition Chromatography	10	1
Blackburn, S., The Determination of Amino Acids by High-Voltage Paper Electrophoresis	13	1
Blow, D. M., see Holmes, K. C.		
Blumenkrantz, Nelly, and Asboe-Hanson, Gustav, Methods for Analysis of Connective-Tissue Macromolecules by Determination of Certain Constituents	24	39
Blumenthal, R., see Loyter, A.		
Bock, Jay L., Recent Developments in Biochemical Nuclear Magnetic Resonance Spectroscopy	31	259
Bodansky, Oscar, see Schwartz, Morton K.		
Bossenmaier, Irene, see Schwartz, Samuel		
Bosshard, Hans Rudolf, Mapping of Contact Areas in Protein-Nucleic Acid and Protein—Protein Complexes by Different Chemical Modification	25	273
Boulton, A. A., see Majer, J. R.		
Boulton, Alan A., The Automated Analysis of Absorbent and Fluorescent Substances Separated on Paper Strips	16	327
Bourne, E. J., see Baker, S. A.		
Brantmark, B. L., see Lindh, N. O.		
Brauser, Bolko, see Sies, Helmut		
Bray, H. G., and Thorpe, W. V., Analysis of Phenolic Compounds of Interest in Metabolism	1	27
Brierley, G. P., see Lessler, M. A.		
Brodersen, R., and Jacobsen, J., Separation and Determination of Bile Pigments	17	31

	VOL.	PAGE
Brodie, Bernard B., see *Udenfriend, Sidney*		
Brooker, Gary, Newer Development in the Determination of Cyclic AMP and Other Cyclic Nucleotides, Adenylate Cyclase, and Phosphodiesterase	22	95
Brzovic, Peter, see *Dunn, Michael*		
Bump, E. A., see *Russo, A.*		
Burtis, Carl A., Tiffany, Thomas O., and Scott, Charles D., The Use of a Centrifugal Fast Analyzer for Biochemical and Immunological Analyses	23	189
Bush, I. E., Advances in Direct Scanning of Paper Chromatograms for Quantitative Estimations	11	149
Bush, I. E., Applications of the R_M Treatment in Chromatographic Analysis	13	357
Erratum	14	497
Calabrese, Lilia, see *Bannister, Joe V.*		
Caldwell, Karin D., see *Giddings, J. Calvin*		
Campbell, Anthony K., Chemiluminescence as an Analytical Tool in Cell Biology and Medicine	31	317
Campbell, I. D., and Dobson, C. M., The Application of High Resolution Nuclear Magnetic Resonance to Biological Systems	25	1
Cartensen, H., Analysis of Adrenal Steroid in Blood by Countercurrent Distribution	9	127
Caster, W. O., A Critical Evaluation of the Gas Chromatographic Technique for Identification and Determination of Fatty Acid Esters, with Particular Reference to the Use of Analog and Digital Computer Methods	17	135
Chambers, Robin E., see *Clamp, John R.*		
Chan, Shirley S., and Austin, Robert H., Laser Photolysis in Biochemistry	30	105
Chance, Britton, see *Maehly, A. C.*		
Chase, Aurin M., The Measurement of Luciferin and Luciferase	8	61
Chinard, Francis P., and Hellerman, Leslie, Determination of Sulfhydryl Groups in Certain Biological Substrates	1	1
Chou, K. C., see *Maggiora*		
Christen, P., and Gehring, H., Detection of Ligand-Induced and Syncatalytic Conformational Changes of Enzymes by Differential Chemical Modification	28	151
Citovsky, V., see *Loyter, A.*		
Clamp, John R., Bhatti, T., and Chambers, R. E., The Determination of Carbohydrate in Biological Materials by Gas-Liquid Chromatography	19	229
Clark, Stanley J., see *Wotiz, Herbert H.*		
Clearly, E. G., see *Jackson, D. S.*		
Clem, Thomas R., Sr., see *Berger, Robert L.*		
Code, Charles F., and McIntyre, Floyd C., Quantitative Determination of Histamine	3	49
Cohn, Waldo E., see *Volkin, Elliot*		
Conzelmann, Ernest, and Sandhoff, Konrad, Activator Proteins for Lysomal Glycolipid Hydrolysis	32	1
Cotlove, Ernest, Determination of Chloride in Biological Materials	12	277
Cox, R. P., see *Degn, H., Cox, R. P., and Lloyd, D.*		
Crabbe, M. James C., Computers in Biochemical Analysis	31	417
Craig, Lyman C., and King, Te Piao, Dialysis	10	175
see also *King, Te Piao*		

	VOL.	PAGE
Crane, F. L., and Dilley, R. A., Determination of Coenzyme Q (Ubiquinone)	11	279
Creech, B. G., see Horning, E. C.		
Creveling, C. R., and Daly, J. W., Assay of Enzymes of Catechol Amines	Supp.	153
Curry, A. S., The Analysis of Basic Nitrogenous Compounds of Toxicological Importance	7	39
Daly, J. W., see Creveling, C. R.		
Davidson, Harold M.., see Fishman, William H.		
Davis, Neil C., and Smith, Emil L., Assay of Proteolytic Enzymes	2	215
Davis, R. J., see Stokstad, E. L. R.		
Davis, Robert P., The Measurement of Carbonic Anhydrase Activity	11	307
deFeijter, Adriaan W., see Wade, Margaret H.		
Dean, H. G., see Whitehead, J. K.		
Degn, H., Cox, R. P., and Lloyd, D., Continuous Measurement of Dissolved Gases in Biochemical Systems with the Quadrupole Mass Spectrometer	31	165
Degn, H., Lundsgaard, J. S., Peterson, L. C., and Ormicki, A., Polarographic Measurement of Steady State Kinetics of Oxygen Uptake by Biochemical Samples	26	47
Dikstein, Shabtay, see Bergmann, Felix		
Dilley, R. A., see Crane, F. L.		
Dinsmore, Howard, see Schwartz, Samuel		
Dische, Zacharias, New Color Reactions for the Determination of Sugars in Polysaccharides	2	313
Dodgson, K. S., and Spencer, B., Assay of Sulfatases	4	211
Douzou, Pierre, The Use of Subzero Temperatures in Biochemistry: Slow Reactions	22	401
Doyle, Matthew J., see Heineman, William R.		
Dunn, Michael, Rapid-Scanning Stopped-Flow Spectroscopy	37	191
Dyer, John N., Use of Periodate Oxidations in Biochemical Analysis	3	111
Edwards, M. A., see Storvick, C. A.		
Eftink, Maurice R., Fluorescence Techniques for Studying Protein Structure	35	127
Elving, P. J., O'Reilly, J. E., and Schmakel, C. O., Polarography and Voltammetry of Nucleosides and Nucleotides and Their Parent Bases as an Analytical and Investigative Tool	21	287
Embree, Norris D., Ames, Sanley R., Lehman, Robert W., and Harris, Philip L., Determination of Vitamin A	4	43
Engel, Lewis L., The Assay of Urinary Neutral 17-Ketosteroids	1	479
Engel, R. W., Salmon, W. D., and Ackerman, C. J., Chemical Estimation of Choline	1	265
Engelman, Karl, see Lovenberg, S. Walter		
Ernster, Lars, see Lindberg, Olov		
Everse, Johannes, Ginsburgh, Charles L., and Kaplan, Nathan O., Immobilized Enzymes in Biochemical Analysis	25	135
Faull, Kym F., and Barchas, Jack D., Negative-Ion Mass Spectrometry, Fused-Silica Capillary Gas Chromatography of Neurotransmitters and Related Compounds	29	325
Felber, J. P., Radioimmunoassay of Polypeptide Hormones and Enzymes	22	1
Fink, Frederick S., see Kersey, Roger C.		
Fisher, Susan R., see Giddings, J. Calvin		
Fishman, William H., Determination of β-Glucuronidases	15	77
Fishman, William H., and Davidson, Harold M., Determination of Serum Acid Phosphatases	4	257

	VOL.	PAGE
Fleck, A., see *Munro, H. N.*		
Forsén, Sture, and *Lindman, Björn,* Ion Bonding in Biological Systems Measured by Nuclear Magnetic Resonance Spectroscopy	27	289
Fraenkel-Conrat, H., Harris, J. Ieuan, and *Levy, A. L.,* Recent Developments in Techniques for Terminal and Sequence Studies in Peptides and Proteins	2	359
Frame, Melinda K., see *Wade, Margaret H.*		
Friedman, Sydney M., Measurement of Sodium and Potassium by Glass Electrodes	10	71
Frisell, Wilhelm R., and *Mackenzie, Cosmo G.,* Determination of Formaldehyde and Serine in Biological Systems	6	63
Gale, Ernest F., Determination of Amino Acids by Use of Bacterial Amino Acid Decarboxylases	4	285
Gardell, Sven, Determination of Hexosamines	6	289
Gaskell, Simon J., Analysis of Steroids by Mass Spectroscopy	29	385
Gershoni, J., Protein Blotting: A Manual	33	1
Gianazza, Elisabetta, see *Righetti, Pier Giorgio*		
Giddings, J. Calvin, Myers, Marcus N., Caldwell, Karin D., and *Fisher, Susan R.,* Analysis of Biological Macromolecules and Particles by Field-Flow Fractionation	26	79
Glusker, J. P., X-ray Crystallography of Proteins	37	1
Gofman, John W., see *Lalla, Oliver F. de*		
Goldberg, Nelson D., and *O'Toole, Ann G.,* Analysis of Cyclic 3',5'-Adenosine Monophosphate and Cyclic 3',5'-Guanosine Monophosphate	20	1
Grabar, Pierre, Immunoelectrophoretic Analysis	7	1
Greenway, R. M., see *Bauld, W. S.*		
Grootveld, M., see *Halliwell, B.*		
Gross, D., see *Whalley, H. C. S. de*		
Grossman, Shlomo, Oestreicher, Guillermo, and *Singer, Thomas P.,* Determination of the Activity of Phospholipases A, C, and D	22	177
Grossman, Shlomo, and *Zakut, Rina,* Determination of the Activity of Lipoxygenase (Lipoxidase)	25	303
Grossowicz, Nathan, and *Ariel, Mira,* Methods for Determination of Lysozyme Activity	29	435
Guilbault, G. G., see *Kauffmann, J.-M.*		
Gupta, M. N., and *Mattiasson, B.,* Unique Applications of Immobilized Proteins in Bioanalytical Systems	36	1
Gutman, Menachem, The pH Jump: Probing of Macromolecules and Solutions by a Laser-Induced, Ultrashort Proton Pulse—Theory and Application in Biochemistry	30	1
Haegele, Klaus D., see *Thénot, Jean-Paul G.*		
Haglund, Herman, Isoelectric Focusing in pH Gradients—A Technique for Fractionation and Characterization of Ampholytes	19	1
Haines, William J., and *Karnemaat, John N.,* Chromatographic Separation of the Steroids of the Adrenal Gland	1	171
Hallett, Maurice B., see *Campbell, Anthony K.*		
Halliwell, B., and *Grootveld, M.,* Methods for the Measurement of Hydroxyl Radicals in Biochemical Systems: Deoxyribose Degradation and Aromatic Hydroxylation	33	59
Halsall, H. Brian, see *Heineman, William R.*		
Hanahan, Donald J., and *Weintraub, Susan T.,* Platelet Activating Factor Insulation, Identification, and Assay	31	195
Hanazato, Yoshio, see *Shiono, Satoru*		

	VOL.	PAGE
Hanessians, Stephen, Mass Spectrometry in the Determination of Structure of Certain Natural Products Containing Sugars	19	105
Harden, Victoria A., see Berger, Robert L.		
Harris, J. Ieuan, see Fraenkel-Conrat, H.		
Harris, Philip L., see Embree, Norris D.		
Heckman, John W., Jr., see Klomparens, Karen L.		
Heineman, William R., Halsall, H. Brian, Wehmeyer, Kenneth R., Doyle, Matthew J., and Wright, D. Scott, Immunoassay with Electrochemical Detection	32	345
Heirwegh, K. P. M., Recent Advances in the Separation and Analysis of Diazo-Positive Bile Pigments	22	205
Hellerman, Leslie, see Chinard, Francis P.		
Hellerqvist, Carl G., Mass Spectrometry of Carbohydrates	34	91
Hermans, Jan, Jr., Methods for the Study of Reversible Denaturation of Proteins and Interpretation of Data	13	81
Hexter, Charles S., see Wilchek, Meir		
Hiromi, Keitaro, Recent Developments in the Stopped-Flow Method for the Study of Fast Reactions	26	137
Hirschbein, L., and Guillen, N., Characterization, Assay, and Use of Isolated Bacterial Nucleoids	28	297
Hjertén, S., see Porath, J.		
Hjertén, Stellan, Free Zone Electrophoresis. Theory, Equipment and Applications	18	55
Hjertén, Stellan, Hydrophobic Interaction Chromatography of Proteins, Nucleic Acids, Viruses, and Cells on Noncharged Amphiphilic Gels	27	89
Hoff-Jorgensen, E., Microbiological Assay of Vitamin B_{12}	1	81
Holman, Ralph T., Measurement of Lipoxidase Activity	2	113
Measurement of Polyunsaturated Acids	4	99
Holmes, K. C., and Blow, D. M., The Use of X-ray Diffraction in the Study of Protein and Nucleic Acid Structure	13	113
Homolka, Jiri, Polarography of Proteins, Analytical Principles and Applications in Biological and Clinical Chemistry	19	435
Horning, E. C., Vanden Heuvel, W. J. A., and Creech, B. G., Separation and Determination of Steroids by Gas Chromatography	11	69
Horvath, C., High-Performance Ion-Exchange Chromatography with Narrow-Bore Columns: Rapid Analysis of Nucleic Acid Constituents at the Subnanomole Level	21	79
Hough, Leslie, Analysis of Mixtures of Sugars by Paper and Cellulose Column Chromatography	1	205
Hughes, Graham J., and Wilson, Kenneth J., High-Performance Liquid Chromatography: Analytic and Preparative Applications in Protein Structure Determination	29	59
Hughes, Thomas R., and Klotz, Irving M., Analysis of Metal-Protein Complexes	3	265
Humphrey, J. H., Long, D. A., and Perry, W. L. M., Biological Standards in Biochemical Analysis	5	65
Hutner, S. H., see Stokstad, E. L. R.		
Jackson, D. S., and Cleary, E. G., The Determination of Collagen and Elastin	15	25
Jacobs, S., The Determination of Amino Acids by Ion Exchange Chromatography	14	177
Jacobs, Stanley, Ultrafilter Membranes in Biochemistry	22	307
Jacobsen, C. F., Léonis, J., Linderstrom-Lang, K., and Ottesen, M., The pH-Stat and Its Use in Biochemistry	4	171
Jacobsen, J., see Brodersen, R.		

	VOL.	PAGE
James, A. T., Qualitative and Quantitative Determination of the Fatty Acids by Gas-Liquid Chromatography	8	1
James, Douglas R., and *Lumry, Rufus W.*, Recent Developments in Control of pH and Similar Variables	29	137
James, Gordon T., Peptide Mapping of Proteins	26	165
Jaques, Louis B., Determination of Heparin and Related Sulfated Mucopolysaccharides	24	203
Jaques, Louis B., and *Bell, Helen J.*, Determination of Heparin	7	253
Jardetzky, C., and *Jardetzky, O.*, Biochemical Applications of Magnetic Resonance	9	235
Jardetzky, O., see *Jardetzky, C.*		
Jenden, Donald J., Measurement of Choline Esters	Supp.	183
Johnson, George, Gel Sieving Electrophores: A Description of Procedures and Analysis of Errors	29	25
Johnson, M. A., see *Sherratt, H. S. A.*		
Johnson, W. Curtis, Jr., Circular Dichroism and Its Empirical Applications to Bipolymers	31	61
Jolicoeur, Carmel, Thermodynamic Flow Methods in Biochemistry: Calorimetry, Densimetry, and Dilatometry	27	171
Jones, Richard T., Automatic Peptide Chromatography	18	205
Josefsson, L. I., and *Lagerstedt, S.*, Characteristics of Ribonuclease and Determination of Its Activity	9	39
Jukes, Thomas H., Assay of Compounds with Folic Acid Activity	2	121
Kabara, J. J., Determination and Localization of Cholesterol	10	263
Kaguni, Jon M., and *Kaguni, Laurie S.*, Enzyme-Labeled Probes for Nucleic Acid Hybridization	36	115
Kaguni, Laurie S., see *Kaguni, Jon M.*		
Kalckar, Herman M., see *Plesner, Paul*		
Kant, Jeffrey A., DNA Restriction Enzymes and RFLPs in Medicine	36	129
Kapeller-Adler, R., Determination of Amine Oxidases	Supp.	35
Kaplan, A., The Determination of Urea, Ammonia, and Urease	17	311
Karnemaat, John N., see *Haines, William J.*		
Kauffmann, J.-M., and *Guilbault, G. G.*, Enzyme Electrode Biosensors: Theory and Applications	36	63
Kearney, Edna B., see *Singer, Thomas P.*		
Keenan, Robert G., see *Saltzman, Bernard E.*		
Kersey, Roger C., and *Fink, Frederick C.*, Microbiological Assay of Antibiotics	1	53
King, Te Piao, and *Craig, Lyman C.*, Countercurrent Distribution	10	201
see also *Craig, Lyman C.*		
Kita, Yoshiko A., see *Arakawa, Tsutomu*		
Kitzinger, Charlotte, and *Benzinger, T. H.*, Principle and Method of Heatburst Microcalorimetry and the Determination of Free Energy, Enthalpy, and Entropy Changes	8	309
Klomparens, Karen L., Transmission Electron Microscopy and Scanning Probe Microscopy	37	73
Klotz, Irving M., see *Hughes, Thomas R.*		
Kobayashi, Yutaka, and *Maudsley, David V.*, Practical Aspects of Liquid-Scintillation Counting	17	55
Kolin, Alexander, Rapid Electrophoresis in Density Gradients Combined with pH and/or Conductivity Gradients	6	259
Kopin, Irwin J., Estimation of Magnitudes of Alternative Metabolic Pathways	11	247
Korn, Edward D., The Assay of Lipoprotein Lipase *in vivo* and *in vitro*	7	145

	VOL.	PAGE
Kuksis, A., New Developments in Determination of Bile Acids and Steroids by Gas Chromatography	14	325
Kunkel, Henry G., Zone Electrophoresis	1	141
Kurnick, N. B., Assay of Deoxyribonuclease Activity	9	1
Kusu, Fumiyo, see Takamura, Kiyoko		
Lagerstedt, S., see Josefsson, L. I.		
Lalla, Oliver F. de, and Gofman, John W., Ultracentrifugal Analysis of Serum Lipoproteins	1	459
Laursen, Richard A., and Machleidt, Werner, Solid-Phase Methods in Protein Sequence Analysis	26	201
Lazarow, Arnold, see Patterson, J. W.		
Leddicotte, George W., Activation Analysis of the Biological Trace Elements	19	345
Lehman, Robert W., Determination of Vitamin E	2	153
see also Embree, Norris D.		
Leloir, Luis F., see Pontis, Horacio G.		
Léonis, J., see Jacobsen, C. F.		
Le Pecq, Jean-Bernard, Use of Ethidium Bromide for Separation and Determination of Nucleic Acids of Various Conformational Forms and Measurement of Their Associated Enzymes	20	41
Lerner, Aaron B., and Wright, M. Ruth, in vitro Frog Skin Assay for Agents that Darken and Lighten Melanocytes	8	295
Lessler, M. A., Adaptation of Polarographic Oxygen Sensors for Biochemical Assays	28	175
Lessler, M. A., and Brierley, G. P., Oxygen Electrode Measurements in Biochemical Analysis	17	1
Levy, A. L., see Fraenkel-Conrat, H.		
Levy, Hilton B., see Webb, Junius M.		
Lichtenberg, D., and Barenholz, Y., Liposomes: Preparation, Characterization, and Preservation	33	337
Lindberg, Olov, and Ernster, Lars, Determination of Organic Phosphorus Compounds by Phosphate Analysis	3	1
Linderstrom-Lang, K., see Jacobsen, C. F.		
Lindh, N. O., and Brantmark, B. L., Preparation and Analysis of Basic Proteins	14	79
Lindman, Björn, see Forsén, Sture		
Lissitzky, Serge, see Roche, Jean		
Lloyd, D., see Degn, H., and Cox, R. P.		
Long, D. A., see Humphrey, J. H.		
Lovenberg, S. Walter, and Engelman, Karl, Serotonin: The Assay of Hydroxyindole Compounds and Their Biosynthetic Enzymes	Supp.	1
Loveridge, B. A., and Smales, A. A., Activation Analysis and Its Application in Biochemistry	5	225
Loyter, A., Citovsky, V., and Blumenthal, R., The Use of Fluorescence Dequenching Measurements to Follow Viral Membrane Fusion Events	33	129
Lumry, Rufus, see Yapel, Anthony F., Jr.		
Lumry, Rufus W., see James, Douglas R.		
Lundquist, Frank, The Determination of Ethyl Alcohol in Blood and Tissues	7	217
Lundsgaard, J. S., see Degn, H.		
McCarthy, W. J., see Winefordner, J. D.		
Machleidt, Werner, see Laursen, Richard A.		
McIntire, Floyd C., see Code, Charles F.		
MacIntyre, Iain, see Alcock, Nancy W.		

	VOL.	PAGE
Mackenzie, Cosmo G., see Frisell, Wilhelm R.		
MacKenzie, S. L., Recent Development in Amino Acid Analysis by Gas-Liquid Chromatography	27	1
McKibbin, John M., The Determination of Inositol, Ethanolamine, and Serine in Lipides	7	111
McLaren, D. S., Read, W. W. C., Awdeh, Z. L., and Tchalian, M., Microdetermination of Vitamin A and Carotenoids in Blood and Tissue	15	1
McPherson, Alexander, The Growth and Preliminary Investigation of Protein and Nuclei Acid Crystals for X-Ray Diffraction Analysis	23	249
Maehly, A. C., and Chance, Britton, The Assay of Catalases and Peroxidases	1	357
Maggiora, G. M., Mao, B., and Chou, K. C., Theoretical and Empirical Approaches to Protein-Structure Prediction and Analysis	35	1
Majer, J. R., and Boulton, A. A., Integrated Ion-Current (IIC) Technique of Quantitative Mass Spectrometric Analysis: Chemical and Biological Application	21	467
Malmström, Bo G., Determination of Zinc in Biological Materials	3	327
Mangold, Helmut K., Schmid, Harald, H. O., and Stahl, Egon, Thin-Layer Chromatography (TLC)	12	393
Mangum, B. W., see Berger, Robert L.		
Mao, B., see Maggiora		
Maret, Arlette, see Beauregard, Guy		
Margoshes, Marvin, and Vallee, Bert L., Flame Photometry and Spectrometry: Principles and Applications	3	353
Mattiasson, B., see Gupta, M. N.		
Maudsley, David V., see Kobayashi, Yutaka		
Mefford, Ivan N., Biomedical Uses of High-Performance Liquid Chromatography with Electrochemical Detection	31	221
Meister, Alton, see Aspen, Anita J.		
Michel, Raymond, see Roche, Jean		
Mickelsen, Olaf, and Yamamoto, Richard S., Methods of Determination of Thiamine	6	191
Miller, Herbert K., Microbiological Assay of Nucleic Acids and Their Derivatives	6	31
Milner, Kelsey, see Ribi, Edgar		
Miwa, I., see Okuda, J.		
Montgomery, Rex, see Smith, Fred		
Muller, Otto H., Polarographic Analysis of Proteins, Amino Acids, and Other Compounds by Means of the Brdicka Reaction	11	329
Munro, H. N., and Fleck, A.., The Determination of Nucleic Acids	14	113
Myers, Marcus N., see Giddings, J. Calvin		
Nakako, Mamiko, see Shiono, Satoru		
Narhi, Linda O., see Arakawa, Tsutomu		
Natelson, Samuel, and Whitford, William R., Determination of Elements by X-Ray Emission Spectrometry	12	1
Nealon, Daniel A., see Lott, John A.		
Neary, Michael P., see Seitz, W. Rudolf		
Neish, William J. P., α-Keno Acid Determinations	5	107
Novelli, G. David, Methods for Determination of Coenzyme A	2	189
Oberleas, Donald, The Determination of Phytate and Inositol Phosphates	20	87
Okuda, J., and Miwa, I., Newer Developments in Enzymatic Determination of D-Glucose and Its Anomers	21	155
Oldham, K. G., Radiometric Methods for Enzyme Assay	21	191

	VOL.	PAGE
Olson, O. E., Palmer, I. S., and Whitehead, E. I., Determination of Selenium in Biological Materials	21	39
O'Neill, Malcolm A., see Selvendran, Robert R.		
O'Reilly, J. E., see Elving, P. J.		
Ormicki, A., see Degn, H.		
Ostreicher, Guillermo, see Grossman, Shlomo		
O'Toole, Ann G., see Goldberg, Nelson D.		
Ottesen, M., see Jacobsen, C. F.		
Ottesen, Martin, Methods for Measurement of Hydrogen Isotope Exchange in Globular Proteins	20	135
Palmer, I. S., see Olson, O. E.		
Parker, Reno, see Ribi, Edgar		
Patterson, J. W., and Lazarow, Arnold, Determination of Glutathione	2	259
Perry, W. L. M., see Humphrey, J. H.		
Persky, Harold, Chemical Determination of Adrenaline and Noradrenaline in Body Fluids and Tissues	2	57
Peterson, L. C., see Degn, H.		
Plesner, Paul, and Kalckar, Herman M., Enzymic Micro Determinations of Uric Acid, Hypoxanthine, Xanthine, Adenine, and Xanthopterine by Ultraviolet Spectrophotometry	3	97
Pontis, Horacio G., and Leloir, Luis F., Measurement of UDP-Enzyme Systems	10	107
Porath, J., and Hjertén, S., Some Recent Developments in Column Electrophoresis in Granular Media	9	193
Porter, Curt C., see Silber, Robert H.		
Potier, Michel, see Beauregard, Guy		
Poulik, M. D., Gel Electrophoresis in Buffers Containing Urea	14	455
Pourfarzaneh, M., Kamel, R. S., Landon, J., and Dawes, C. C., Use of Magnetizable Particles in Solid Phase Immunoassay	28	267
Raaflaub, Jurg, Applications of Metal Buffers and Metal Indicators in Biochemistry	3	301
Radin, Norman S., Glycolipide Determination	6	163
Ramwell, P. W., see Shaw, Jane E.		
Read, W. W. C., see McLaren, D. S.		
Ribi, Edgar, Parker, Reno, and Milner, Kelsey, Microparticulate Gel Chromatography Accelerated by Centrifugal Force and Pressure	22	355
Righetti, Pier Giorgio, and Gianazza, Elisabetta, Isoelectric Focusing in Immobilized pH Gradients: Theory and Newer Technology	32	215
Robins, Eli, The Measurement of Phenylalanine and Tyrosine in Blood	17	287
Robins, S. P., Analysis of Crosslinking Components in Collagen and Elastin	28	329
Roche, Jean, Lassitzky, Serge, and Michel, Raymond, Chromatographic Analysis of Radioactive Iodine Compounds from the Thyroid Gland and Body Fluids	1	243
Roche, Jean, Michel, Raymond, and Lassitzky, Serge, Analysis of Natural Radioactive Iodine Compounds by Chromatographic and Electrophoretic Methods	12	143
Roe, Joseph H., Chemical Determinations of Ascorbic, Dehydroascorbic, and Diketogulonic Acids	1	115
Rosenkrantz, Harris, Analysis of Steroids by Infrared Spectrometry	2	1
Infrared Analysis of Vitamins, Hormones, and Coenzymes	5	407
Roth, Marc, Fluorimetric Assay of Enzymes	17	189
Russo, A., and Bump, E. A., Detection and Quantitation of Biological Sulfhydryls	33	165
Salmon, W. D., see Engel, R. W.		

	VOL.	PAGE
Saltzman, Bernard E., and Keenan, Robert G., Microdetermination of Cobalt in Biological Materials	5	181
Salvayre, Robert, see Beauregard, Guy		
Sandhoff, Konrad, see Conzelmann, Ernest		
Schayer, Richard W., Determination of Histidine Decarboxylase Activity	16	273
Determination of Histidine Decarboxylase	Supp.	99
Schindler, Melvin, see Wade, Margaret H.		
Schmakel, C. O., see Elving, P. J.		
Schmid, Harald H. O., see Mangold, Helmut K.		
Schram, Karl H., Mass Spectrometry of Nucleic Acid Components	34	203
Schubert, Jack, Measurement of Complex Ion Stability by the Use of Ion Exchange Resins	3	247
Schuberth, Jan, see Sorbo, S. Bo		
Schulten, Hans-Rolf, Field Desorption Mass Spectrometry and Its Application in Biochemical Analysis	24	313
Schwartz, Morton K., and Bodansky, Oscar, Automated Methods for Determination of Enzyme Activity	11	211
Schwartz, Morton K., and Bodansky, Oscar, Utilization of Automation for Studies of Enzyme Kinetics	16	183
Schwartz, Samuel, Berg, Marie H., Bossenmaier, Irene, and Dinsmore, Howard, Determination of Porphyrins in Biological Materials	8	221
Scott, Charles D., see Burtis, Carl A.		
Scott, J. E., Aliphatic Ammonium Salts in the Assay of Acidic Polysaccharides from Tissues	8	145
Seaman, G. R., see Stokstad, E. L. R.		
Sebald, Walter, see Werner, Sigurd		
Seiler, N., Use of the Dansyl Reaction in Biochemical Analysis	18	259
Seitz, W. Rudolf, and Neary, Michael P., Recent Advances in Bioluminescence and Chemiluminescence Assay	23	161
Selvendran, Robert R., and O'Neill, Malcolm A., Isolation and Analysis of Cell Walls from Plant Material	32	25
Shaw, Jane E., and Ramwell, P. W., Separation, Identification, and Estimation of Prostaglandins	17	325
Sherratt, H. S. A., Watmough, N. J., Johnson, M. A., and Turnbull, D. M., Methods for Study of Normal and Abnormal Skeletal Muscle Mitochondria	33	243
Shibata, Kazuo, Spectrophotometry of Opaque Biological Materials: Reflection Methods	9	217
Spectrophotometry of Translucent Biological Materials: Opal Glass Transmission Method	7	77
Shiono, Satoru, Honazato, Yoshio, and Nakako, Mamiko, Advances in Enzymatically Coupled Field Effect Transistors	36	151
Shore, P. A., Determination of Histamine	Supp.	89
Sies, Helmut, and Brauser, Bolko, Analysis of Cellular Electron Transport Systems in Liver and Other Organs by Absorbance and Fluorescence Techniques	26	285
Silber, Robert H., Fluorimetric Analysis of Corticoids	14	63
Silber, Robert H., and Porter, Curt C., Determination of 17,21-Dihydroxy-20-Ketosteroids in Urine and Plasma	4	139
Singer, Thomas P., see Grossman, Shlomo		
Singer, Thomas P., Determination of the Activity of Succinate NADH, Choline, and α-Glycerophosphate Dehydrogenases	22	123
Singer, Thomas P., and Kearney, Edna B., Determination of Succinic Dehydrogenase Activity	4	307
Sjovall, Jan, Separation and Determination of Bile Acids	12	97

	VOL.	PAGE
Skeggs, Helen R., Microbiological Assay of Vitamin B_{12}	14	53
Smales, A. A., see Loveridge, B. A.		
Smith, Emil L., see Davis, Neil C.		
Smith, Fred, and Montgomery, Rex, End Group Analysis of Polysaccharides	3	153
Smith, Lucile, Spectrophotometric Assay of Cytochrome c Oxidase	2	427
Sorbo, S. Bo, and Schuberth, Jan, Measurements of Choline Acetylase	Supp.	275
Spencer, B., see Dodgson, K. S.		
Sperry, Warren M., Lipid Analysis	2	83
Spink, Charles H., and Wadsö, Ingemar, Calorimetry as an Analytical Tool in Biochemistry and Biology	23	1
Stahl, Egon, see Mangold, Helmut K.		
St. John, P. A., see Winefordner, J. D.		
Stokstad, E. L. R., Seaman, G. R., Davis, R. J., and Hutner, S. H., Assay of Thioctic Acid	3	23
Storvick, C. A., Benson, E. M., Edwards, M. A., and Woodring, M. J., Chemical and Microbiological Determination of Vitamin B_6	12	183
Stott, R. A., see Walker, M. R.		
Strehler, Bernard L., Bioluminescence Assay: Principles and Practice	16	99
Strehler, B. L., and Totter, J. R., Determination of ATP and Related Compounds: Firefly Luminescence and Other Methods	1	341
Stults, John T., Peptide Sequencing by Mass Spectrometry	34	145
Swartz, Harold M., and Swartz, Sharon M., Biochemical and Biophysical Applications of Spin Resonance	29	207
Swartz, Sharon M., see Swartz, Harold M.		
Sweetman, Brian, see Hellerqvist, Carl G.		
Takamura, Kiyoko, and Kusu, Fumiyo, Electro-Optical Reflection Methods for Studying Bioactive Substances at Electrode-Solution Interfaces—An Approach to Biosurface Behavior	32	155
Talalay, Paul, Enzymic Analysis of Steroid Hormones	8	119
Tchalian, M., see McLaren, D. S.		
Thénot, Jean-Paul G., and Haegele, Klaus D., Analysis of Morphine and Related Analgesics by Gas Phase Methods	24	1
Thiers, Ralph E., Contamination of Trace Element Analysis and Its Control	5	273
Thorpe, G. H. G., see Walker, M. R.		
Thorpe, W. V., see Bray, H. G.		
Tiffany, Thomas O., see Burtis, Carl A.		
Tinoco, Ignacio, Jr., Application of Optical Rotatory Dispersion and Circular Dichroism to the Study of Biopolymers	18	81
Tolksdorf, Sibylle, The in vitro Determination of Hyaluronidase	1	425
Totter, J. R., see Strehler, B. L.		
Travers, Franck, see Barman, Thomas E.		
Treadwell, C. R., see Vahouny, George V.		
Tulp, Abraham, Density Gradient Electrophoresis of Mammalian Cells	30	141
Turnbull, D. M., see Sherratt, H. S. A.		
Udenfriend, Sidney, Weissbach, Herbert, and Brodie, Bernard B., Assay of Serotonin and Related Metabolites, Enzymes, and Drugs	6	95
Ushakov, A. N., see Vaver, V. A.		
Vahouny, George V., and Treadwell, C. R., Enzymatic Synthesis and Hydrolysis of Cholesterol Esters	16	219
Vallee, Bert L., see Margoshes, Marvin		
Vanden Heuvel, W. J. A., see Horning, E. C.		
Van Pilsum, John F., Determination of Creatinine and Related Guanidinium Compounds	7	193

	VOL.	PAGE
Vaver, V. A., and *Ushakov, A. N.,* High Temperature Gas-Liquid Chromatography in Lipid Analysis	26	327
Venkateswarlu, P., Determination of Fluorine in Biological Materials	24	93
Vessey, D. A., see *Zakim, D.*		
Vestling, Carl S., Determination of Dissociation Constants for Two-Substrate Enzyme Systems	10	137
Volkin, Elliot, and *Cohn, Waldo E.,* Estimation of Nucleic Acids	1	287
Vollenweider, H. J., Visual Biochemistry: New Insight into Structure and Function of the Genome	28	201
Wacker, Warren, E. C., see *Amador, Elias*		
Wade, Margaret H., Quantitative Fluorescence Imaging Techniques for the Study of Organization and Signaling Mechanisms in Cells	37	117
Wadsö, Ingemar, see *Spink, Charles H.*		
Waldemann-Meyer, H., Mobility Determination by Zone Electrophoresis at Constant Current	13	47
Walker, M. R., Stott, R. A., and *Thorpe, G. H. G.,* Enzyme-Labeled Antibodies in Bioassays	36	179
Walter, Bert, Fundamentals of Dry Reagent Chemistries: The Role of Enzymes	36	35
Wang, C. H., Radiorespirometry	15	311
Watmough, M. A., see *Sherratt, H. S. A.*		
Webb, Junius M., and *Levy, Hilton B.,* New Developments in the Chemical Determination of Nucleic Acids	6	1
Weeks, Ian, see *Campbell, Anthony K.*		
Wehmeyer, Kenneth R., see *Heineman, William R.*		
Weil-Malherbe, H., The Estimation of Total (Free + Conjugated) Catecholamines and Some Catecholamine Metabolites in Human Urine	16	293
Determination of Catechol Amines	Supp.	119
Weinstein, Boris, Separation and Determination of Amino Acids and Peptides by Gas-Liquid Chromatography	14	203
Weintraub, Susan T., see *Hanahan, Donald J.*		
Weissbach, Herbert, see *Udenfriend, Sidney*		

CUMULATIVE SUBJECT INDEX, VOLUMES 1-37 AND SUPPLEMENTAL VOLUME

	VOL.	PAGE
Absorbent and Fluorescent Substances, The Automated Analysis of, Separated on Paper Strips (Boulton)	16	327
Activation Analysis and Its Application in Biochemistry (Loveridge and Smales)	5	225
Activation Analysis of Biological Trace Elements (Leddicotte)	19	345
Adenine, Enzymic Micro Determination, by Ultraviolet Spectrophotometry (Plesner and Kalckar)	3	97
Adrenal Gland, Steroids of, Chromatographic Separation (Haines and Karnemaat)	1	171
Adrenalin, Chemical Determination, in Body Fluids and Tissues (Persky)	2	57
Adrenal Steroids in Blood, Analysis of, by Countercurrent Distribution (Carstensen)	9	127
Affinity Chromatography, The Purification of Biologically Active Compounds by Aliphatic Ammonium Salts in the Assay of Acidic Polysaccharides from Tissues (Scott)	8	145
Alternative Metabolic Pathways, Estimation of Magnitudes of (Kopin)	11	247
Amine Oxidases, Determination of (Kapeller-Adler)	Supp.	35
Amino Acid Analysis by Gas-Liquid Chromatography, Recent Developments in, (MacKenzie)	27	1
Amino Acids, Analysis by Means of Brdicka Reaction (Müller)	11	329
Amino Acids, Determination by High-Voltage Paper Electrophoresis (Blackburn)	13	1
Amino Acids, Determination by Ion Exchange Chromatography (Jacobs)	14	177
Amino Acids, Determination by Use of Bacterial Amino Acid Decarboxylases (Gale)	4	285
Amino Acids, Separation and Determination by Gas-Liquid Chromatography (Weinstein)	14	203
Ammonium Salts, Aliphatic, in the Assay of Acidic Polysaccharides from Tissues Scott	8	145
Ampholytes, A Technique of Fractionation and Characterization of Isoelectric Focusing in—pH Gradients (Haglund)	19	1
Analgesics, Analysis by Gas Phase Methods (Thénot and Haegele)	24	1
Antibiotics, Microbiological Assay (Kersey and Fink)	1	153
Application of High Resolution Nuclear Magnetic Resonance in Biological Systems (Campbell and Dobson)	25	1
Ascorbic Acid, Chemical Determination (Roe)	1	115
Atomic, Absorption Spectroscopy, Analysis of Biological Materials by (Willis)	11	1
Automated Enzyme Assays (Lott and Wealon)	37	143
ATP, Determination of Firefly Luminescence (Strehler and Totter)	1	341
Avidin-Biotin, Use of, as Tool in Molecular Biology (Bayer and Wilchek)	26	1
Bacterial Amino Acid Decarboxylases in Determination of Amino Acids (Gale)	4	285
Basic Proteins, Preparation and Analysis of (Lindh and Brantmark)	14	79
Bile Acids, Newer Developments in the Gas Chromatographic Determination of (Kuksis)	14	325

313

	VOL.	PAGE
Bile Acids, Separation and Determination of (Sjöval)	12	97
Bile Pigments, Separation and Determination of (Brodersen and Jacobsen)	17	31
Bioanalytical Systems, Unique Applications of Immobilized Proteins in (Gupta and Mattiasson)	36	1
Bioassays, Enzyme-Labeled Antibodies in (Walker, Stott, and Thorpe)	36	179
Biochemical Anslysis, Computers in (Crabbe)	31	179
Biochemical Applications of Magnetic Resonance (Jardetzky and Jardetzky)	9	235
Biochemical Nuclear Magnetic Resonance Spectroscopy, Recent Developments in (Bock)	31	259
Biochemistry, Historical Development and Newer Means of Temperature Measurement in (Berger, Clem, Harden, and Mangum)	30	269
Biochemistry, Laser Photolysis in (Chan and Austin)	30	105
Biological Materials, Analysis by Atomic Absorption Spectroscopy (Willis)	11	1
Biological Materials, Determination of Nitrogen in (Jacobs)	13	241
Biological Materials, Determination of Porphyrins in (Schwartz, Berg, Bossenmaier, and Dinsmore)	8	221
Biological Materials, Determination of Zinc in (Malmstrom)	3	327
Biological Materials, Methods for Estimating Magnesium in (Alcock and MacIntyre)	14	1
Biological Materials, Microdetermination of Cobalt in (Saltzman and Keenan)	5	181
Biological Materials, Opaque, Spectrophotometry of; Reflection Methods (Shibata)	9	217
Biological Materials, Translucent, Spectrophotometry of; Opal Glass Methods (Shibata)	7	77
Biological Standards in Biochemical Analysis (Humphrey, Long, and Perry)	5	65
Biological Systems, Determination of Serine in (Frisell and Mackenzie)	6	63
Biological Systems, Ion Binding in, Measured by Nuclear Magnetic Resonance Spectroscopy (Forsén and Lindman)	27	289
Biological Trace Elements, Activation Analysis of (Leddicotte)	19	345
Bioluminescence Assay: Principles and Practice (Strehler)	16	99
Bioluminescence and Chemiluminescence Assay, Recent Advances in	23	161
Blood, Analysis of Adrenal Steroids in, by Countercurrent Distribution (Cartensen)	9	127
Blood, Determination of Ethyl Alcohol in (Lindquist)	7	217
Body Fluids, Chemical Determination of Adrenaline and Noradrenaline in (Persky)	2	57
Body Fluids, Chromatographic Analysis of Radioactive Iodine Compounds from (Roche, Lissitzky, and Michel)	1	243
Buffers, Containing Urea, Gel Electrophoresis in (Poulik)	14	455
Calorimetry as an Analytical Tool in Biochemistry and Biology (Spink and Wadsö)	23	1
Carbohydrate, The Determination of, in Biological Materials by Gas-Liquid Chromatography (Clamp, Bhatti, and Chambers)	19	229
Carbohydrate Derivatives, Separation of, by Gas-Liquid Partition Chromatography (Bishop)	10	1
Carbohydrate Structure, Use of Infrared Analysis in Determination of (Baker, Bourne, and Whiffen)	3	213
Carbonic Anhydrose Activity, Measurements of (Davis)	11	307
Carotene, Determination of (Bickoff)	4	1
(Creveling and Daly)	Supp.	153
Catalases, Assay of (Maehly and Chance)	1	357
Catechol Amine Biosynthesis and Metabolism, Assay of Enzymes of Catecholamines and Catecholamine Metabolites, Estimation of Total (Free + Conjugated), in Human Urine (Weil-Malherbe)	16	293

CUMULATIVE SUBJECT INDEX, VOLUMES 1-37 AND SUPPLEMENTAL VOLUME 315

	VOL.	PAGE
Catechol Amines, Determination of (Weil-Malherbe)	Supp.	119
Cell Biology and Medicine, Chemiluminescence as an Analytical Tool in (Campbell, Hallett, and Weeks)	31	317
Cell Particles and Macromolecules, Partition Methods for Fractionation of (Albertsson)	10	229
Cellular Electron Transport Systems in Liver and Other Organs, Analysis of, by Absorbance and Fluorescence Techniques (Sies and Brauser)	26	285
Cellulose Column Chromatography, Analysis of Mixtures of Sugars by (Hough)	1	205
Cell Walls, Isolation and Analysis of, from Plant Material (Selvendran and O'Neill)	32	25
Centrifugal Fast Analyzer for Biochemical and Immunological Analysis, The Use of (Burtis, Tiffany, and Scott)	23	189
Centrifugation, Preparative Zonal (Anderson)	15	271
Chemiluminescence as an Analytical Tool in Cell Biology and Medicine (Campbell, Hallet, and Weeks)	31	317
Chloride in Biological Materials, Determination of (Cotlove)	12	277
Cholesterol, Determination and Microscopic Localization of (Kabara)	10	263
Cholesterol Esters, Enzymatic Synthesis and Hydrolysis of (Vahouny and Treadwell)	16	219
Choline, Chemical Estimation of (Engel, Salmon, and Ackerman)	1	265
Choline Acetylase, Measurements of (Sorbo and Schuberth)	Supp.	275
Choline Esters, Measurement of (Jenden)	Supp.	183
Cholinesterases, Assay Methods of (Augustinsson)	5	1
Cholinesterases, Determination of (Augustinsson)	Supp.	217
Chromatographic Analysis, Applications of the R_M Treatment in (Bush)	13	357
Chromatographic Analysis, Applications of the R_M Treatment in, Erratum (Bush)	14	497
Chromatographic Analysis of Radioactive Iodine Compounds from the Thyroid Gland and Body Fluids (Roche, Lissitzky, and Michel)	1	243
Chromatographic and Electrophoretic Methods, Analysis of Natural Radioactive Iodine Compounds by (Roche, Michel and Lissitzky)	12	143
Chromatographic Separation of Steroids of the Adrenal Gland (Haines and Karnemaat)	1	171
Chromatography, Gas, in Determination of Bile Acids and Steroids (Kuksis)	14	325
Chromatography, Gas, Separation and Determination of Steroids by (Horning, Vanden Heuvel, and Creech)	11	69
Chromatography, Gas-Liquid, Determination of the Fatty Acids by (James)	8	1
Chromatography, Gas-Liquid, Separation and Determination of Amino Acids and Peptides by (Weinstein)	14	203
Chromatography, Gas-Liquid Partition, Separation of Carbohydrate Derivatives by (Bishop)	10	1
Chromatography, High-Performance Liquid: Analytic and Preparative Application in Protein Structure Determination (Hughes and Wilson)	29	59
Chromatography, High-Temperature Gas-Liquid, in Lipid Analysis (Vaver and Ushakov)	26	327
Chromatography, Ion Exchange, Determination of Amino Acids by (Jacobs)	14	177
Chromatography, Paper and Cellulose Column, Analysis of Mixture of Sugars by (Hough)	1	205
Chromatography, of Proteins, Nucleic Acids, Viruses, and Cells on Noncharged Amphiphilic Gels, Hydrophobic Interaction (Hjertén)	27	89
Chromatography, Thin-Layer (TLC) (Mangold, Schmid, and Stahl)	12	393
Circular Dichroism and Its Empirical Application to Biopolymers (Johnson)	31	61
Cobalt, Microdetermination of, in Biological Materials (Saltzman and Keenan)	5	181
Coenzyme A, Methods for Determination (Novelli)	2	189

	VOL.	PAGE
Coenzyme Q, Determination of (Crane and Dilley)	11	279
Coenzyme, Infrared Analysis of (Rosenkrantz)	5	407
Collagen and Elastin, Analysis of the Crosslinking Components in (Robins)	28	329
Collagen and Elastin, The Determination of (Jackson and Cleary)	15	25
Color Reactions, New, for Determination of Sugars in Polysaccharides (Dische)	2	313
Column Electrophoresis in Granular Media, Some Recent Developments (Porath and Hjertén)	9	193
Complexes, Metal Protein (Hughes and Klotz)	3	265
Complex Ion Solubility, Measurement by Use of Ion Exchange Resins (Schubert)	3	247
Connective-Tissue Macromolecules, Analysis by Determination of Certain Constituents (Blumenkrantz and Asboe-Hansen)	24	39
Contamination in Trace Element Analysis and Its Control (Thiers)	5	273
Corticoids, Fluorimetric Analysis of (Silber)	14	63
Countercurrent Distribution (King and Craig)	10	201
Countercurrent Distribution, Analysis of Adrenal Steroids in Blood by (Cartensen)	9	127
Creatinine and Related Guanidinium Compounds, Determination of (Van Pilsum)	7	193
Current, Constant, Mobility Determination by Zone Electrophoresis at (Waldmann-Meyer)	13	47
Cyclic 3',5'-Adenosine Monophosphate and Cyclic 3',5'-Guanosine Monophosphate, Analysis of (Goldberg and O'Toole)	20	1
Cyclic AMP and Other Cyclic Nucleotides, Adenylate Cyclase, and Phosphodiesterase, Newer Developments in the Determination of (Brooker)	22	95
Cyclochrome c Oxidase, Spectrophotometric Assay of (Smith)	2	427
Dansyl Reaction, Use of the, in Biochemical Analysis (Seiler)	18	259
Dehydroascorbic Acid, Chemical Determination of (Roe)	1	115
Dehydrogenases, Determination of the Activity of Succinate, NADH, Choline α-Glycerophosphate (Singer)	22	123
Denaturation, Reversible, of Proteins, Methods of Study and Interpretation of Data for (Hermans, Jr.)	13	81
Density Gradients, Rapid Electrophoresis in (Kolin)	6	259
Deoxyribonuclease Activity, Assay of (Kurnick)	9	1
Diagnosis, Enzymatic Methods of (Amador and Wacker)	13	265
Dialysis (Craig and King)	10	175
Diazo-Positive Bile Pigments, Recent Advances in the Separation and Analysis of (Heirwegh)	22	205
Diffraction, X-ray, in the Study of Protein and Nucleic Acid Structure (Holmes and Blow)	13	113
17,21-Dihydroxy-20-Ketosteroids, Determination in Urine and Plasma (Silber and Porter)	9	139
Diketogulonic Acid, Chemical Determination of (Roe)	1	115
Dissociation Constants, Determination of, for Two-Substrate Enzyme Systems (Vestling)	10	137
Dissolved Gases in Biochemical Systems with the Quadrupole Mass Spectrometer, Continuous Measurement of (Degn, Cox and Lloyd)	31	165
DNA Restriction Enzymes and RFLPs in Medicine (Kant)	36	129
Dry Reagent Chemistries, Fundamentals of: The Role of Enzymes (Walter)	36	35
Electrochemical Detection, Biomedical Uses of High-Performance Liquid Chromatography with (Mefford)	31	221
Electrochemical Detection, Immunoassay with (Heineman, Halsall, Wehmeyer, Doyle, and Wright)	32	345
Electron Probe Microanalyzer, An Introduction to, and Its Application to Biochemistry (Anderson)	15	147
Electron Spin Resonance, Biochemical and Biophysical Application of (Swartz and Swartz)	29	207

	VOL.	PAGE
Electro-Optical Reflection Methods for Studying Bioactive Substances at Electrode-Solution Interfaces—An Approach to Biosurface Behavior (Takamura and Kusu)	32	155
Electrophoresis, Free Zone, Theory, Equipment, and Applications (Hjertén)	18	55
Electrophoresis, Gel, in Buffers Containing Urea (Poulik)	14	455
Electrophoresis, Gel Sieving: A Description of Procedures and Analysis of Errors (Johnson)	29	25
Electrophoresis, Paper, Determination of Amino Acids at High-Voltage by (Blackburn)	13	1
Electrophoresis, Rapid, in Density Gradients Combined with pH and/or Conductivity Gradients (Kolin)	6	259
Electrophoresis in Granular Media, Column, Some Recent Developments (Porath and Hjertén)	9	193
Electrophoresis Zone (Kunkel)	1	141
Electrophoresis Zone, Constant Current Mobility Determination by (Waldmann-Meyer)	13	47
Electrophoretic Methods, Analysis of Natural Radioactive Iodine Compounds by (Roche, Michel, and Lissitzky)	12	143
Elements, Determination of, by X-Ray Emission Spectrometry (Natelson and Whitford)	12	1
Enthalpy and Entropy Changes, Determination by Heatburst Microcalorimetry (Kitzinger and Benzinger)	8	309
Enzymatically Coupled Field Effect Transistors, Advances in (Shiono, Honazato, and Nakako)	36	151
Enzymatic Methods, in Diagnosis (Amador and Wacker)	13	265
Enzyme Activity, Automated Methods for Determination of (Schwartz and Bodansky)	11	211
Enzyme Assay, Ratiometric Methods of (Oldham)	21	191
Enzyme Electrode Biosensors: Theory and Applications (Kauffmann and Guilbault)	36	63
Enzyme Kinetics, Utilization of Automation for Studies of (Schwartz and Bodansky)	16	183
Enzymes, Assay of in Catechol Amine Biosynthesis and Metabolism (Creveling and Daly)	Supp.	153
Enzymes, Detection of Ligand-Induced and Syncatalytic Conformational Changes of, by Differential Chemical Modification (Christen and Gehring)	28	151
Enzymes, Fluorimetric Assay of (Roth)	17	189
Enzymes, Immobilized, in Biochemical Analysis (Everse, Ginsburgh, and Kaplan)	25	135
Enzymes, Proteolytic Assay of (Davis and Smith)	2	215
Enzymes, Related to Serotonin, Assay of (Udenfriend, Weissbach, and Brodie)	6	95
Enzyme Systems, Two Substrate, Determination of Dissociation Constants for (Vestling)	10	137
Enzymic Analysis of Steroid Hormones (Talalay)	8	119
Enzymic Determination of D-Glucose and Its Anomers, New Developments in (Okuda and Miwa)	21	155
Estrogens, Chemicals Determination of, in Human Urine (Bauld and Greenway)	5	337
Ethanolamine, Determination of, in Lipids (McKibbin)	7	111
Fast Reactions in Biochemistry, the Rapid-Flow-Quench Method in the Study of (Barman and Travers)	31	1
Fatty Acid Esters, A Critical Evaluation of the Gas Chromatographic Technique for Identification and Determination of, with Particular Reference to the Use of Analog and Digital Computer Methods (Caster)	17	135
Fatty Acids, Determination by Gas-Liquid Chromatography (James)	8	1
Field Desorption Mass Spectrometry: Application in Biochemical Analysis (Schulten)	24	313

	VOL.	PAGE
Field-Flow Fractionation, Analysis of Biological Macromolecules and Particles by (Giddings, Myers, Caldwell, and Fisher)	26	79
Firefly Luminescence, Determination of ATP by (Strehler and Totter)	1	341
Flame Photometry, Principles and Applications (Margoshes and Vallee)	3	353
Flavins, Chemical Determination of (Yagi)	10	319
Fluids, Body, Chemical Determination of Adrenaline and Noradrenaline in (Persky)	2	57
Fluids, Body, Chromatographic Analysis of Radioactive Iodine Compounds from (Roche, Lissitzky, and Michel)	1	243
Fluorescence Dequenching Measurements, The Use of, to Follow Viral Membrane Fusion Events (Loyter, Citovsky, and Blumenthal)	33	129
Fluorescence Techniques, for Studying Protein Structure (Eftink)	35	127
Fluorimetric Analysis of Corticoids (Silber)	14	63
Fluorine, Determination in Biological Materials (Venkateswarlu)	24	93
Folic Acid Activity, Assay of Compounds with (Jukes)	2	121
Formaldehyde, Determination of, in Biological Systems (Frisell and Mackenzie)	6	63
Fractionation of Cell Particles and Macromolecules, Partition Methods for (Albertsson)	10	229
Free Energy Changes, Determination by Heatburst Microcalorimetry (Kitzinger and Benzinger)	8	309
Frog Skin Assay for Agents that Darken and Lighten Melanocytes (Lerner and Wright)	8	295
Gas-Liquid Chromatography, The Determination in Carbohydrates and Biological Materials (Clamp, Bhatti, and Chambers)	19	229
Gel Electrophoresis in Buffers Containing Urea (Poulik)	14	455
β-*Glucuronidases, Determination of* (Fishman)	15	77
UDP-Glucuronyltransferase, Glucose-6 Phosphatase, and Other Tightly-Bound Microsomal Enzymes, Techniques for the Characterization of (Zakin and Vessey)	21	1
Glutamic and Aspartic Acids and Their Amides, Determination of (Balis)	20	103
Glutathione, Determination of (Patterson and Lazarow)	2	259
Glycolipid Determination (Radin)	6	163
Glycoproteins, Serum, Determination of (Winzler)	2	279
Gradients, Density, Rapid Electrophoresis in (Kolin)	6	259
Heatburst Microcalorimetry, Principle and Methods of, and Determination of Free Energy, Enthalpy, and Entropy Changes (Kitzinger and Benzinger)	8	309
Heparin, Determination of (Jaques)	24	203
Heparin, Determination of (Jaques and Bell)	7	253
Hexosamines, Determination of (Gardell)	6	289
High-Performance Ion-Exchange Chromatography with Narrow-Bore Columns: Rapid Analysis of Nucleic Acid Constituents at the Subnanomole Level (Horvath)	21	79
Histamine, Determination of (Shore)	Supp.	89
Histamine, Quantitative Determination of (Code and McIntire)	3	49
Histidine, Decarboxylase, Determination of (Schayer)	Supp.	99
Histidine, Decarboxylase Activity, Determination of (Schayer)	16	273
Hormones, Infrared Analysis of (Rosenkrantz)	5	407
Hormones, Plant, Analysis of (Bentley)	9	75
Hormones, Steroid, Enzymic Analysis of (Talalay)	8	119
Hyaluronidase, in vitro Determination (Tolksdorf)	1	425
Hydrogen Exchange Data, Acquisition and Interpretation of, from Peptides, Polymers, and Proteins (Barksdale and Rosenberg)	28	1
Hydrogen Isotope Exchange in Globular Proteins, Methods for Measurement (Ottesen)	20	135
Hydrophobic Interaction Chromatography of Proteins, Nucleic Acids, and Cells on Noncharged Amphiphilic Gels (Hjertén)	27	89

CUMULATIVE SUBJECT INDEX, VOLUMES 1-37 AND SUPPLEMENTAL VOLUME

	VOL.	PAGE
Hydroxyl Radicals, Methods for the Measurement of in Biochemical Systems: Deoxyribose Degradation and Aromatic Hydroxylation (Halliwell and Grootveld)	33	59
Hypoxanthine, Enzymic Micro Determination, by Ultraviolet Spectrophotometry (Plesner and Kalckar)	3	97
Immunoassay of Plasma Insulin (Yalow and Berson)	12	69
Immunoassay with Electrochemical Detection, see Electrochemical Detection, Immunoelectrophoretic Analysis (Garbar)	7	1
Immunological Techniques for Studies on the Biogenesis of Mitochondrial Membrane Proteins (Werner and Sebald)	27	109
Infrared Analysis, Use of, in the Determination of Carbohydrate Structure (Baker, Bourne, and Whiffen)	3	213
Infrared Analysis of Vitamins, Hormones, and Coenzymes (Rosenkrantz)	5	407
Infrared Spectrometry, Analysis of Steroids by (Rosenkrantz)	2	1
Inositol, Determination of, in Lipides (McKibbin)	7	111
Iodine, in Biological Material, Determination of (Binnerts)	22	251
Iodine Compound, Radioactive, from Thyroid Gland and Body Fluids, Chromatographic Analysis (Roche, Lissitzky, and Michel)	1	243
Iodine Compounds, Natural Radioactive, Analysis by Chromatographic and Electrophoretic Methods (Roche, Michel, and Lissitzky)	12	143
Ion Binding in Biological Systems Measured by Nuclear Magnetic Resonance Spectroscopy (Forsen and Lindman)	27	289
Ion Exchange Resins, Measurement of Complex Ion Stability by Use of (Schubert)	3	247
Isoelectric Focusing in Immobilized pH Gradients: Theory and Newer Technology (Righetti and Gianazza)	32	215
Isolated Bacterial Nucleoids, Characterization, Assay, and Use of (Hirschbein and Guillen)	28	297
Isotope Derivative Method in Biochemical Analysis, The (Whitehead and Dean)	16	1
Ketose, Determination, in Plant Products (de Whalley and Gross)	1	307
α-Keto Acid Determinations (Neish)	5	107
17-Ketosteroids, Urinary Neural, Assay of (Engel)	1	459
Lipase, Lipoprotein, Assay of, in vivo and in vitro (Korn)	7	145
Lipide Analysis (Sperry)	2	83
Lipides, Determination of Inositol, Ethanolamine, and Serine in (McKibbin)	7	111
Lipid Transfer Activity, Quantitation of (Wetterau and Zilversmit)	30	199
Lipoprotein Lipase, Assay of, in vivo and in vitro (Korn)	7	145
Lipoproteins, Serum, Ultracentrifugal Analysis (de Lalla and Gofman)	1	459
Liposomes: Preparation, Characterization, and Preservation (Lichtenberg and Bareneholz)	33	337
Lipoxidase Activity, Measurement of (Holman)	2	113
Lipoxygenase (Lipoxidase), Determination of the Activity of (Grossman and Zakut)	25	303
Liquid-Scintillation Counting, Practical Aspects of (Kobayashi and Maudsley)	17	55
Luciferin and Luciferase, Measurement of (Chase)	8	61
Lysosomal Glycolipid Hydrolysis, Activator Proteins for (Conzelmann and Sandhoff)	32	1
Lysozyme Activity, Methods for Determination of (Grossowicz and Ariel)	29	435
Magnesium Estimation, in Biological Materials (Alcock and MacIntyre)	14	1
Magnetic Resonance, Biochemical Applications of (Jardetzky and Jardetzky)	9	235
Mammalian Cells, Density Gradient Electrophoresis of (Tulp)	30	141
Mass Spectrometry, Analysis of Steroids by (Gasdell)	29	385
Mass Spectrometry, Field Desorption: Application in Biochemical Analysis (Schulten)	24	313

	VOL.	PAGE
Mass Spectrometry of Carbohydrates (Hellerqvist and Sweetman)	34	91
Mass Spectrometry in the Determination of Structure of Certain Natural Products Containing Sugars (Hanessian)	19	105
Mass Spectrometry: An Introduction (Anderegg)	34	1
Mass Spectrometry of Nucleic Acid Components (Schram)	34	203
Mass Spectrometry in Pharmacology (Abramson)	34	289
Melanocytes, Darkening and Lightening, Frog Skin Assay for (Lerner and Wright)	8	295
Metabolic Pathways, Alternative, Estimation of Magnitudes of (Kopin)	11	247
Metabolism, Analysis of Phenolic Compounds of Interest in (Bray and Thorpe)	1	27
Metal Buffers, Applications, in Biochemistry (Raaflaub)	3	301
Metal Indicators, Applications, in Biochemistry (Raaflaub)	3	301
Metal-Protein Complexes, Analysis of (Hughes and Klotz)	3	265
Microbiological Assay of Antibiotics (Kersey and Fink)	1	53
Microbiological Assay of Vitamin B_{12} (Hoff-Jorgensen)	1	81
Microbiological Assay of Vitamin B_{12} (Skeggs)	14	53
Microbiological Determination of Vitamin B_6 (Storvick, Benson, Edwards, and Woodring)	12	183
Microparticulate Gel Chromatography Accelerated by Centrifugal Force and Pressure (Ribi, Parker, and Milner)	22	355
Mobility, Determination by Zone Electrophoresis at Constant Current (Waldmann-Meyer)	13	47
Molecular Size, Estimation of, and Molecular Weights of Biological Compounds by Gel Filtration (Andrews)	18	1
The Use of Monoclonal Antibodies and Limited Proteolysis in Elucidation of Structure-Function Relationships in Proteins (Wilson)	35	207
Morphine, and Related Analgesics, Analysis by Gas Phase Methods (Thénot and Haegele)	24	1
Mucopolysaccharides, Sulfated, Determination of (Jaques)	24	203
Negative-Ion Mass Spectrometry, Fused-Silica Capillary Gas Chromatography of Neurotransmitters and Related Compounds (Faull and Barchas)	29	325
Neuraminic (Sialic) Acids, Isolation and Determination of (Whitehouse and Zilliken)	8	199
Nitrogen Determination in Biological Materials (Jacobs)	13	241
Nitrogenous Compounds, Basic, of Toxicological Importance, Analysis of (Curry)	7	39
Noradrenaline, Chemical Determination, in Body Fluids and Tissues (Persky)	2	57
Nucleic Acid, Structure, X-ray Diffraction in the Study of (Holmes and Blow)	13	113
Nucleic Acid Hybridization, Enzyme-Labeled Probes for (Kaguni and Kaguni)	36	115
Nucleic Acids, Chemical Determination of (Webb and Levy)	6	1
Nucleic Acids, The Determination of (Munro and Fleck)	14	113
Nucleic Acids, Estimation (Volkin and Cohn)	1	287
Nucleic Acids and Their Derivatives, Microbiological Assay of (Miller)	6	31
Nucleic Acids of Various Conformational Forms and Measurement of Their Associated Enzymes, Use of Ethidium Bromide for Separation and Determination of (Le Pecq)	20	41
Nucleosides and Nucleotides and Their Parent Bases as an Analytical and Investigative Tool, Polarography and Voltammetry of (Elving, O'Reilly, and Schmakel)	21	287
Optical Rotatory Dispersion, Application of, and Circular Dichroism to the Study of Biopolymers (Tinoco, Jr.)	18	81
Organic Phosphorus Compounds, Determination of, by Phosphate Analysis (Lindberg and Ernster)	3	1
Oxidations, Periodate, Use of, in Biochemical Analysis (Dyer)	3	111
Oxygen Electrode Measurements in Biochemical Analysis (Lessler and Brierly)	17	1

	VOL.	PAGE
Paper Chromatograms, for Analysis of Mixture of Sugars (Hough)	1	205
Paper Chromatograms, Direct Scanning of, for Quantitative Estimations (Bush)	11	149
Partition Methods for Fractionation of Cell Particles and Macromolecules (Albertson)	10	229
Peptide Chromatography, Automatic (Jones)	18	205
Peptide Mapping of Proteins (James)	26	165
Peptides, Terminal and Sequence Studies in, Recent Developments in Techniques for (Fraenkel-Conrat, Harris, and Levy)	2	359
Peptides and Amino Acids in Normal Human Urine, Separation and Quantitation of (Lou and Hamilton)	25	203
Peptide Sequencing by Mass Spectrometry (Stultz)	34	145
Peptides Separation and Determination, by Gas-Liquid Chromatography (Weinstein)	14	203
Peroxidases, Assay of (Maehly and Chance)	1	357
Peroxidate Oxidations, Use of, in Biochemical Analysis (Dyer)	3	111
Phase Partition, Interaction Between Biomolecules Studied by (Albertsson)	29	1
Phase Partition—A Method for Purification and Analysis of Cell Organelles and Membrane Vesicles (Albertsson, Andersson, Larsson, and Akerlund)	28	115
Phenolic Compounds of Interest in Metabolism (Bray and Thorpe)	1	27
Phenylalanine and Tyrosine in Blood, The Measurement of (Robins)	17	287
pH Gradients, Isoelectric Focusing in—A Technique for Fractionation and Characterization of Ampholytes (Haglund)	19	1
pH Jump, The: Macromolecules and Solutions by a Laser-Induced, Ultrashort, Proton Pulse, Probing of—Theory and Application in Biochemistry (Gutman)	30	1
pH and Similar Variable, Recent Developments in Control of (James and Lumry)	29	137
pH-Stat and Its Use in Biochemistry (Jacobsen, Léonis, Linderstrøm-Lang, and Ottesen)	4	171
Phosphate Analysis, Determination of Organic Phosphorus Compound by (Lindberg and Ernster)	3	1
Phospholipases, A, C, and D, Determination of the Activity of (Grossman, Oestreicher, and Singer)	22	177
Phosphorimetry, as an Analytical Approach in Biochemistry (Winefordner, McCarthy, and St. John)	15	369
Phosphorus Compounds, Organic, Determination of, by Phosphate Analyses (Lindberg and Ernster)	3	1
Photometry, Flame, Principles and Applications of (Margoshes and Vallee)	3	353
Phytate and Inositol Phosphates, the Determination of (Oberleas)	20	87
Plant Hormones, Analysis of (Bentley)	9	75
Plasma, Determination of 17,21-Dihydroxy-20-Ketosteroids in (Silber and Porter)	4	139
Plasma Insulin, Immunoassay of (Yalow and Berson)	12	69
Platelet-Activating Factor Isolation, Identification, and Assay (Hanahan and Weintraub)	31	195
Polarographic Analysis of Proteins, Amino Acids, and Other Compounds by Means of the Brdicka Reaction (Müller)	11	329
Polarographic Oxygen Sensors, Adaptation of, for Biochemical Assays (Lessler)	28	175
Polysaccharides, Acidic, from Tissues, Aliphatic Ammonium Salts in the Assay of (Scott)	8	145
Polysaccharides, End Group Analysis of (Smith and Montgomery)	3	153
Polysaccharides, Sugars in, New Color Reactions for Determination of (Dische)	2	313
Polyunsaturated Fatty Acids, Measurement of (Holman)	4	99
Porphyrins in Biological Materials, Determination of (Schwartz, Berg, Bossenmaier, and Dinsmore)	8	221
Prostaglandins, Separation, Identification, and Estimation of (Shaw and Ramwell)	17	325

	VOL.	PAGE
Protein, Structure, X-ray Diffraction in the Study of (Holmes and Blow)	13	113
Protein, Terminal and Sequence Studies in Recent Developments in Techniques for (Fraenkel-Conrat, Harris, and Levy)	2	359
Protein Blotting: A Manual (Gershoni)	33	1
Protein-Ligand Interaction as a Method to Study Surface Properties (Arakawa, Kita, and Narhi)	35	87
Protein-Nucleic Acid and Protein-Protein Complexes by Differential Chemical Modification, Mapping of Contact Areas (Bosshard)	25	273
Proteins, Analysis by Means of Brdicka Reaction (Müller)	11	329
Proteins; Basic, Preparation and Analysis of (Lindh and Brantmark)	14	79
Proteins, Mitochondrial Membrane, Immunological Techniques for Studies on the Biogenesis of (Werner and Sebald)	27	109
Proteins, Polarography of, Analytical Principles and Applications in Biological and Clinical Chemistry (Homolka)	19	435
Proteins, Reversible Denaturation of, Methods of Study and Interpretation of Data for (Hermans, Jr.)	13	81
Protein Sequence Analysis, Solid Phase Methods in (Laursen and Machleidt)	26	201
Proteolytic Enzymes, Assay of (David and Smith)	2	215
Purification of Biologically Active Compounds by Affinity Chromatography, The Purines, New Methods for Purification and Separation of (Bergmann and Dikstein)	6	79
Quantitative Fluorescence Imaging Techniques for the Study of Organization and Signaling Mechanisms in Cells (Wade, deFeijter, Frame, and Schindler)	37	117
Quantitative Mass Spectrometric Analysis: Chemical and Biological Applications Integrated Ion-Current (IIC) Technique of (Majer and Boulton)	21	467
Radiation Inactivation Method as a Tool to Study Structure-Function Relationships in Proteins (Beauregard Maret, Salvayre, and Potier)	32	313
Radioactive Iodine Compounds, from Thyroid Gland and Body Fluids, Chromatographic Analysis of (Roche, Lissitzky, and Michel)	1	243
Radioimmunoassay of Polypeptide Hormones and Enzymes (Felber)	22	1
Radiorespirometry (Wang)	15	311
Raffinose Determination in Plant Products (de Whalley and Gross)	1	307
Rapid-Scanning Stopped-Flow Spectroscopy (Dunn and Brzovic)	37	191
Resins, Ion Exchange, Measurement of Complex Ion Stability, by Use of (Schubert)	3	247
Resonance, Magnetic, Biochemical Applications of (Jardetzky and Jardetzky)	9	235
Ribonuclease, Characterization of, and Determination of Its Activity (Josefsson and Lagerstedt)	9	39
R_M Treatment, Applications in Chromatographic Analysis (Bush)	13	357
R_M Treatment, Applications in Chromatographic Analysis Erratum (Bush)	14	497
Selenium in Biological Materials, Determination of (Olson, Palmer, and Whitehead)	21	39
Serine, Determination of, in Biological Systems (Frisell and Mackenzie)	6	63
Serine, Determination of, in Lipides (McKibbin)	7	111
Serotonin: The Assay of Hydroxyindole Compounds and Their Biosynthetic Enzymes (Lovenberg and Engelman)	Supp.	1
Serotonin and Related Metabolites, Enzymes, and Drugs, Assay of (Udenfriend, Weissbach, and Brodie)	6	95
Serum Acid Phosphatases, Determinations (Fishman and Davidson)	4	257
Serum Glycoproteins, Determinations of (Winzler)	2	279
Serum Lipoproteins, Ultracentrifugal Analysis of (de Lalla and Gofman)	1	459
-SH Groups in Proteins, Determinations of (Benesch and Benesch)	10	43
Sialic Acids, see Neuraminic Acids		

	VOL.	PAGE
Skeletal Muscle Mitochondria, Methods for Study of Normal and Abnormal (Sherratt, Watmough, Johnson, and Turnball)	33	243
Sodium and Potassium, Measurements of, by Glass Electrodes (Friedman)	10	71
Solid Phase Immunoassay, Use of Magnetizable Particles in (Pourfarzaneh, Kamel, Landon, and Dawes)	28	267
Spectrometry, Infrared, Analysis of Steroids by (Rosenkrantz)	2	1
Spectrometry, Principles and Applications (Margoshes and Vallee)	3	353
Spectrometry, X-ray Emission, Determination of Elements (Natelson and Whitford)	12	1
Spectrophotometric Assay of Cytochrome c Oxidase (Smith)	2	427
Spectrophotometric Oxyhemoglobin Method, Measurement of Oxygen Consumption by (Bârzu)	30	227
Spectrophotometry, Ultraviolet Enzymic Micro Determination of Uric Acid, Hypoxanthine, Xanthine, Adenine, and Xanthopterine by (Plesner and Kalckar)	6	97
Spectrophotometry of Opaque Biological Materials; Reflection Methods (Shibata)	9	217
Spectrophotometry of Translucent Biological Materials; Opal Glass Method (Shibata)	7	77
Standards, Biological, in Biochemical Analysis (Humphrey, Long, and Perry)	5	65
Steady State Kinetics of Oxygen Uptake by Biological Samples, Polarographic Measurement of (Degn, Lundsgaard, Peterson and Ormicki)	26	47
Steroid Hormones, Enzymic Analysis of (Talalay)	8	119
Steroids, Adrenal, in Blood, Analysis by Countercurrent Distribution (Carstensen)	9	227
Steroids, Analysis by Infrared Spectrometry (Rosenkrantz)	2	1
Steroids, Newer Developments in the Analysis of, by Gas-Chromatography (Wotiz and Clark)	18	339
Steroids, Newer Developments in the Gas Chromatographic Determination of (Kuksis)	14	325
Steroids, Separation and Determination, by Gas Chromatography (Horing, Vanden Heuvel, and Creech)	11	69
Steroids of the Adrenal Gland, Chromatographic Separation (Haines and Karnemaat)	1	171
Stopped-Flow Method, Recent Development in, for the Study of Fast Reactions (Hiromi)	26	137
Subzero Temperatures in Biochemistry: Slow Reactions, The Use of (Douzou)	22	401
Succinic Dehydrogenase Activity, Determination of (Singer and Kearney)	4	307
Sugars, Analysis of Mixtures, by Paper and Cellulose Column Chromatography (Hough)	1	205
Sugars, the Determination of Structure of Certain Natural Products Containing Sugars (Hanessian)	19	105
Sugars, in Polysaccharides, Determination, New Color Reactions for (Dische)	2	313
Sulfatases, Assay (Dodgson and Spencer)	4	211
Sulhydryl Groups, Determination in Biological Substances (Chinard and Hellerman)	1	1
Sulfhydryls, Detection and Quantitation of Biological (Russo and Bump)	33	165
Superoxide Dismutase Assays for (Bannister and Calabrese)	32	279
Temperature-Jump Method for Measuring the Rate of Fast Reactions, A Practical Guide to (Yapel and Lumry)	20	169
Theoretical and Empirical Approaches to Protein-Structure Prediction and Analysis (Maggiora, Mao, and Chou)	35	1
Thermodynamic Flow Methods in Biochemistry: Calorimetry, Densimetry and Dilatometry (Jolicoeur)	27	171
Thiamine, Methods for the Determination of (Mickelsen and Yamamoto)	6	191

	VOL.	PAGE
Thioctic Acid, Assay of (Stockstad, Seaman David, and Hutner)	3	23
Thyroid Gland, Chromatographic Analysis of Radioactive Iodine Compounds from (Roche, Lissitzky, and Michel)	1	243
Tissues, Aliphatic Ammonium Salts in the Assay of Acidic Polysaccharides from (Scott)	8	145
Tissues, Body, Chemical Determination of Adrenaline and Noradrenaline in (Persky)	2	57
Tissues, Determination of Ethyl Alcohol in (Lundquist)	7	217
Trace Element Analysis, Contamination in, and Its Control (Thiers)	5	273
Transaminase, Determination of (Aspen and Meister)	6	131
Transmission Electron Microscopy and Scanning Probe Microscopy (Klomparens and Heckman)	37	73
Ubiquinone, Determination of (Crane and Dilley)	11	279
UDP-Enzyme Systems, Measurements of (Pontis and Leloir)	10	107
Ultracentrifugal Analysis of Serum Lipoproteins (de Lalla and Gofman)	1	459
Ultrafilter Membranes in Biochemistry (Jacobs)	22	307
Ultraviolet Spectrophotometry, Enzyme Micro Determinations of Uric Acid, Hypoxanthine, Xanthine, Adenine, and Xanthopterine by (Plesner and Kalckar)	3	97
Urea, Ammonia, and Urease, The Determination of (Kaplan)	17	311
Urea, Gel Electropohoresis in Buffers Containing (Poulik)	14	455
Uric Acid, Enzymic Micro Determinations, by Ultraviolet Spectrophotometry (Plesner and Kalckar)	3	97
Urinary Neutral 17-Ketosteroids, Assay of (Engel)	1	479
Urine, Determination of 17,21-Dihydroxy-2-Ketosteroids in (Silber and Porter)	4	139
Urine, Human, Chemical Determination of Estrogens in (Bauld and Greenway)	5	337
Visual Biochemistry: New Insight Into Structure and Function of the Genome (Vollenweider)	28	201
Vitamin A, Determination of (Embree, Ames, Lehman, and Harris)	4	43
Vitamin A and Carotenoids, in Blood and tissue, Microdetermination of (McLaren, Read, Awdeh, and Tchalian)	15	51
Vitamin B_6, Chemical and Microbiological Determination of (Storvick, Benson, Edwards, and Woodring)	12	183
Vitamin B_{12}, Microbiological Assay of (Skeggs)	14	53
Vitamin E Determination (Lehman)	2	153
Vitamins, Infrared Analysis of (Rosenkrantz)	5	407
Xanthine, Enzymic Micro Determination by Ultraviolet Spectrophotometry (Plesner and Kalckar)	3	97
Xanthopterine, Enzymic Micro Determinations, Ultraviolet Spectrophotometry (Plesner and Kalckar)	3	97
X-Ray Diffraction, in the Study of Protein and Nucleic Acid Structure (Holmes and Blow)	13	113
X-Ray Crystallography of Proteins (Glusker)	37	1
X-Ray Diffraction Analysis, The Growth and Preliminary Investigation of Protein Nucleic Acid Crystals for	23	249
X-Ray Emission Spectrometry, Determination of Elements by (Natelson and Whitford)	12	1
Zinc, Determination of, in Biological Materials (Malmstrom)	3	327
Zone Electrophoresis (Kunkel)	1	141
Zone Electrophoresis, at Constant Current, Mobility Determination by (Waldmann-Meyer)	13	47